Linking Aquatic Exposure and Effects

Risk Assessment of Pesticides

Other Titles from the Society of Environmental Toxicology and Chemistry (SETAC)

Veterinary Medicines in the Environment
Crane, Boxall, Barrett
2008

Relevance of Ambient Water Quality Criteria for Ephemeral and Effluent-dependent Watercourses of the Arid Western United States
Gensemer, Meyerhof, Ramage, Curley
2008

Extrapolation Practice for Ecotoxicological Effect Characterization of Chemicals
Solomon, Brock, de Zwart, Dyev, Posthumm, Richards, editors
2008

Environmental Life Cycle Costing
Hunkeler, Lichtenvort, Rebitzer, editors
2008

Valuation of Ecological Resources: Integration of Ecology and Socioeconomics in Environmental Decision Making
Stahl, Kapustka, Munns, Bruins, editors
2007

Genomics in Regulatory Ecotoxicology: Applications and Challenges
Ankley, Miracle, Perkins, Daston, editors
2007

Population-Level Ecological Risk Assessment
Barnthouse, Munns, Sorensen, editors
2007

Effects of Water Chemistry on Bioavailability and Toxicity of Waterborne Cadmium, Copper, Nickel, Lead, and Zinc on Freshwater Organisms
Meyer, Clearwater, Doser, Rogaczewski, Hansen
2007

Ecosystem Responses to Mercury Contamination: Indicators of Change
Harris, Krabbenhoft, Mason, Murray, Reash, Saltman, editors
2007

For information about SETAC publications, including SETAC's international journals, Environmental Toxicology and Chemistry and Integrated Environmental Assessment and Management, contact the SETAC Administratice Office nearest you:

SETAC Office
1010 North 12th Avenue
Pensacola, FL 32501-3367 USA
T 850 469 1500   F 850 469 9778
E setac@setac.org

SETAC Office
Avenue de la Toison d'Or 67
B-1060 Brussells, Belguim
T 32 2 772 72 81   F 32 2 770 53 86
E setac@setaceu.org

www.setac.org
Environmental Quality Through Science®

Linking
Aquatic Exposure
and Effects

Risk Assessment of Pesticides

Editors:
Theo C.M. Brock
Anne Alix
Colin D. Brown
Ettore Capri
Bernhard F.F. Gottesbüren
Fred Heimbach
Chris M. Lythgo
Ralf Schulz
Martin Streloke

EU and SETAC Europe
Workshop ELINK
Bari, Italy, and Wageningen,
Netherlands

Coordinating Editor of SETAC Books
Joseph W. Gorsuch
Gorsuch Environmental Management Services, Inc.
Webster, New York, USA

CRC Press
Taylor & Francis Group
Boca Raton London New York

CRC Press is an imprint of the
Taylor & Francis Group, an **informa** business

Information contained herein does not necessarily reflect the policy or views of the Society of Environmental Toxicology and Chemistry (SETAC). Mention of commercial or noncommercial products and services does not imply endorsement or affiliation by the author or SETAC.

CRC Press
Taylor & Francis Group
6000 Broken Sound Parkway NW, Suite 300
Boca Raton, FL 33487-2742

First issued in paperback 2017

© 2010 by Taylor & Francis Group, LLC
CRC Press is an imprint of Taylor & Francis Group, an Informa business

No claim to original U.S. Government works

ISBN 13: 978-1-138-11264-3 (pbk)
ISBN 13: 978-1-4398-1347-8 (hbk)

Library of Congress Cataloging-in-Publication Data

European Union Workshop on Linking Aquatic Exposure and Effects in the Registration
   Procedure of Plant Protection Products (2007 : Bari, Italy, and Wageningen, Netherlands)
   Linking aquatic exposure and effects : risk assessment of pesticides / editors, Theo C. M.
   Brock ... [et al.]
      p. cm.
   Includes bibliographical references and index.
   ISBN 978-1-4398-1347-8 (hardcover : alk. paper)
   1. Pesticides--Risk assessment--Congresses. 2.
Water--Pollution--Toxicology--Congresses. 3. Aquatic ecology--Congresses. I. Brock, Theo
C. M. II. Title. IIIl. Series.

   QH545.P4E97 2007
   628.1′6842--dc22
                                                                                    2009035209

Visit the Taylor & Francis Web site at
http://www.taylorandfrancis.com

and the CRC Press Web site at
http://www.crcpress.com

and the SETAC Web site at
www.setac.org

SETAC Publications

Books published by the Society of Environmental Toxicology and Chemistry (SETAC) provide in-depth reviews and critical appraisals on scientific subjects relevant to understanding the impacts of chemicals and technology on the environment. The books explore topics reviewed and recommended by the Publications Advisory Council and approved by the SETAC North America, Latin America, or Asia/Pacific Board of Directors; the SETAC Europe Council; or the SETAC World Council for their importance, timeliness, and contribution to multidisciplinary approaches to solving environmental problems. The diversity and breadth of subjects covered in the series reflect the wide range of disciplines encompassed by environmental toxicology, environmental chemistry, and hazard and risk assessment, and life-cycle assessment. SETAC books attempt to present the reader with authoritative coverage of the literature, as well as paradigms, methodologies, and controversies; research needs; and new developments specific to the featured topics. The books are generally peer reviewed for SETAC by acknowledged experts.

SETAC publications, which include Technical Issue Papers (TIPs), workshops summaries, newsletter (SETAC Globe), and journals (*Environmental Toxicology and Chemistry* and *Integrated Environmental Assessment and Management*), are useful to environmental scientists in research, research management, chemical manufacturing and regulation, risk assessment, and education, as well as to students considering or preparing for careers in these areas. The publications provide information for keeping abreast of recent developments in familiar subject areas and for rapid introduction to principles and approaches in new subject areas.

SETAC recognizes and thanks the past coordinating editors of SETAC books:

A.S. Green, International Zinc Association
Durham, North Carolina, USA

C.G. Ingersoll, Columbia Environmental Research Center
US Geological Survey, Columbia, Missouri, USA

T.W. La Point, Institute of Applied Sciences
University of North Texas, Denton, Texas, USA

B.T. Walton, US Environmental Protection Agency
Research Triangle Park, North Carolina, USA

C.H. Ward, Department of Environmental Sciences and Engineering
Rice University, Houston, Texas, USA

Contents

PART I   *Guidance on Linking Aquatic Exposure and Effects in the Registration Procedure of Plant Protection Products*

Acknowledgments

This book presents the proceedings of a workshop convened by the Society of Environmental Toxicology and Chemistry Europe (SETAC Europe) in Bari, Italy (14 to 16 March 2007), and Wageningen, the Netherlands (19 to 21 September 2007). The 53 scientists involved in this workshop represented 12 European countries and the United States. They offered expertise in aquatic exposure assessment, aquatic ecotoxicology, risk assessment, and risk management of plant protection products.

The workshop was made possible by the time and expertise offered by all workshop participants (Appendix 1) and by the generous financial support of several sponsors (Appendix 2). Based on the discussions during the two workshop meetings, we (the organizing committee) drafted guidance chapters on linking exposure and effects in the aquatic risk assessment procedure for plant protection products (Part I of this book, Chapters 1 to 11). All workshop participants were invited to comment on earlier versions of these guidance chapters, and we would like to thank those participants who provided us with useful and constructive comments.

Besides the input provided by workshop participants, the writing of Chapter 14 was supported by Thierry Caquet and Virginie Ducrot of the French National Institute on Research in Agronomy (INRA), France. Gertie Arts, Dick Belgers, and Caroline van Rhenen of Alterra (Wageningen UR, the Netherlands), and Jeremy Biggs of Oxford Brookes University (United Kingdom) contributed to Chapter 15.

We thank the Instituto Agronomico Mediterraneo di Bari (IAMB) and Wageningen UR for the hospitality offered during the workshop meetings and the Università Cattolica del Sacro Cuore (UCSC) for hosting the ELINK (European Union Workshop on Linking Aquatic Exposure and Effects in the Registration Procedure of Plant Protection Products) Web site (http://www.elink-info.org/) and for the financial support to publish this book.

We also thank Lorraine Maltby for her inspiring enthusiasm when acting as workshop rapporteur during the Bari meeting; Dave Arnold for editing previous versions of this book; Keith Solomon, who independently reviewed the ELINK proceedings on behalf of SETAC; Joe Gorsuch, who is the coordinating editor of SETAC books, and Mimi Meredith of SETAC for their editorial support.

Theo Brock
Anne Alix
Colin Brown
Ettore Capri
Bernhard Gottesbüren
Fred Heimbach
Chris Lythgo
Ralf Schulz
Martin Streloke

About the Editors

Theo C.M. Brock, after finishing his PhD in 1985 at the University of Nijmegen, held several positions as an aquatic and wetland ecologist at the University of Amsterdam, the Wageningen Agricultural University, and the University of Utrecht. Since 1991, he has been employed at Alterra (Wageningen University and Research Centre; formerly SC-DLO) to scientifically underpin the ecological risk assessment procedures for pesticides and other chemicals. He is author or coauthor of more than 140 scientific papers and reports, covering subjects on ecology of macrophyte-dominated freshwater ecosystems, vegetation dynamics of wetlands and wetland forests, aquatic ecotoxicology, and ecological risk assessment and risk management of pesticides.

Brock has been secretary and president of the Netherlands Society of Aquatic Ecology and council member and president of the Society of Environmental Toxicology and Chemistry (SETAC) in Europe. Currently, he is associate editor of the SETAC journal *Integrated Environmental Assessment and Management* (IEAM). He participated in the organizing committee of EU/SETAC workshops on higher-tier risk assessment (e.g., Higher-Tier Aquatic Risk Assessment for Pesticides [HARAP], Community-Level Aquatic System Studies–Interpretation Criteria [CLASSIC]). He also chaired the ELINK (European Union Workshop on Linking Aquatic Exposure and Effects in the Registration Procedure of Plant Protection Products) workshop.

Anne Alix, after receiving a PhD in ecotoxicology conducted on an integrated pest management problem, worked as an environmental risk assessor at Novartis in France investigating the environmental fate and ecotoxicological effects of plant protection products (PPPs) in support of applications for national registration.

In December 2000, she joined the French National Institute on Research in Agronomy (INRA) and started to work at the office in charge of the scientific evaluation of PPPs for the Ministry of Agriculture. In this office, she evaluated the dossiers submitted by companies in a European context (application for inclusion

of a pesticide substance in Annex I of Directive 91/414/EC) and in a national context (application for placing the product on the market). She participated in working groups in charge of proposing and updating guidance documents for the assessment of pesticides and in the working group in charge of Directive 91/414/EC revision.

Since July 2006, Alix has worked at the Plant and Environment Directorate (Direction du Végétal et de l'Environnement) of the French Agency on the Safety of Food (AFSSA), as the director of the unit in charge of environment and ecotoxicology evaluations. The Plant and Environment Directorate is in charge of the evaluation of PPPs and fertilizers and is involved in the development of risk assessment tools at the national level.

Colin D. Brown is professor of ecochemistry at the University of York, United Kingdom. He has 20 years of experience researching the fate and behavior of organic contaminants in the environment. Current research interests focus on methodological development for ecological risk assessment, including probabilistic techniques and applications of landscape analysis, toxicokinetic–toxicodynamic modeling, transport of solutes from soil to water, kinetics of chemical sorption to soil, and development of mathematical models to predict environmental exposure to organic contaminants. Brown is a member of the UK Advisory Committee for Pesticides and chairs its Environmental Panel. He chairs the BioResources Group of the Society of Chemical Industry and chaired the DG-Sanco FOCUS (Forum for the Co-ordination of Pesticide Fate Models and Their Use) Work Group on Landscape and Mitigation Factors in Ecological Risk Assessment.

Ettore Capri (PhD) has been associate professor at the Catholic University in Milan in pesticide risk assessment and food contaminants since 2002. His research work concerns the agricultural chemistry and mainly the environmental fate and risk assessment of pollutants such as biocides, nitrates, persistent organic pollutants (POPs), and trace elements. He has published about 130 scientific papers and communications and has edited 4 national books and 5 international monographs.

Bernhard F.F. Gottesbüren specializes in the evaluation of pesticide behavior in soil and aquatic systems, estimation of pesticide parameters from field and lysimeter studies (inverse modeling), and prediction of pesticide persistence, accumulation, and leaching. He received his PhD in 1991 at the University of Hannover based on research performed at the Federal Biolicial Institute for Agriculture and Forestry (BBA) in Braunschweig on degradation of pesticides in soils under laboratory and field conditions and prediction and on side effects of herbicides on rotational crops. In the period 1992 to 1995, he was a scientist at the Institute of Radioagronomy (Research Centre Jülich), and here his research focused on the evaluation of pesticide leaching models and the validation of these models with lysimeter data. Since 1995, he has been a scientific researcher and environmental fate assessor at BASF SE. Currently, he is team leader of BASF's environmental fate modeling team.

He was an organizing committee member of the European concerted COST Action 66 to test pesticide leaching models on the European level and of the ELINK workshop. Furthermore, he was a member of the FOCUS Workgroup for Groundwater Scenarios, developing 9 EU scenarios for assessment of pesticide leaching for EU registration (1997 to 2000), member of the FOCUS Groundwater Workgroup developing guidance on assessment of pesticide leaching to groundwater for national and EU registration (2004 to 2007), and a member of the FOCUS Version Control group.

In 2007, he was a participant at the SETAC Europe Workshop, "Semi-Field Methods for the Environmental Risk Assessment of Pesticides in Soil (PERAS)."

Fred Heimbach works as a consultant scientist at RIFCon GmbH in Leichlingen, Germany. He obtained his MSc degree and PhD in conducted research on marine insects at the Institute of Zoology, Physiological Ecology, at the University of Cologne.

From 1979 to 2007, Heimbach worked at Bayer CropScience in Monheim, Germany, on the side effects of pesticides on nontarget organisms. In addition to his work, he gave lectures on ecotoxicology at the University of Cologne. Heimbach has researched the development of single-species toxicity tests for both terrestrial and aquatic organisms and has worked with microcosms and mesocosms in the development of multispecies tests for these organisms. As an active member of European and international working groups, he participated in the development of suitable test methods and risk assessment of pesticides and other chemicals for their potential side effects on nontarget organisms.

For several years, Heimbach served on the SETAC Europe Council and the SETAC World Council, and he has been an active member of the organizing committees of several European workshops on specific aspects of the ecotoxicology of pesticides.

Chris M. Lythgo works in the pesticide risk assessment peer review unit (PRAPeR) at the European Food Safety Authority (EFSA) in Parma, Italy. He is leader of a team responsible for peer reviewing the environmental exposure assessments carried out by the EU member state competent authorities (which have responsibility for the regulatory regime for plant protection products) that underpin EU-level decision making. He obtained his BSc (honors) degree in agricultural science (crops) at the University of Leeds and MSc through research on bioreactor systems and bacteria-mediated biotransformations at the University of Manchester Institute of Science and Technology (UMIST).

Between 1998 and 2005, Lythgo worked at the Pesticide Safety Directorate (PSD) in York, the UK competent authority for the authorization of plant protection products, carrying out environmental exposure assessments. Between 1994 and 1997, at the same organization, he carried out consumer exposure assessments, assessments for the setting of maximum residue levels (MRLs) in food, and assessments of data regarding chemical identity and quality of plant protection products. Between 1991 and 1993, he was employed as a chemical analyst in a UK government laboratory (the Central Science Laboratory) in Harpenden; prior to this, he worked in the food manufacturing sector at Weetabix Limited (1986 to 1990).

Ralf Schulz is a professor for environmental sciences and the head of the Department for Environmental Sciences at University Koblenz-Landau/Germany. Between 2002 and 2004, he worked as an aquatic team leader at Syngenta AG, Bracknell/UK. His main area of expertise is exposure, effects, and risk mitigation strategies for agricultural pesticides.

Martin Streloke is an aquatic ecotoxicologist with special interests in regulatory risk assessment and management. In 1991, he received a PhD in zoology from Hanover University, and in 1992 he joined the Biologische Bundesanstalt für Land und Forstwirtschaft (BBA) in Braunschweig, Germany, where he worked in the field of regulatory risk assessment and management for aquatic organisms. In 2002, he joined the German Federal Office of Consumer  Protection and Food Safety (BVL), the responsible authority for the authorization of PPPs, where he is responsible for risk management issues (e.g., risk mitigation measures to protect the environment) and head of the environmental unit. He was a member of the organizing committees of HARAP, CLASSIC, Workshop on Risk Assessment, Risk Mitigation Measures in the Context of the Authorization of Plant Protection Products (WORMM), and the Effects of Pesticides in the Field (EPiF) workshops dealing with the development of improved test methods, higher-tier risk assessments, realistic risk management tools, and monitoring studies. He assists in the development of EU documents on regulation for PPPs (e.g., Guidance Document on Aquatic Organisms).

Part I

Guidance on Linking Aquatic Exposure and Effects in the Registration Procedure of Plant Protection Products

The guidance chapters in Part I of this book (Chapters 1 to 11) have been produced by the members of the organizing committee of the ELINK workshop. The content of these guidance chapters is based on the input of workshop participants, on further information provided by the ELINK work groups, and on information found in the scientific literature.

The ELINK organizing committee invited all workshop participants to comment on an earlier version of these chapters with the aim of reflecting as far as possible consensus regarding the conclusions and recommendations that are presented.

Primary Authors
    Anne Alix, Theo Brock, Colin Brown, Ettore Capri, Bernhard Gottesbüren,
        Fred Heimbach, Chris Lythgo, Ralf Schulz, and Martin Streloke

Contributors
    Alf Aagaard, Paulien Adriaanse, Elena Alonso Prados, Dave Arnold, Roman
        Ashauer, Jos Boesten, Eric Bruns, Peter Campbell, Peter Dohmen,
        Mark Douglas, Igor Dubus, Gunilla Ericson, Antonio Finizio, Gerhard
        Görlitz, Mick Hamer, Paul Hendley, James Hingston, Udo Hommen,

Karen Howard, Andreas Huber, Katja Knauer, Igor Kondzielski, Jenny Kreuger, Roland Kubiak, Laurent Lagadic, Matthias Liess, Neil Mackay, Lorraine Maltby, Steve Maund, Gary Mitchel, Aidan Moody, Michael Neumann, Steve Norman, Apolonia Novillo Villajos, Jo O'Leary Quinn, Werner Pol, Toni Ratte, Elena Redolfi, Helmut Schäfer, Paul Sweeney, Csaba Szentes, Paul Van den Brink, Peter Van Vliet, and Jörn Wogram

1 Executive Summary and Recommendations
European Union Workshop on Linking Aquatic Exposure and Effects in the Registration Procedure of Plant Protection Products (ELINK)

CONTENTS

1.1 WORKSHOP OBJECTIVE

The principle objective of European Union Workshop on Linking Aquatic Exposure and Effects in the Registration Procedure of Plant Protection Products (ELINK) was to bring together specialists in aquatic exposure and effects assessment to improve guidance on linking exposure and effects in the aquatic risk assessment procedure for pesticides under the European Plant Protection Products Directive 91/414/EEC (EC, 1997).

1.2   REASON FOR ELINK

Although several European guidance documents deal with aquatic exposure or effect assessments of pesticides, the interaction between these fields of expertise has received relatively little attention in these documents. Moreover, current procedures for aquatic risk assessment have not been able to adequately address some of the uncertainties arising from time-variable surface water exposure profiles that are more often the rule than the exception in the field. Hence, there was a need to provide a forum for balanced, open, and constructive communication among and between fate and effects experts to seek a better understanding of each other's issues and to find a way forward in addressing the uncertainties posed by time-variable exposures.

1.3   WORKSHOP STRUCTURE AND APPROACH

The ELINK workshop was the first workshop organized in Europe in which fate and effects experts played an equal part. The participants consisted of 53 invited scientists from different EU member states and the United States, representing government, business, and academia. The ELINK workshop comprised two meetings, one in Bari, Italy (14 to 16 March 2007), and the other in Wageningen, the Netherlands (19 to 21 September 2007). The Bari meeting focused on problem formulation and on the identification and prioritizing of the most relevant topics. Six ELINK working groups were formed during the meeting, dealing with 1) communication and education, 2) characterizing the exposure profile, 3) key parameters for risk assessment, 4) interaction between fate and effect experts, 5) extrapolation tools, and 6) characteristics of water bodies. The Wageningen meeting focused on development of actual guidance based on the input of the different working groups, and decision schemes for acute and chronic risk assessment were produced to support this guidance.

In the postworkshop period, members of the organizing committee proceeded with improving these decision schemes and came to the conclusion that one combined scheme for acute and chronic risk assessment is sufficient to present the new insight gained on linking exposure and effects (see Figure 2.1 in Chapter 2). In addition, they drafted a guidance chapter around this scheme (Chapters 2 to 11), based on workshop discussions, further information provided by the working groups (Chapters 12 to 16), and information found in the scientific literature.

1.4   RECOMMENDED APPROACHES FOR LINKING EXPOSURE AND EFFECTS

1.4.1   Generalized Exposure Profiles and Tiered Exposure Assessment

Available monitoring data, for certain moderately persistent and water-soluble compounds, suggest that Forum for the Co-ordination of Pesticide Fate Models and Their Use (FOCUS) simulations are able to reproduce the general characteristics of measured pesticide concentration profiles in the water column of edge-of-field water bodies such as streams and ditches. Hence, the implications of such exposure profiles in relation to ecotoxicology need to be considered in the risk assessment process. Whether the simulated exposure patterns in these edge-of-field surface waters are generally valid for compounds that differ in physicochemical properties needs to be confirmed.

RECOMMENDATION 1

The implications of simulated time-variable exposure profiles in relation to ecotoxicology need to be considered in the aquatic risk assessment process for pesticides. To underpin the validity of long-term exposure profiles simulated by the FOCUS scenarios and models, further experimental and monitoring work is recommended in different types of edge-of-field surface waters and on a range of compounds and conditions.

Time-variable exposure profiles of pesticides in surface water may vary in types of ecosystem, in periods of the year, and in the fate properties of substances. To categorize the complex exposure profiles simulated for stream, ditch, and pond environments, metrics that allow a description of different generalized exposure profiles are proposed. These metrics, or parameters, which describe exposure characteristics, can be used to delineate exposure regimes for higher-tier effects studies.

Key exposure characteristic parameters are as follows:

- height of peak concentrations
- area-under-the-curve concentrations
- duration of peak exposure
- interval between peaks
- height of a possible long-term background concentration
- frequency of peaks

A further parameter to characterize the exposure pattern is

- single first-order DT50 (period required for 50% disappearance) or other kinetic parameters to describe peak decline

Parameters supportive for risk assessments are the following:

- entry route of peak (link to mitigation measures)
- time of exposure above the acute and chronic threshold levels for effects (link to risk assessment)

RECOMMENDATION 2

Key exposure parameters should be used to construct generalized exposure profiles for edge-of-field surface waters (streams, ditches, ponds) and for pesticides that differ in physicochemical properties and use patterns to support problem formulation and to provide information for the exposure profiles that must be simulated in ecotoxicological effects studies.

For an appropriate assessment of risk from exposure profiles characterized by repeated pulsed exposures, it is (in the first instance) important to determine whether the pulses are toxicologically independent of each other. To demonstrate the toxicological independence of pulsed exposures, either specially designed pulsed exposure toxicity tests or toxicokinetic and toxicodynamic (TK/TD) models for the relevant organisms and pesticides are required. When evidence can be provided that different pulsed exposures are toxicologically independent, it may, for certain risk assessments, be important also to demonstrate their ecological independence. Such demonstration will, for example, be necessary when recovery of populations is taken into account. Peaks may be considered ecologically independent if peak intervals are greater than the relevant recovery time of the sensitive population of concern (which may depend on properties of the affected species or populations and on landscape characteristics).

1.4.2 Peak or Time-Weighted Average Concentrations

One key aspect of FOCUS simulations is that they give a range of predicted environmental concentrations (PECs) that vary both temporally and spatially as a consequence of the pattern of use of a pesticide and the geoclimatic situation. These exposure concentrations are then compared with the short-term and long-term effect assessment endpoints derived from lower- and higher-tier effects assessment procedures. In this book, we refer to these measures of effect as regulatory acceptable concentrations (RACs) and to the interface between the exposure and effects assessment as the ecotoxicologically relevant concentration (ERC). There are two distinctly different exposure estimates in pesticide risk assessment: 1) exposure estimates that relate to exposure in the field (i.e., PEC) and 2) exposure estimates that relate to exposure in ecotoxicological studies (to derive the RAC). In first-tier toxicity tests with standard species performed according to Organization for Economic Cooperation and Development (OECD) guidelines, the exposure regime is fixed. It is in the higher-tier tests in particular that refined exposure regimes may be adopted and that a clear idea of the ERC is required. In the risk assessment, the ERC needs to be consistently applied so that field exposure and test concentration profiles can be compared as readily as possible. Important considerations in determining the ERC include these questions:

- In which environmental compartment (e.g., water or sediment) do the organisms at risk live, and what constitutes the bioavailable fraction?

- What is the influence of time-variable exposure patterns on the type and magnitude of the effects?
- Is it appropriate to use the maximum (peak) concentration over time or some time-weighted average (TWA) concentration to derive the PEC and the RAC?

Generally, PEC_{max} values are used in acute risk assessments, while in chronic risk assessments, the PEC_{max} and, under certain conditions, a TWA PEC may be used. However, there are situations in which the use of a TWA concentration in risk assessment is not appropriate.

SITUATIONS IN WHICH A TWA IS NOT APPROPRIATE

- In risk assessments that use RACs derived from 1) laboratory toxicity tests with algae or 2) effect studies in which the exposure is not maintained and loss of the active substance in the test system other than uptake by the test organism is fast
- When the effect endpoint in the chronic test (used to derive the RAC) is based on a developmental process during a specific sensitive life-cycle stage and when it cannot be excluded that the exposure will occur when the sensitive stage is present
- When there is evidence of an endocrine disruption effect (unless the mechanism for the effect is clearly understood and it is proven that long-term exposure is required to elicit the effect)
- When the effect endpoint in the chronic test (used to derive the RAC) is based on mortality occurring early in the test (e.g., in the first 96 hours) or if the acute-to-chronic ratio (acute EC50 [median effective concentration that affects 50% of the test organisms] or LC50 [median lethal concentration]–chronic no observed effect concentration [NOEC]) based on immobility or mortality is <10
- If latency of effects has been demonstrated or might be expected due to mode of action of the pesticide or by appropriate other data

When it is possible to use a TWA concentration approach, it is proposed to use a TWA PEC of 7 days as a default if no specific information is available on the relation between exposure pattern and time to onset of effects for the relevant life stages of the organisms that triggered the chronic risk. It may be scientifically justified to lengthen or shorten the default 7-day TWA period when the information on time to onset of effect is made available. However, the time window for the TWA PEC should never be 1) longer than the duration of the ecotoxicological test that triggered the risk or 2) longer than the duration of the life stage of highest ecotoxicological concern in this test. These proposed criteria and triggers are based on current knowledge, expert judgment, and pragmatism. Further research is recommended to scientifically underpin the criteria that can be used to decide whether the TWA approach is appropriate and to set the appropriate TWA time window.

RECOMMENDATION 3

Ecotoxicologists must determine, based on knowledge of ecotoxicological data, whether the TWA concentration approach is appropriate to use in chronic risk assessment and on which time window to base the TWA. The time window of the TWA PEC should be equal to or smaller than the length of the relevant chronic toxicity test (or life stage of highest ecotoxicological concern) that triggered the risk. For invertebrates and fish (and possibly also macrophytes), a default 7-day TWA time window is proposed if the TWA concentration approach is deemed appropriate and no further information on the relation between exposure pattern and time to onset of effects is provided.

1.4.3 SPECIES SENSITIVITY DISTRIBUTION APPROACH

Species sensitivity distributions (SSDs) have the advantage of making more use of the available laboratory toxicity data for a larger array of species. SSDs can be based on either acute or chronic toxicity data. They enable estimates to be made of the proportion of the species affected at different concentrations (e.g., HC5 or HC1, the hazardous concentration to 5% or 1% of the tested taxa, respectively), and they can be shown together with confidence limits that demonstrate the sampling uncertainty due to the limited number of species tested. For example, the lower-limit HC5 based on NOECs protects 95% of the tested taxa with 95% certainty. In pesticide risk assessment, it is critical that the SSD is constructed with toxicity data from the relevant taxonomic group, which may be a specific group such as arthropods for insecticides, or a wider array of species, such as in the case of a fungicide with a narcotic mode of action.

Empirical data reveal that the median HC5 value (derived from SSDs constructed with acute toxicity data) divided by an appropriate uncertainty factor (UF; which may be 1 for single short-term pulse exposures to certain compounds) may be used as the acute RAC. In addition, the corresponding lower-limit HC5 or the median HC1 may be used as the RAC for repeated pulsed exposures that are toxicologically not independent. In the risk assessment, these acute RACs should be higher than the PEC_{max}.

Fewer empirical data are available for pesticides that allow a proper evaluation of HC5 values derived from SSDs constructed with chronic toxicity data. More research is required that enables the construction of chronic SSDs for compounds that differ in their mode of action and the comparison of the chronic HC5 values with threshold levels of microcosm or mesocosm studies with the same compounds that simulate a chronic exposure regime. However, the available data for a limited number of compounds indicate that the median chronic HC5 value divided by an appropriate UF (which may be 1), or the lower limit chronic HC5, may be used as the RAC for long-term exposure. When the SSD approach is used in the chronic risk assessment, either the PEC_{max} or the appropriate TWA PECs should be lower than the RAC derived from the chronic SSD.

RECOMMENDATION 4

The HC5 or HC1 values calculated from SSDs that were constructed with acute toxicity data from the relevant taxonomic group (considering the toxic mode of action of the pesticide) may be used to derive RACs for single or repeated pulse exposure regimes. In the risk assessment, these RACs should be higher than the PEC_{max}. More research is required to scientifically underpin whether the chronic HC5, based on SSDs constructed with chronic toxicity data from the relevant taxonomic groups, can be used generally as the chronic RAC.

1.4.4 USE OF REFINED EXPOSURE SINGLE-SPECIES AND POPULATION STUDIES

If the TWA approach cannot be applied in the risk assessment, refined exposure studies may provide useful information under more realistic conditions at higher tiers. In designing refined exposure studies with standard or additional species, parameters such as mode of action, time to onset of effect, and sensitive life stages should be considered. The refined exposure regime tested should be guided by the generalized exposure regime that is derived from the relevant predicted (modeled) field exposure profiles.

If standard test species or populations are tested in such refined exposure studies, a reduction of the UF is not justified when the RAC is derived. However, a higher toxicity value from the refined exposure study with standard test species and the application of the appropriate UF (e.g., 10 to derive the chronic RAC) may change the overall risk assessment. Refined exposure studies may also be used to put other higher-tier data into perspective.

This approach warrants "confirmation and validation," for example, by comparing the results with threshold concentrations (or RACs) derived from appropriate micro/mesocosm experiments that simulate similar exposure regimes with the same compounds.

RECOMMENDATION 5

If the TWA approach cannot be applied in the risk assessment (all acute and several chronic risk assessments), refined ecotoxicological exposure studies may be a higher-tier option. Exposure conditions in refined exposure single-species or population studies should be guided by species properties (e.g., sensitive life stages and life cycle characteristics) and by the generalized exposure regime that is derived from the relevant predicted field exposure profiles.

1.4.5 TOXICOKINETIC AND TOXICODYNAMIC MODELING

The TK/TD models describe the processes that link exposure to effects in an individual organism. Toxicokinetics consider the time course of concentrations within an organism in relation to concentrations in the external medium. Toxicodynamics describe the time course of damage and repair to the target organisms based on specific patterns of exposure to the test compound. Several models are available, ranging from generalized models to those specific to a particular mode of action. To date, the modeling tools available have not been extensively used in pesticide regulatory risk assessment and primarily concern a research activity.

The development of TK/TD models for the aquatic risk assessment of pesticides relies on controlled experiments; methods are available, and requirements seem realistic. At present, TK/TD modeling is most highly developed for aquatic invertebrates. Because these are relatively simple animals, the assumption of uniform internal concentration appears to hold, and the TK can be simulated with a single-compartment first-order model. A major constraint to current models is that they apply when the duration of exposure is less than the duration of the life of the test species, and they generally assume negligible growth and negligible change in lipid content during the period of exposure. In addition, species- and compound-specific parameterizations of the models are necessary. The parameterization facilitates a better understanding of the causes for the distribution of species sensitivities toward pesticides that differ in toxic mode of action.

RECOMMENDATION 6

The TK/TD models are promising tools in the future risk assessment for time-variable exposure regimes of pesticides. The primary areas for research and development are 1) broader application and evaluation of models currently available to explore the feasibility of extrapolation of parameters and model concepts to other species and compounds, 2) development of models for sublethal endpoints in aquatic invertebrates, 3) development of the models to account for aquatic invertebrates with short generation times and rapid growth of individuals (e.g., *Daphnia*), and 4) development of approaches for use in fish and aquatic plants.

1.4.6 MODEL ECOSYSTEM APPROACH

The selection of an exposure regime in a micro/mesocosm study should be guided by information from both lower-tier effects studies and the most relevant modeled exposure profile (e.g., using FOCUS tools) based on the use pattern of the compound.

If the expected and relevant field exposure regime is characterized by a single pulse (e.g., due to drift application) or by repeated pulses that are both toxicologically and ecologically independent, a single application experimental design is an appropriate exposure regime to study in the micro/mesocosm experiment. Not only should the PEC_{max} guide the treatment levels but also the pulse duration should either be equal to or larger in the micro/mesocosm experiment than that predicted for the field or be easy to extrapolate concentration–response relationships for shorter peaks to that of broader peaks.

If the expected field exposure profile consists of several short pulses and the time frame between pulses is relatively short (e.g., <7 to 14 days) so that pulses are probably toxicologically dependent, it is logical to adopt a repeated exposure regime in the micro/mesocosm study. Views of the workshop participants were divided with respect to the number of applications really necessary to address risks of repeated exposures. A pragmatic option would be the selection of a more or less regular multiple-exposure regime (e.g., weekly application) on the basis of the predicted field exposure concentrations for the most relevant exposure scenarios. The number of applications has to be considered carefully in relation to the expected biological effects — but should be as low as possible — guided by the duration and responses observed in the toxicity tests that triggered the micro/mesocosm study and by biological information of the species potentially at risk.

If the expected field exposure profile triggers concerns due to long-term exposure, a realistic long-term exposure regime covering the duration of the sensitive life stages of organisms for which the risks were triggered in the lower tiers should be adopted in the micro/mesocosm study, or appropriate models should be applied for spatiotemporal extrapolation of the test results.

The short-term or long-term RAC representative for the threshold level of toxic effects may be derived by applying an appropriate UF to the Effect class 1 to 2 concentration from the micro/mesocosm experiment. This is a test concentration, for which no (Effect class 1) to slight (Effect class 2) treatment-related effects are observed on the most sensitive measurement endpoint (for a more detailed description, see Chapter 8). The size of this UF (which may be 1) should depend on (among other considerations) the relevance of the tested assemblage for the species potentially at risk, the other higher-tier information available (e.g., toxicity data for additional test species and other micro/mesocosm experiments), and known variability in Effect class 1 and 2 concentrations for related compounds with a similar toxic mode of action.

If population recovery as observed in the micro/mesocosm study is considered in the RAC, for example, by applying a UF for spatiotemporal extrapolation to an Effect class 3 concentration, an appropriate risk assessment can be performed only by also plotting the RAC representative for the threshold level for direct toxic effects on the predicted field exposure profiles or a justified constructed generalized field exposure profile. An Effect class 3 concentration is the test concentration for which pronounced but short-term treatment-related effects are observed on the most sensitive measurement endpoints (for a more detailed description, see Chapter 8). If, in the appropriate field scenarios, the pulses are lower than the RAC value based on population recovery (e.g., derived from Effect class 3 concentration) but higher than the threshold level for direct toxic effects (e.g., RAC derived from Effect class 1 concentration), the interval between successive peaks should be carefully considered to evaluate the duration of the total period of possible effects. Population or community models may be the appropriate tools to evaluate whether the peaks are dependent or independent from an ecological point of view. These models may be based on empirical data for related compounds with a similar fate profile and toxic mode of action and on ecological knowledge of the sensitive species of concern.

To date, most micro/mesocosm experiments with pesticides have studied single- or repeated-pulse exposures. More effort is required for micro/mesocosm experiments simulating a more-or-less-constant exposure regime to get insight into ecological risks of worst-case long-term exposures.

RECOMMENDATION 7

Under the condition that a realistic (worst-case) exposure regime is studied and the test systems contain enough representatives of sensitive taxonomic groups, Effect class 1 and 2 concentrations can be used to derive an RAC representative for the threshold level of toxic effects, for example, by applying an appropriate UF (which may be 1). Such a micro/mesocosm experiment might also be used to derive an RAC that addresses population recovery (e.g., based on Effect class 3 concentrations) by applying an appropriate UF or modeling approach for spatiotemporal extrapolation. If an RAC based on population recovery is used in the risk assessment, the intervals between pulse exposures above the threshold level of toxic effects should be considered to estimate the duration of the total period of possible effects.

1.4.7 ECOLOGICAL MODELS IN PESTICIDE RISK ASSESSMENT

Ecological modeling may be used to extrapolate from either one or a number of situations with measured deterministic data to other (new) situations. This extrapolation is usually done to reduce experimental effort, to predict outcomes of untested situations, and to check to which extent ecological complexity has been understood by comparing model outputs and real data.

Within the context of pesticide risk assessment, and linking aquatic exposure and effects, ecological models may be used to refine the acute or chronic effect thresholds for defined populations and communities (at levels of biological organization higher than that of the individual covered by TK/TD modeling). In addition, the spatiotemporal extrapolation of results of focused studies (e.g., mesocosms) with respect to interactions between functional groups (to predict possible indirect effects) and recovery rates of populations, communities, and ecological processes (e.g., nutrient cycling) can be performed with ecological models.

The following ecological modeling tools deserve further attention and improvement in terms of their use in pesticide risk assessment:

- *Population models.* These models generally describe the dynamics of one population. In pesticide risk assessment, population models for standard test species and for focal species of edge-of-field freshwater ecosystems (e.g., *Gammarus, Asellus*) have been developed to predict effects on populations of individuals that differ in age and developmental state. Particularly when external recovery of sensitive populations in streams, ditches, and ponds is addressed, spatially explicit population models are a promising tool.

* *Community, food web, or ecosystem models.* These models aim to describe the interactions of populations or functional groups within communities, the transfer of contaminants within food webs (e.g., for pesticides with a high bioconcentration factor [BCF]), and the interaction of communities with their abiotic environment. Except for models that predict the transfer of pesticides with a high BCF through the food web, community and food web models are primarily used in research and, until now, generally not in pesticide risk assessment for regulatory purposes.

* *Empirical models.* These models aim to use existing data (e.g., database of micro/mesocosm studies with pesticides) for extrapolation without mechanistic understanding. These "data-mining" models may be used for the ecological interpretation of chemical monitoring data.

RECOMMENDATION 8

Ecological modeling is generally considered a helpful and promising tool in pesticide risk assessment. For example, modeling approaches have been used successfully for the spatiotemporal extrapolation of recovery rates as observed in micro/mesocosm experiments. However, because many of the existing examples of ecological modeling in pesticide risk assessment are still snapshots, guidance is needed regarding when and how to use ecological models, particularly concerning the validity of model input data and model parameterization. Quality control of the modeling results is needed, for example, through verification of model results using independent (experimental or monitoring) ecological data sets.

1.4.8   Ecological Field Data

Ecological field data allow the description of the physical, chemical, and biological characteristics of the aquatic ecosystems of concern. These data are a prerequisite to obtaining characteristics of the water bodies that we intend to protect by the risk assessment performed. Collecting these data is important to developing and improving the tiered risk assessment approach by incorporating realistic ecosystem properties in the scenarios and tools for both the exposure and the effects assessment.

By their nature, ecological data sets may be particularly useful in higher-tier risk assessments derived from either model ecosystems or ecosystem models. But, ecological data may also be useful at earlier steps of the risk assessment because field monitoring data may provide useful information about the relevant ecosystem properties that affect pesticide fate (e.g., pH, light, organic matter content), the relevant species to be tested (e.g., to construct SSDs), or the ecologically relevant experimental conditions for refined fate and effects studies.

Additional ecological data sets are needed. It is proposed that the generation of these ecological field data sets should follow a FOCUS-based distinction of streams, ditches, and ponds, and that these data should be collected in different European regions. The generated data should be sufficiently representative for the geographical area of concern to allow a proper risk assessment. Furthermore, monitoring data are needed as a reality check for the risk predictions. Work sharing is strongly recommended to ensure the harmonization of data collection and of data use. Integrating surface water data collected from agricultural landscapes into a European database is recommended.

RECOMMENDATION 9

Ecological data sets (including physical, chemical, and biological properties) for ditches, streams, and ponds in agricultural landscapes should be collected for different geoclimatic zones in Europe to be used in a generic way as a "handbook of ecological scenarios" to which fate and effect experts can refer in performing and evaluating refined exposure, effect, and risk assessments.

2 Introduction to the Guidance on Linking Aquatic Exposure and Effects in the Risk Assessment for Plant Protection Products

CONTENTS

2.1 INTRODUCTION AND REGULATORY BACKGROUND IN EUROPE

The aquatic risk assessment procedure for pesticides under the Plant Protection Products Directive 91/414/EEC (EC, 1997) consists of 2 parts: 1) exposure assessment, which is the domain of environmental fate and behavior and exposure modeling, and 2) effects assessment, which is in the domain of ecotoxicology. Guidance on aquatic effects assessment for pesticides is provided in various workshop reports such as HARAP (Higher-Tier Aquatic Risk Assessment for Pesticides; Campbell et al. 1999), CLASSIC (Community-Level Aquatic System Studies–Interpretation Criteria; Giddings et al. 2002), EPiF (Effects of Pesticides in the Field; Liess et al. 2005), and AERA (Van den Brink, Maltby, et al. 2007), the SANCO (Santé de consommateurs) guidance document on aquatic ecotoxicology (EC 2002), and some recent opinions of the Scientific Panel on Plant Health, Plant Protection Products, and Their Residues (European Food Safety Authority [EFSA] 2005a, 2005b, 2006a, 2006b). Guidance on exposure assessments at the EU level became available in the form of FOCUS (Forum for the Co-ordination of Pesticide Fate Models and Their Use) models and scenarios (FOCUS 2001, which primarily prescribes steps 1 to 3) and in the FOCUS degradation kinetics report (FOCUS 2006). In addition, options for risk mitigation measures and step 4 exposure calculations are described in the FOCUS landscape and mitigation report (FOCUS 2007a, 2007b). Although such guidance provides valuable information and, in some cases, tools, to aid the conduct of regulatory aquatic risk assessments, it is considered that the issue of the linkage of effects (on aquatic organisms) with sometimes complex pesticide exposure patterns needs to be addressed at a fundamental level (EFSA 2005a; Boesten et al. 2007).

Until recently, most higher-tier ecotoxicological experiments with pesticides only addressed the impact of spray drift. With the advent of the FOCUS surface water tools and documents, procedures no longer address exposure only from drift but also from drainage and runoff. Hence, a much wider range of exposure patterns has to be considered in the aquatic risk assessment. A question therefore arises regarding whether the exposure regime in higher-tier ecotoxicological studies (e.g., microcosm and mesocosm experiments) is sufficiently challenging in view of the complex patterns of exposure that occur in the field as indicated by FOCUS approaches. On the other hand, it can be disputed whether the current FOCUS scenarios are sufficiently geared to the ecological properties of the aquatic systems they intend to simulate and, consequently, predict realistic exposure regimes for the habitats (e.g., littoral or pelagic zones in freshwater ecosystems) of aquatic populations actually at risk.

It is, however, simply not feasible (or desirable for prompt decision making) to study all types of possible exposure regimes in the effect assessment for a wide array of aquatic species and communities or to confirm the predicted exposure concentrations for all pesticides in several types of surface waters in European member states (e.g., by expensive chemical monitoring programs). Nevertheless, for a proper aquatic risk assessment some issues need to be addressed. Important questions that relate to the appropriate linking of exposure and effects are as follows:

- What is our confidence in the predictive value of our current exposure and effect assessment procedure, and how do fate and effect experts communicate this with each other?
- Are there types of exposure patterns that can be distilled from the current modeling procedures based on FOCUS scenarios and chemical monitoring programs, and can these types of exposure patterns be used to better design the exposure in ecotoxicological experiments?
- Do we have appropriate data sets and tools to extrapolate the effects of time-variable exposures, and what is the appropriate ecotoxicologically relevant concentration to use in the risk assessment?

The information in Part I of this book (Chapters 1 to 11) is intended to provide guidance on linking exposure and effects when assessing the environmental risks of time-variable concentrations of pesticides in surface waters. This guidance is based on the output of two 2.5-day ELINK workshops organized in Bari, Italy (14 to 16

March 2007), and Wageningen, the Netherlands (19 to 21 September 2007) and is drafted by the members of the organizing committee. ELINK is the acronym for EU Workshop on Linking Aquatic Exposure and Effects in the Registration Procedure of Plant Protection Products.

The main aim of the ELINK workshops was to bring together specialists in aquatic exposure and effects assessment and risk assessors and risk managers of governmental authorities to improve guidance on linking exposure and effects in the risk assessment procedure for pesticides within the Plant Protection Products Directive. The meeting in Bari focused on problem formulation and the identification and prioritizing of relevant topics needed for guidance recommendations.

The following 6 themes (priorities) were identified:

- Theme 1: Communication and education
- Theme 2: Characterizing the exposure profile
- Theme 3: Key parameters for the risk assessment
- Theme 4: Interaction between fate and effect experts
- Theme 5: Extrapolation tools
- Theme 6: Characteristics of water bodies

Experts were assigned to working groups to address issues within each theme and, having identified key criteria to address (problem formulation), were given the task of producing reports for discussion at the Wageningen meeting. The main goals of the Wageningen meeting were to produce guidance on 1) how to improve the communication between fate and effect experts and 2) how to improve the linking of exposure and effects in the risk assessment of pesticides. During the final session of the Wageningen meeting, draft decision schemes for acute and chronic risk assessment were produced that incorporated the insights of how to link exposure and effects when addressing time-variable exposure regimes of plant protection products. In the postworkshop period, the members of the organizing committee proceeded with improving these decision schemes (among others based on input of the 6 ELINK working groups and information in the scientific literature). Finally, we came to the conclusion that 1 combined decision scheme for acute and chronic risk assessment is sufficient to present the new insight gained on linking exposure to effects (see Section 2.2). In addition, the members of the ELINK organizing committee drafted guidance (Part I, Chapters 1 to 11 of this book) around this final decision scheme for acute and chronic risk assessment, of which this text is the introduction.

The guidance drafted in Chapters 1 to 11 may be considered as part of ongoing developments in aquatic risk assessment. The decision scheme proposed (see Section 2.2) should be revisited and improved when more insight becomes available (by means of modeling and monitoring) on the characteristics of the exposure profiles of pesticides that differ in use patterns and fate properties and on the spatiotemporal extrapolation of effects by means of novel ecotoxicological and ecological modeling tools. Nevertheless, the success of the ELINK workshops should not be underestimated as this was the first meeting in Europe in which fate and effects experts were brought together to play an equal part in problem solving. We sensed an awareness of the need to improve the communication among and between fate and effects experts so that both groups gained a better understanding of each other's issues. It is hoped that the ELINK workshops will pave the way for the 2 disciplines to work more closely to address the increasing complexity of environmental risk assessment, and that this guidance chapter will be a "building block" to achieve this.

It should be noted that similar, and certainly no less important, building blocks to improve the communication between fate and effect experts are provided by the different working group reports incorporated in Part II of this ELINK workshop document (Chapters 12 to 16). The ELINK working group Communication and Education produced "A Novice's Guide to FOCUS Surface Water" (Chapter 12), which is a simplified but very useful guide to exposure assessment using FOCUS surface water modeling approaches. In Chapter 13, a template for information exchange between fate and effects experts is provided as output of the ELINK working group Interaction between Fate and Effect Experts. The ELINK working group Extrapolation Tools produced an overview of approaches and models that can be used to extrapolate the ecotoxicological effects of time-variable exposure regimes to support the risk assessment (Chapter 14). Chapter 15 presents the report of the ELINK working group Characteristics of Water Bodies. The ecological characterization of edge-to-field surface waters is important to inform both the exposure and effect assessment of plant protection products, particularly when adopting scenarios for these surface waters and when designing and evaluating higher-tier tests.

The information provided in Part II of this book was useful when drafting the guidance chapters (Part I). Nevertheless, it should be noted that the views expressed in the working group reports of Part II are those of the authors and were not commented by all ELINK participants. The information in Chapters 1 to 11, however, aims to reflect the consensus view of all ELINK workshop participants. They were all invited to comment on an earlier draft of these guidance chapters, and we received many useful and constructive comments. Nevertheless, we are aware that individual workshop participants expressed other views on a few items. In these cases, a consensus decision was made by the members of the organizing committee of the ELINK workshop, who are also the authors of Chapters 1 to 11 and the editors of this book.

2.2   INTRODUCTION TO THE DECISION SCHEME FOR AQUATIC RISK ASSESSMENT

The proposed decision scheme for acute and chronic aquatic risk assessment of time-variable exposure regimes for plant protection products is presented in Figure 2.1. For your convenience, this decision scheme is also presented in Appendix 4 at the end of this book as a foldout page so that it can be referenced from several locations without paging back and forth. This section aims to 1) introduce a few important concepts that underlie this decision scheme and 2) explain the different decision boxes in this decision scheme (Figure 2.1) by providing short comments and referring to other guidance chapters where more detailed information and guidance can be found.

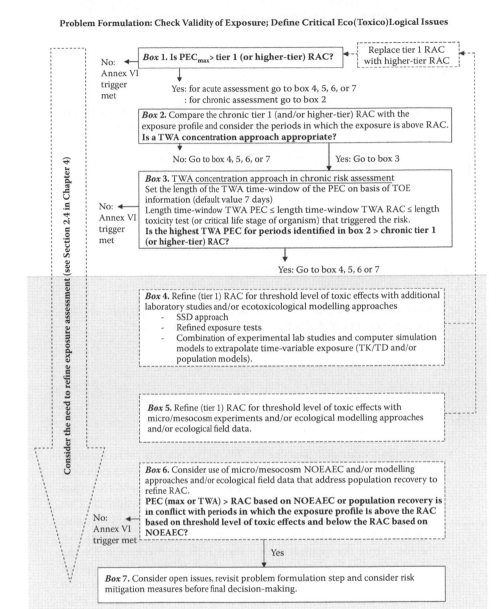

FIGURE 2.1 ELINK decision scheme for acute and chronic aquatic risk assessment for plant protection products. (For explanatory notes, see the following Boxes 1 to 7.)

EXPLANATORY NOTES FOR FIGURE 2.1

BOX 1

- The PEC_{max} can be the highest actual concentration from a range of exposure profiles from different scenarios or from a single-exposure profile (e.g., $PEC_{global\ max}$ from FOCUS step 3) (see Chapters 3 and 4).
- A summary of the current first-tier effects endpoints for acute and chronic risk assessment is presented in Tables 3.1 and 3.2, respectively (Chapter 3).

BOX 2

- Situations for which a TWA approach is not appropriate in the chronic risk assessment are described in Section 3.4 of Chapter 3.

BOX 3

- The default value of 7 days for the TWA PEC is a general recommendation if no specific information about the relation between exposure pattern and time to onset of effect (TOE) for the (relevant life stages of the) organisms that trigger the chronic risk are available. This default value helps to avoid lengthy discussion about the time window for TWA calculations. On the one hand, it is very unlikely that the risks from compounds with a short TOE are overlooked. On the other hand, simply using the duration of the effects test that triggered the chronic risk might be too liberal if the TOE is short. Information on criteria to determine the length of the time window of the TWA PEC and the corresponding TWA RAC is given in Section 3.4 of Chapter 3.
- Usually, the highest TWA PEC for the relevant time period, either from each individual scenario or water body or from the highest of the range of available pertinent exposure profiles, should be used for risk assessment (Chapters 3 and 4).

BOX 4

- Information on the SSD approach in the aquatic risk assessment can be found in Chapter 5. To date, SSDs for plant protection products have been generated mainly with toxicity values from acute tests. Often, not enough chronic NOEC or chronic EC10 (effective concentration that effects 10% of test organisms) values are available to construct a chronic SSD, partly due to the fact that test methods for nonstandard test species are lacking. The HC5 or HC1 (hazardous concentration to 5% or 1% of the tested taxa, respectively) from the distribution curve may be used for setting the RAC (Section 5.4 of Chapter 5).
- For further information on refined exposure tests with the relevant (standard) test organisms or populations (or sensitive life stages), refer to Chapter 6. The exposure regime tested should be guided by exposure predictions (e.g., maximum duration of pulse exposure) or the relevant generalized exposure profile (Chapter 4).
- If the adopted approach comprises a combination of experimental laboratory studies and computer simulation models to extrapolate time-variable exposure, further information on toxicokinetic and toxicodynamic (TK/TD) and population models can be found in Chapters 7 and 9, respectively.

BOX 5

- Information on the model ecosystem approach can be found in Chapter 8 and on linking the threshold concentration for effects to exposure predictions in Section 8.4 of that chapter.
- In the micro/mesocosm experiment, the tested exposure regime should have been guided by exposure predictions or the relevant generalized exposure profile (Chapter 4 and Section 8.3 of Chapter 8). Note that especially due to the different intended uses it might only be possible to test a representative exposure pattern. Furthermore, importance of exposure routes and different mitigation options depends on the regional context.
- Consider if micro/mesocosms contain the relevant populations and community potentially at risk (compare with ecological scenario derived from ecological field data; Chapter 10).
- Ecological models (e.g., population and food web models) may be used for spatiotemporal extrapolation of the threshold levels for toxic effects derived from micro/mesocosm tests (Chapter 9).

BOX 6

- Information on factors that affect recovery of sensitive populations in micro/mesocosm experiments and how to link exposure and effects when recovery is taken into account is presented in Sections 8.5 to 8.7 of Chapter 8.
- In the micro/mesocosm experiment, the tested exposure regime should have been guided by exposure predictions or the relevant generalized exposure profile (Chapter 4 and Section 8.3 of Chapter 8).
- Consider if micro/mesocosms contain the relevant populations and community potentially at risk (compare with ecological scenario derived from ecological field data; Chapter 10).
- Consider if external recovery is important for potentially sensitive populations; if yes, apply modeling (e.g., metapopulation models) or experimentation with a spatial component to refine RAC (see Section 8.6 and Chapters 9 and 10).
- Compare relevant RAC from boxes 1 to 5 and the RAC based on the NOEAEC with exposure profile and evaluate differences in terms of field relevance (see Section 8.7 and Chapter 10).

BOX 7

- Even if the triggers are not met in the procedures described in Boxes 1 to 6 (indicating unacceptable risks), there might be further tools available to refine the risk assessments, for example, the combination of probabilistic exposure and effects assessment approaches. Since these methods, and the implications for linking exposure and effects, were not discussed during the ELINK workshops, no specific guidance on these methods is given here.

2.2.1 REGULATORY ACCEPTABLE CONCENTRATION

The regulatory acceptable concentration (RAC) is the effects assessment endpoint, expressed in terms of a permissible concentration in the environment, that is directly used in the risk assessment by comparing it with the appropriate field exposure estimate (e.g., PEC_{max}). Usually, the RAC is derived by applying an uncertainty factor to the relevant measurement endpoint for the most sensitive species (e.g., EC50 [median effective concentration that affects 50% of the test organisms] and NOEC [no observed effect concentration] from single-species toxicity tests or Effect class concentrations from micro/mesocosm experiments) (for explanation, see Section 8.2 in Chapter 8) or by means of an appropriate (statistical) extrapolation procedure (e.g., SSD [species sensitivity distribution] approach). In boxes 1 to 5 of the decision scheme (Figure 2.1), the RAC represents an estimation of the "threshold concentration for toxic effects" that does not consider ecological recovery of sensitive species. In box 6 of the decision scheme, the RAC is based on the no observed ecologically adverse effect concentration (NOEAEC), in which short-term effects on sensitive measurement endpoints are considered acceptable (e.g., based on Effect class 3 concentrations from micro/mesocosm experiments).

2.2.2 TIME-WEIGHTED AVERAGE CONCENTRATION APPROACH

Effects of plant protection products on aquatic organisms may be similar when exposed for a short time to a greater concentration or for a longer time to a smaller concentration, a phenomenon referred to as "reciprocity" (Giesy and Graney 1989). Reciprocity relates to Haber's law, which assumes that toxicity depends on the product of concentration and time. For example, a 2-day exposure at 10 µg/L may cause the same effects as a 1-day exposure at 20 µg/L or a 12-hour exposure at 40 µg/L, an example of linear reciprocity. Linear reciprocity is the basis of the time-weighted average (TWA) approach, by which exposure concentration is integrated over time (area under the curve, AUC) and then divided by the duration of the toxicity test. When this approach is applied, different exposure patterns with the same AUC are assumed to have the same effects.

Theoretically, reciprocity should only apply where both uptake or elimination of a compound into the test organism (toxicokinetics, TK) and damage or repair processes (toxicodynamics, TD) have reached steady state (Rozman and Doull 2000). In practice, evidence either for or against reciprocity is scarce. Parsons and Surgeoner (1991) studied toxicity of permethrin, fenitrothion, carbaryl, and carbofuran to mosquito larvae (*Aedis aegypti*) and observed the same effects from two 1-hour exposures as from a single 2-hour exposure. Naddy et al. (2000) found that reciprocity held for effects of chlorpyrifos on *Daphnia magna* at large exposure concentrations over short durations, but that reciprocity was not observed for smaller concentrations and longer exposure periods. Repair of damage caused by chlorpyrifos is slow, and the evidence for reciprocity in the work of Naddy et al. thus conflicts with theory. No reciprocity for chlorpyrifos was observed by Ashauer et al. (2007b) when extrapolating from short- to long-term exposures to *Gammarus pulex*. These authors found that the TWA approach based on an acute toxicity test greatly underestimated mortality in longer-term exposure studies, whereas it overestimated mortality caused by pentachlorophenol. In long-term toxicity tests with *Gammarus pulex*, however, Ashauer et al. demonstrated that the TWA concentration approach can be used to extrapolate results of a chronic pulse test to other chronic exposures for both chlorpyrifos and pentachlorophenol. This observation supports the use of the TWA concentration approach in chronic risk assessments.

Based on the paucity of available data, it should be noted that the inference of reciprocity to underpin the TWA approach is a simplifying approximation, and that its degree of appropriateness will depend on the chemical, the endpoint, and the exposure scenario investigated. In particular, the reciprocity relationship may break down for pulses of very short duration. Generally, the TWA approach is not used in acute risk assessment, whereas the inherent approximation has been deemed acceptable in many situations in chronic risk assessment. The longer duration of chronic tests implies a greater probability that TK and TD will approach steady state by the end of the study period. Consequently, in the ELINK decision scheme (Figure 2.1) the TWA approach is adopted in the chronic risk assessment only, at least when certain conditions are met.

2.2.3 HOW TO USE THE DECISION SCHEME

The decision scheme (Figure 2.1) is centered around a "refine exposure assessment" arrow and several risk assessment boxes (1 to 7). The decision boxes 1 to 3 (white area in Figure 2.1) all relate to the first-tier risk assessment approach, for which guidance can be found in Chapter 3. More detailed guidance for the refine exposure assessment arrow is provided in Chapter 4. Guidance for the higher-tier risk assessment boxes 4 to 6 (shaded area in Figure 2.1) can be found in Chapters 5 to 10. The aim of these chapters is not only to clarify the decision scheme but also to improve the communication between fate and effects experts by providing essential background information on different tiered risk assessment procedures currently in use and on promising novel risk assessment tools that may be used in the near future.

The boxes in the decision scheme (Figure 2.1) contain criteria that help the user to decide on the next appropriate step. In the explanatory notes (see their specific text box in this chapter), further information is given on these criteria and on the chapters where further guidance can be found. Whenever a box number is referred to in the text of Chapters 3 to 11, it relates to the box in the decision scheme.

For a proper interpretation of the decision scheme (Figure 2.1), please consider the following comments:

- In general, both acute and chronic risks need to be assessed. The assessment that results in the highest risk will be most crucial for decision making. Due to considerable overlap in approaches and to ease understanding, a combined scheme for acute and chronic risk assessment is presented in Figure 2.1.
- The decision-making scheme is considered relevant for the aquatic risk assessment of the normal agricultural use of the majority of plant protection products; adjustments might be needed for special cases.
- Setting of uncertainty factors to derive RACs on the basis of results of higher-tier ecotoxicological studies was not under the remit of ELINK. Therefore, no advice on the size or numerical values of possible uncertainty factors is given in the boxes of the decision scheme.
- Different methods are available to refine the exposure assessment. Especially, step 4 of FOCUS delivers such methods, but there are also approaches that are specific for certain member states. The downward arrow on the left hand in the scheme stands for all these approaches. Further information can be found in Chapter 4.
- Risk mitigation measures might be introduced at every stage of the scheme if appropriate.

3 Tier 1 Aquatic Risk Assessment for Plant Protection Products in Europe

CONTENTS

3.1   BACKGROUND

The characterization of risk to aquatic organisms is carried out by comparing effects endpoints of indicator aquatic species to exposure concentrations that might be expected in aquatic systems (both the water column and sediment). In the European Union, when carrying out these assessments for plant protection products, effects endpoints are divided by the exposure concentrations, which can be expressed as a "toxicity–exposure ratio" (TER). For the European Union, the guidance document that currently outlines the detail for this aquatic risk characterization is the aquatic guidance document SANCO/3268/2001 revision 4 (EC 2002). This document provides guidance on how to use the data provided to address the EU data requirements for the authorization of plant protection products to formulate a risk characterization or risk assessment. Supporting this guidance document are other guidance documents (Forum for the Co-ordination of Pesticide Fate Models and Their Use [FOCUS] 2001, 2007a, 2007b) and national guidance provided by member state regulatory authorities responsible for the registration of plant protection products. FOCUS guidance documents provide detailed procedures for calculating predicted environmental concentrations (PECs) of pesticides in water and sediment to which nontarget aquatic organisms may be exposed. The output of these calculations is a range of pesticide concentrations that vary both temporally and spatially as a consequence of the pattern of use of a plant protection product and the geoclimatic situation in the area of use. The principal aim of ELINK (EU Workshop on Linking Aquatic Exposure and Effects in the Registration Procedure of Plant Protection Products) was to develop guidance on how to link the effects thresholds with the range of varying patterns of exposure concentrations that occur in water bodies in the agricultural landscape.

In the European Union, the uniform principles for decision making on product authorizations (currently Annex VI of Directive 91/414/EEC) and the aquatic guidance document (EC 2002) define effects endpoints and TER triggers (acute and chronic) for a base set of indicator species. Where TERs exceed the triggers defined in Annex VI, the risk is considered to be low, and further refinement of the risk being assessed is not required. This situation is regarded as a tier 1 risk assessment. Higher tiers of assessment are required if TERs are lower than those defined in Annex VI. This chapter of the ELINK guidance provides a brief summary of the procedure for carrying out this tier 1 assessment and provides the details that support the nonshaded upper portion of the decision scheme outlined in Figure 2.1 (see Chapter 2). Figure 2.1 reflects what was agreed at the workshop. In the context of the tier 1 risk assessment, the pertinent uncertainty factor is the TER trigger for the base set of indicator species, which are currently (under 91/414/EEC) as summarized in Tables 3.1 (acute risk assessment) and 3.2 (chronic risk assessment).

3.2   EXPOSURE CONCENTRATIONS: PREDICTED ENVIRONMENTAL CONCENTRATIONS IN SURFACE WATER AND SEDIMENT

Further discussion on the calculation of the exposure concentrations PECsw and PECsed (predicted environmental concentration in surface water and predicted environmental concentration in sediment, respectively) and the profiles that result is provided in Chapter 4. For first-tier assessments, PECs are calculated utilizing base set physicochemical data required by legislation on the chemical substance that is being assessed[1] combined with data on the use pattern of the substance in a variety of agro(geo)climatic scenarios. PECs are calculated in a series of steps, each with an increasing level of refinement of exposure. At Step 1, a very simple definition of the environment is used in the calculation procedure, and the assumptions made are accepted as representing conservative values that would cover a very wide range of geoclimatic conditions. At higher steps (in total 4 steps are defined by EU-level guidance), more realistic (but still simplified) scenario definitions are used that encompass narrower ranges of geoclimatic conditions. Consequently, a number of sets of PECs are produced at these higher steps, so a number of risk characterizations that represent different combinations of soil, climate, and landscape definitions become available. In the context of the ELINK assessment scheme (Figure 2.1), all risk assessments that use officially[2] defined scenarios without modification, standard agreed or defined calculation methods, and the base set[3] of physicochemical properties for PEC calculations are considered as tier 1 risk assessments. Following

TABLE 3.1

Summary of Current[a] Acute (Short-Term) EU Ecotoxicology Testing Requirements and Current EU Endpoints and Triggers Pertinent for Tier 1 Assessment

Test/Indicator Species	Encompassed Organisms Exposure Regime(s) Possible in the Test	Effects Endpoint	TER Trigger or Divisor for Effect Threshold
Rainbow trout (*Oncorhynchus mykiss*)	Aquatic vertebrates (e.g., fish and amphibians) Static, semistatic, or flow through possible	LC50	100
Warm-water fish species	Aquatic vertebrates (e.g., fish and amphibians) Static, semistatic, or flow through possible	LC50	100
Daphnia magna	Aquatic invertebrates Usually static although flow through may be necessary for volatile test substances	EC50	100
First-instar *Chironomus riparius* (48-hour water only study) Only required for insecticides for which toxicity to *Daphnia* is low (*Daphnia* EC50 >1 mg/L, NOEC > 0.1 mg/L) or for insect growth regulators	Aquatic invertebrates Usually static although flow through may be necessary for volatile test substances	EC50	100

[a] Directive 91/414/EEC, SANCO/3268/2001 revision 4 (European Union 2002).

EU-level assessment, this means all FOCUS scenarios defined at Steps 1 to 3. Step 4 calculations that maintain Step 3 scenario definitions but introduce exposure mitigation in agreed standardized approaches[4] may also be considered as tier 1 risk assessments. Although this is not in complete agreement with the terminology used in FOCUSsw guidance, it reflects EU-level regulatory practice. At the member-state level, this includes national officially defined scenarios and calculation tools or methods that are published or otherwise made available to applicants in an official way. Note, however, that authorities may publish exposure scenarios that they choose to define as not appropriate for use in tier 1 risk assessments. Any exposure concentration that is calculated on the basis of additional (to the base set) properties of the chemical or uses scenario definitions that are not published or otherwise defined as an official standard scenario by a regulatory authority appropriate for use in a first-tier risk assessment is considered as higher tier (not tier 1).

From the environmental exposure perspective, tier 1 risk assessments for a formulated (marketed) product as opposed to the active substance require a simplified approach as there will only be exposure to the coformulants from aerial deposition of spray drift from a single application.[5] When a risk assessment is carried out using effect thresholds for a formulation, in a tier 1 risk assessment, the TER should only be calculated using an initial PEC from a single application and should not include any input resulting from runoff or drainage from the soil compartment. When a formulation has a lower effect threshold than the active substance or substances contained in the formulation, by 1 order of magnitude or greater in acute tests, appropriate chronic effects tests may be required.[6] In this situation, the exposure assessment required to complete the chronic risk assessment would be expected to be higher tier as fate and behavior data on the pertinent coformulants would be needed for such an assessment.

The EU risk assessments also need to take account of situations for which groundwater becomes surface water. The PEC groundwater (PECgw) calculated in the regulatory exposure assessment for metabolites can be used directly for calculating the TER to characterize both acute and chronic risk.[7] In EU exposure assessments, these groundwater PECs represent annual average concentrations. Consequently, when the PECsw, which identifies low risk, is greater than the PECgw for the same chemical moiety (or moieties), separate calculations specifically using PECgw are not necessary.

3.3   ACUTE (SHORT-TERM) RISK ASSESSMENT

The following notes support the acute risk assessment according to the scheme presented in Figure 2.1 (see Chapter 2 and Appendix 4):

Box 1 in Figure 2.1: Acute TER = EC50 or LC50/PEC_{max}

where
   Acute TER = Acute (short-term) toxicity exposure ratio
   EC50 = Effective concentration for 50% of individuals in a test group
   LC50 = Lethal concentration for 50% of individuals in a test group
   PEC_{max} = The highest predicted environmental concentration for a defined environmental scenario
   TER > Trigger (100) = Low risk due to short-term exposure: Perform chronic risk assessment (revisit decision scheme; go to Section 3.4).
   TER < Trigger (100) = Higher-tier acute risk assessment triggered (see Boxes 4, 5, 6, or 7 in Figure 2.1)

Or, using ELINK "effects threshold" (equal to the regulatory acceptable concentration [RAC] based on assessments not considering ecological recovery) terminology,

   If PEC_{max} < EC50/100 or LC50/100 = Low risk
   If PEC_{max} > EC50/100 or LC50/100 = Higher-tier assessment triggered

A summary of current (91/414/EEC) tests and endpoints is shown in Table 3.1.

3.4   CHRONIC (LONGER-TERM) RISK ASSESSMENT

The following notes support the chronic risk assessment outlined in Figure 2.1 (see Chapter 2 and Appendix 4).

Box 1 in Figure 2.1: Chronic TER = NOEC or EC50/PEC$_{max}$

where
Chronic TER = Chronic (long-term) toxicity exposure ratio
NOEC = No observed effect concentration

When determining the EC50 for algae and *Lemna* sp., where the exposure concentration in the test system is not maintained, if the sterile aqueous hydrolysis or aqueous photolysis DT50 (period required for 50% disappearance) is (are) more than 3.1 times, or DT90 is (are) more than 10.3 times, the test duration[8] and the test substance did not partition to test vessel walls, then the EC50 should be calculated based on the nominal or initial measured concentration. If these criteria are not met, then the EC50 should be calculated based on a mean measured concentration over the study duration.

TER > Trigger (10) = Low risk

TER < Trigger (10), move to Box 2 of the decision scheme (Figure 2.1)

Or, using ELINK effects threshold terminology,

If PEC$_{max}$ < NOEC/10 or EC50/10 = Low risk

If PEC$_{max}$ > NOEC/10 or EC50/10 = Move to Box 2 in Figure 2.1 (see also the box in this chapter discussing Box 2).

BOX 2

Consider if the use of a time-weighted average (TWA) PEC is appropriate or not.
Use the following checklist to decide:

- Do not use with algal study endpoints at tier 1. Plot the chronic effects threshold onto the exposure profile associated with each pertinent scenario, conclude that TWA concentrations are not appropriate at tier 1, and move to a higher-tier risk assessment (choice of Box 4 or Box 5 in Figure 2.1). At higher tiers, a TWA PEC approach can only be justified when time-to-effect information can be elucidated for both algae and macrophytes.

- For effects studies in which the exposure is not maintained (often the case in tier 1 tests to determine the EC50 for *Lemna* or the 20- to 28-day *Chironomus* test to determine a NOEC): Was the sterile aqueous hydrolysis and/or aqueous photolysis DT50 greater than 3.1 times or DT_{90} greater than 10.3 times the test duration, and did the test substance not partition to test vessel walls? If yes, move to Box 3 in Figure 2.1. If no, plot the chronic effects threshold that should be expressed on a mean measured concentration basis onto the exposure profile associated with each pertinent scenario, conclude that TWA concentrations are not appropriate at tier 1, and move to a higher-tier risk assessment (choice of Box 4 or Box 5 if it will be necessary to use a peak and not TWA PEC whenever the higher-tier assessment utilizes a test with a static exposure regime).
- Is the effect endpoint in the chronic test from which the NOEC is derived based on a developmental process during a specific sensitive life-cycle stage of the organism, or is there any evidence of an endocrine disruption effect? If yes, plot the chronic effects threshold onto the exposure profile associated with each pertinent scenario and move to higher-tier risk assessment (choice of Box 4 or Box 5 in Figure 2.1). Do not use a TWA PEC even at higher tiers unless there is very strong justification based on good evidence why this may be acceptable (e.g., the exposure never occurs when the sensitive life-cycle stage occurs).
- Is the effect endpoint in the chronic test from which the NOEC is derived based on mortality occurring early (in first 96 hours) in the study, or is the acute-to-chronic ratio (EC50 or LC50/NOEC) less than 10 when the endpoint is mortality? If yes, plot the chronic effects threshold onto the exposure profile associated with each pertinent scenario and move to a higher-tier risk assessment (choice of Box 4 or Box 5 in Figure 2.1). Do not use a TWA PEC even at higher tiers.
- If latency of effects[9] has either been demonstrated or might be expected due to the mode of action, plot the chronic effects threshold onto the exposure profile associated with each pertinent scenario and move to higher-tier risk assessment (choice of Box 4 or Box 5 in Figure 2.1). Do not use a TWA PEC even at higher tiers.

Box 3 in Figure 2.1: Default procedure: Chronic

TER = NOEC or EC50/PEC TWA of 7 days

TER > Trigger (10) = Low risk

TER < Trigger (10) = Higher-tier assessment triggered

(choice of Box 4 or Box 5 in Figure 2.1)

If a 7-day TWA was used for the PEC and the duration of the chronic standard toxicity test is longer than 7 days, consider lengthening the TWA period of the PEC to the time to the observed effect if scientifically justified (e.g., on the basis of time-to-onset-of-effect [TOE] information in the chronic test). (Note that the observations necessary to determine TOE will often not be made in standard guideline chronic tests; hence, this procedure will usually be considered higher tier.) The length of the TWA period should never be longer than the study duration or the relevant chronic toxicity test that triggered the risk (e.g., in tier 1 tests, 72 hours for algae, 7 days for *Lemna*, 21 days for *Daphnia*, 20 to 28 days for *Chironomus*, and 28 days for fish). If TER is greater than trigger (10), then it is low risk; if TER is less than trigger (10), then higher-tier assessment is triggered (choice of Box 4 or Box 5 in Figure 2.1).

3.4.1 ADDITIONAL CRITERIA, SCREENING FOR BIOCONCENTRATION POTENTIAL

If the log Pow (octanol-water partitioning coefficient) of a substance is less than 3, a fish bioconcentration study is not required. A fish bioconcentration study to determine a bioconcentration factor (BCF) is always required for substances that have a log Pow of 3 or greater unless it can be justified that exposure leading to bioconcentration cannot occur. This is only likely when either the substance is not stable in water and is effectively degraded in sediment or its use pattern dictates that it will not enter a water course. To make the case that a pattern of use and substance property combination would justify not doing the fish bioconcentration study, the fate and behavior aquatic exposure assessment (i.e., PEC profile, for example; at the EU level, FOCUS output profiles) would need to indicate just 1 or 2 short exposure events per year in the water column (sum of event durations is less than 10 days) and that the number and duration of peaks in the sediment are few and of short duration (again sum of event durations per year is less than 10 days, with sediment concentrations predicted to be zero between events). To be clear, the partition of a substance into the sediment cannot be an argument to conclude that there is low potential of bioconcentration in aquatic species since some substances may accumulate in living organisms even when a sediment phase is present in the system (particularly if they live or feed in the sediment). As a consequence, the dissipation behavior of the substance as a whole must be considered, not just what happens in the water column.

1. If a substance either is or would be classified as "not readily biodegradable"[10] and has a log Pow of 3 or greater and a BCF in fish of less than 100, then a low risk of bioconcentration can be concluded at the first tier of assessment. However, if with these properties, the BCF of 100 or greater, a higher risk is indicated with a need to address possible biomagnification in a higher-tier assessment. Go to Box 5 in Figure 2.1 (also see Section 9.2.2 of Chapter 9 and Section 14.2.4 of Chapter 14).

2. If a substance either is or would be classified as "readily biodegradable"[11] and has a log Pow of 3 or greater and BCF less than 1000, then a low risk of bioconcentration is indicated at the first tier of assessment. However, if with these properties the BCF is 1000 or more, a higher risk is indicated with a need to address possible biomagnification in a higher-tier assessment. Go to Box 5 in Figure 2.1 (also see Section 9.2.2 of Chapter 9 and Section 14.2.4 of Chapter 14).

A summary of chronic tests and current EU endpoints is shown in Table 3.2.

TABLE 3.2
Summary of Current[a] Chronic (Long-Term) EU Ecotoxicology Testing
Requirements and Current EU Endpoints and Triggers Pertinent
for Tier 1 Assessment

Test/Indicator Species	Encompassed Organisms Exposure Regime(s) Possible in the Test	Effects Endpoint	TER Trigger or Divisor for Effect Threshold
Rainbow trout (*Oncorhynchus mykiss*) 28-day test (21-day test may sometimes be accepted) with endpoints for survival, growth, and behavior reported Note: not required if satisfactory ELS (early life-stage toxicity) or FLC (full life-cycle toxicity) tests are available	Aquatic vertebrates (e.g., fish and amphibians) Semistatic or flow through possible	NOEC	10
Bioconcentration in fish Not required if log Pow is < 3	All aquatic organisms Flow through	BCF uptake and depuration rate constants	Not a TER trigger but if substance is "not readily biodegradable" and BCF > 100, higher-tier assessment triggered[b]
Fish ELS test Required when BCF is > 100 or acute LC50 < 0.1 mg/L; not required if a satisfactory fish FLC test is available	Aquatic vertebrates (egg, fish, and amphibians) Semistatic or flow through possible	NOEC	10
Fish FLC test; only required if BCF > 1000 with low depuration (<95% in 14 days) and the test substance is stable in water or sediment (DT90 > 100 days). In addition, it should also be considered if studies (e.g., in mammalian toxicology) indicate effects on reproduction or the endocrine system. If this is indicated, the study is required.	Aquatic vertebrates (egg, fish, and amphibians) Semistatic or flow through possible	NOEC	10

TABLE 3.2 (CONTINUED)

Summary of Current[a] Chronic (Long-Term) EU Ecotoxicology Testing Requirements and Current EU Endpoints and Triggers Pertinent for Tier 1 Assessment

Test/Indicator Species	Encompassed Organisms Exposure Regime(s) Possible in the Test	Effects Endpoint	TER Trigger or Divisor for Effect Threshold
Chironomus riparius (20- to 28-day study with sediment, sediment or water spiked design possible[c]) Only required if substance partitions to sediment (>10% AR in sediment in lab; sediment water study at or after 14 days) and NOEC *Daphnia* < 0.1 mg/L or where partitioning is less than this if FOCUS step 2 PECsw gives a TER to *Daphnia* < 100 acute or < 10 chronic	Sediment dwellers Water spiked test static or semistatic, exceptionally flow through possible Sediment spiked test static	NOEC	10
Daphnia	Aquatic invertebrates Semistatic or flow through possible	NOEC	10
Chironomus riparius (water spiked 20- to 28-day study with sediment present[d]) Only required if 48-hour EC50 (insect) < 1/10 48-hour EC50 (*Daphnia*) or if TER (insect acute < 100) or for insect growth regulators	Aquatic invertebrates Static or semistatic possible	NOEC	10
Green algae	Plants Algae Static	EbC50 (median effective concentration on biomass) or ErC50 (median effective concentration on growth rate) for period of exponential growth	10
Blue-green algae or diatoms Only required for herbicides and plant growth regulators	Plants Algae Static	EbC50 or ErC50 for period of exponential growth	10

(continued)

TABLE 3.2 (CONTINUED)
Summary of Current[a] Chronic (Long-Term) EU Ecotoxicology Testing Requirements and Current EU Endpoints and Triggers Pertinent for Tier 1 Assessment

Test/Indicator Species	Encompassed Organisms Exposure Regime(s) Possible in the Test	Effects Endpoint	TER Trigger or Divisor for Effect Threshold
Lemna sp. Only required for herbicides and plant growth regulators	Plants Macrophytes Static, semistatic or flow through possible	ErC50 or EbC50, frond number EC50, or other endpoint EC50	10

[a] Directive 91/414/EEC, SANCO/3268/2001 revision 4 (EC 2002).
[b] In addition, if the substance is "readily biodegradable" and the BCF > 1000, higher-tier assessment is triggered.
[c] TER should be calculated by comparing the NOEC expressed as a sediment concentration (usually measured average concentration over the study duration or time to effect if this is shorter) with the pertinent PECsed for each pertinent scenario (i.e., both in units of µg/kg dw (dry weight) or mg/kg dw). However, in some older water spiked studies in which sediment concentrations were not measured, this will not be possible. In this situation, when calculating TER the PECsed should be expressed as an equivalent water concentration taking into account the mass of sediment and water volume in the study design of the available ecotoxicology test; then, this concentration can be compared to the NOEC from this study expressed as a water concentration (i.e., both expressed in units of µg/L or mg/L). The PECsw for the water column in the scenario (in µg/L or mg/L) should not be used to calculate this TER, which addresses sediment dweller risk.
[d] TER should be calculated by comparing the NOEC expressed as a water concentration (usually measured average concentration over the study duration or time to effect if this is shorter) with the pertinent PECsw for each relevant scenario (i.e., both in units of µg/L or mg/L) to calculate this TER, which addresses aquatic invertebrate risk.

NOTES

1. The chemical may be the active substance, a breakdown or transformation product of the active substance, or for some assessments the formulated product.
2. Defined by regulatory authorities of governments as appropriate to be used at the first tier or in agreement with EU guidance documents.
3. Those specified as required by the EU plant protection product data requirements (i.e., currently Sections 2 and 7 of Annex II, Part A of Directive 91/414/EEC).
4. Currently, as outlined in FOCUS (2007a), at the EU level these are restricted to where spray drift reductions are implemented following the procedures defined by FOCUS and using the spray drift calculator tool provided in SWASH (Surface Water Scenarios Help software) for drift reductions up to a maximum of 95% compared to drift at the base distance defined by FOCUS; runoff reductions for which both substance mass loss and runoff water volume are reduced (including the defined treated proportion of the upstream

catchment) up to a maximum of 90% but with greater uncertainty for less strongly adsorbed substances (see Sections 3.5.2 and 3.5.3 of FOCUS 2007a and European Food Safety Authority [EFSA] 2006b, which can be found appended to FOCUS 2007a); and where drainage inputs (both substance mass loss from soil and drainage water volume) are reduced by a maximum of 90%.

5. There would be chromatographic-type separation and differential degradation rates of the different formulation components in the soil compartment and receiving water body.

6. Section 2.5.3 SANCO/3268/2001 revision 4 (final) (EC 2002).

7. According to SANCO/3268/2001 revision 4 (EC 2002), a further dilution factor of 10 may be taken into account, but this is not appropriate in arid regions when all flow in surface water bodies originates from groundwater sources.

8. Time equivalent to an 80% measured initial value that SANCO/3268/2001 revision 4 states would allow a nominal initial dosed concentration to be used to define the effects endpoint assuming a first-order rate of breakdown.

9. Delayed responses are observed in acute studies after exposure to a stressor ends or in studies with long-term exposure regimes, observations over the whole life cycle or observations of offspring indicate a delayed response.

10. As defined by the Organization for Economic Cooperation and Development (OECD), "ready biodegradability" test guidelines (No. 301 A-F or the Headspace Test [TG 310]).

11. As defined by OECD, "ready biodegradability" test guidelines (No. 301 A-F or the Headspace Test [TG 310]).

4 Generalized Exposure Regimes and Tiered Exposure Assessment for Plant Protection Products

CONTENTS

4.1   INTRODUCTION

In the ELINK (EU Workshop on Linking Aquatic Exposure and Effects in the
Registration Procedure of Plant Protection Products) risk assessment scheme
(Figure 2.1; see Chapter 2 and Appendix 4), the large arrow on the left contains
the text "consider the need to refine exposure assessment." In fact, this arrow refers
to a tiered exposure assessment scheme. The output of this tiered exposure assessment scheme needs to be a realistic (worst-case) and ecotoxicologically relevant
concentration (ERC) or exposure curve for the environmental compartment of
concern (e.g., water, interstitial water, or sediment) that can be fed into the different boxes of the decision scheme. There are general and specific options to
revisit and refine exposure assessments. The general options can be applied at
any point in the aquatic risk assessment and are intended to be more realistic and
thus less conservative with respect to exposure. Refined exposure may also aim
to provide a more tailored exposure assessment as laid down in the Forum for the
Co-ordination of Pesticide Fate Models and Their Use (FOCUS; 2007a, 2007b).
These general and specific options for refined exposure assessment may result in
predicted environmental concentrations (PECs) that are below the lower-tier regulatory acceptable concentration (RAC) and may hence negate the need for higher-
tier effects assessment.

A crucial step in the risk assessment is to define which type of concentration
from the exposure assessment is needed as the "exposure input" to the effect tiers.
The choice should be based on ecotoxicological considerations because this should
provide the best correlation with ecotoxicological effects. This type of concentration
is defined as the ERC. For aquatic organisms, the ERC could, for example, be the
maximum concentration over time, a time-weighted average (TWA) concentration
(for a certain time interval) in the ambient water, or it could be a peak or TWA of
the concentration in the organism of concern (critical body burden concept). For
sediment-dwelling organisms, the ERC could, for example, be the peak or TWA
concentration in the water column, the pore water, the bulk sediment, or even the
organism. In this context, there are 2 distinct exposure estimates that may be used

for pesticide risk assessment: 1) exposure estimates that relate to exposure in the field (i.e., PEC) and 2) exposure estimates that relate to exposure in ecotoxicological studies (test concentration profile). The ERC needs to be consistently applied so that field exposure and test concentration profiles can be compared as readily as possible (Boesten et al. 2007).

From a historical perspective and for identifying potential intrinsic hazards, lower-tier ecotoxicological tests have been designed with the purpose of maintaining the applied exposure concentration, if necessary, through renewal of the test substance or using flow-through systems. Such regimes allow the derivation of worst-case concentration–response relationships to determine the inherent toxicological properties of a compound. The Organisation for Economic Co-operation and Development (OECD) test guidelines often require this type of exposure, and it does not fall under the remit of ELINK to amend this. However, in reality, time-variable exposure patterns of pesticides are often the rule rather than the exception, certainly in edge-of-field surface waters. For that reason, it is appropriate to consider the surface water exposure pattern in the refined steps of the risk assessment. Time-varying exposure in the field may be simulated with various levels of complexity, and these estimates not only may be used directly in the risk assessment but also may serve as a basis to simulate more realistic time-variable exposure regimes in ecotoxicological studies. It is, however, neither practically nor for animal welfare or financial reasons feasible to study all possible exposure regimes. A pragmatic way forward would be to identify generalized exposure regimes typical of the different FOCUS surface water scenarios (ponds, streams, and ditches) that can be used as a basis for exposure regimes in ecotoxicological studies.

4.2   TIERED FOCUS APPROACH CURRENTLY USED IN THE EXPOSURE ASSESSMENT FOR PESTICIDES AT THE EU LEVEL

This section aims to provide information on the tiered approach currently used in the exposure assessment for pesticides and to describe and discuss different cost-effective options that can be used. It also presents approaches to derive generalized exposure regimes.

One of the options for refinement of the exposure profile is to introduce 1 or more mitigation measures to reduce overall levels of exposure (FOCUS 2007a, 2007b). Problem formulation needs to consider that the inclusion of mitigation measures can also significantly alter the pattern of exposure. For example, reduction of exposure via spray drift using a no-spray buffer zone or a vegetated buffer zone to mitigate surface runoff may greatly reduce the significance of short-term peaks in exposure that typically arise from spray drift and runoff but not those resulting from drainage. Restricting applications of a mobile herbicide to exclude autumn treatment may reduce the overall importance of exposure via drain flow and influence the timing of peaks in exposure.

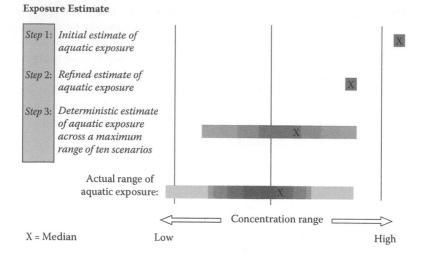

FIGURE 4.1   Conceptual relationship between FOCUS surface water Steps 1, 2, and 3.

The FOCUS Surface Water Scenarios Working Group (FOCUS 2001) has implemented 3 sequential steps for modeling aquatic exposure to pesticides, all of which fit into Tier 1 risk assessment:

Step 1: a simple spreadsheet calculation intended to provide conservative aquatic concentration estimates somewhat higher than would actually be observed

Step 2: a refined spreadsheet calculation intended to represent the high end of actual aquatic exposures

Step 3: mechanistic modeling of drift, drainage, runoff, and erosion coupled with aquatic fate for scenarios designed to capture a range of realistic worst-case conditions for the European agricultural area

This sequence of steps is represented schematically in Figure 4.1. Higher-tier exposure assessments are conducted at Step 4.

Step 1 and 2 calculations include a number of conservative, simplifying assumptions, such as assuming 2% to 10% instantaneous aquatic loading from runoff or drainage as well as a single set of fixed dimensions for the receiving water body. Such assumptions make it possible to determine an initial concentration in a water body using simple algebraic equations. However, due to the conservative nature of these assumptions, the estimated concentrations from Steps 1 and 2 are likely to be higher than or at the very top end of the distribution of actual environmental concentrations. In practice, the Step 1 to 2 calculator is likely to provide PECsw (PEC in surface water) values leading to acceptable risk for active substances and metabolites with low aquatic toxicity.

The more mechanistic calculations performed at Step 3 are an attempt to generate realistic worst-case aquatic concentrations (as defined by the FOCUS Surface Water Scenarios Work Group). However, it should be recognized that even Step 3 calculations include a number of conservative assumptions, such as the use of minimal buffer zones between crop and water for evaluation of spray drift and delivery of edge-of-field runoff directly into surface water, ignoring photolysis as a pesticide degradation process, and adopting worst-case environmental default values for the PEC predictions in the ditch, stream, and pond scenarios (e.g., low organic matter content in sediment and no macrophyte biomass). As a result of these assumptions, the predicted concentrations are expected to be toward the top end of the distribution of concentrations that would be observed across the usage area; thus, these scenarios are assumed to generate "reasonable worst-case" aquatic concentrations.

The goal of performing Step 4 surface water modeling is either to provide a more accurate estimate of exposure concentration likely under actual usage conditions or to evaluate the influence on exposure of 1 or more mitigation options. The proposed refinements can be divided into 3 types of change:

1) Relatively straightforward changes to individual model parameters to alter chemical properties, application rates, application dates, loadings from different emission routes, or system properties that affect degradation or dissipation, such as hydrology, sediment organic matter content, and macrophyte biomass
2) Changes to the modeling to incorporate the use of a risk mitigation measure
3) More complex refinements that might involve creation of new scenarios, use of chemical monitoring data, or the application of probabilistic approaches or distributed catchment models

The FOCUS working group on landscape and mitigation factors in ecological risk assessment (FOCUS 2007a, 2007b) has produced detailed documentation to support step 4 exposure modeling, and this material is not repeated here. Some refinements applied at Step 4 are standardized and have been used so routinely that they are generally included in Tier 1 risk assessment; an example of this is the inclusion of mitigation for spray drift using no-spray buffer zones. Other Step 4 refinements involve relatively new approaches (e.g., catchment-scale modeling) or may be developed specifically for a single product (e.g., new scenarios). In such instances, the exposure estimate needs to be evaluated on a case-by-case basis, and the risk assessment has clearly moved beyond Tier 1. Assessments at the catchment scale are currently not used under the umbrella of 91/414/EEC, which has its focus on risk assessment in edge-of-field surface waters.

The spatial scale of exposure assessment based on the FOCUS surface water scenarios is clearly defined as being edge of field for 3 types of water body (streams, ditches, and ponds) with standard dimensions (FOCUS 2001). It is generally anticipated that Step 4 estimates will also relate to small edge-of-field water bodies. However, catchment modeling and monitoring offer the possibility to assess exposure in multiple water bodies and thus extend the spatial scale of the analysis (FOCUS 2007a, 2007b). Using output from catchment exposure models to evaluate unacceptable effects, however, also requires that the protection goals be defined at a larger spatiotemporal scale.

4.2.1   WHAT TYPES OF EXPOSURE OCCUR IN THE FIELD?

4.2.1.1   Exposure in Flowing Waters

A review was undertaken of research and monitoring studies that characterized concentrations of isoproturon in surface waters (Ashauer and Brown 2007). Evidence from field studies at multiple scales suggests that exposure of nontarget aquatic organisms to pesticides normally occurs as pulses or at fluctuating concentrations. An example is given in Figure 4.2. Constant concentrations over time must be considered as an exception that only occurs for special cases of compound properties in combination with special rainfall patterns and hydrology of the water body. Pesticide concentrations in ditches or streams vary due to physicochemical substance properties, hydrological dilution, dissipation and sorption of the compound, and time-varying input that is driven by application timing, rainfall patterns, crop growth stage, and landscape characteristics such as soil type, slope, and connectivity to surface water.

Ashauer and Brown (2007) also provided a number of examples at the edge-of-field scale that recorded 1) pulses of chlorpyrifos from spray drift and runoff in surface waters adjacent to Italian vineyards and 2) drainflow-related pesticide concentrations in drainage outlets or ditches. The large number of drainage studies (covering a variety of locations and soils) demonstrates the wide range of pesticide sorption and degradation characteristics that are associated with multiple peaks of pesticide in water bodies. River-monitoring data show that the pulsed/time-varying patterns of exposure are conserved as the scale of observation increases. Monitoring and field mobility studies have tended to focus to a large extent on relatively mobile compounds (primarily herbicides) that exhibit larger exposure concentrations in surface waters. There are a number of studies that included more strongly sorbed

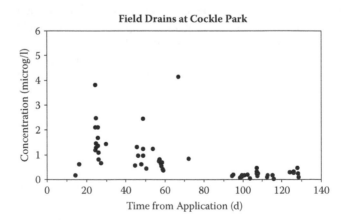

FIGURE 4.2   Concentrations of isoproturon measured in a research program in the United Kingdom.

compounds (e.g., pendimethalin, trifluralin, diflufenican) and demonstrated similar temporal variation in exposure in the water column, albeit at smaller concentrations (Ashauer and Brown 2007).

These findings were confirmed by evaluating almost 4000 samples taken from 4 sites in Ohio over a 10-year time period (Crawford 2004). Concentrations of strongly sorbing compounds in surface waters showed an extremely high temporal variation, as levels that are above the limit of quantification often only occur for very short time periods, that is, for total time spans of a few days per year. It was shown, for example, that the insecticide chlorpyrifos only occurred in about 2.5% of all the samples taken, while less-adsorbing compounds like atrazine or metolachlor appeared in 75 to 89% of the samples.

4.2.1.2 Exposure in (Semi)Static Waters

Monitoring data in Europe are almost exclusively available for flowing water bodies. Patterns of exposure in static waters are expected to be less temporally dynamic. However, slowly flowing waters (e.g., in drainage ditches) are generally most intimately associated with arable agriculture and often drive the risk assessment due to their relatively small size and thus low potential for dilution of pesticide residues.

4.2.2 Do Simulated Exposure Profiles Mimic Field Observations?

Comparisons have been undertaken between exposure simulated with the FOCUS SWS (surface water scenarios) and field observations for isoproturon and sulfosulfuron in the United Kingdom (Ashauer and Brown 2007). Data sets for concentrations of pesticides in drainflow, streamflow, and 2 UK rivers were subjected to time series analysis to quantify their main characteristics. FOCUS surface water simulations were undertaken and analyzed to determine similarities and discrepancies with measured behavior. The analyses focused on the herbicides isoproturon and sulfosulfuron due to availability of detailed concentration profiles for these compounds. Measured data showed that characteristics of time series varied considerably for the same compound at different locations and in different years. Nevertheless, comparison between isoproturon concentrations measured at edge of field and in small streams with simulated data showed broad correspondence with respect to interval between peaks, peak duration, and to a lesser extent, the decrease in peak concentrations for successive pulses. Edge-of-field measurements and simulations both differed from measured concentrations in large rivers, presumably because there was considerable integration of inputs and processes at the larger scale.

Evidence reported by Ashauer and Brown (2007) was exclusively for the United Kingdom. A further analysis was undertaken within ELINK for isoproturon concentrations in the Nil catchment in Belgium. The Nil catchment (3200 ha) is significantly larger than any FOCUS scenario (the stream scenario has a 100-ha upstream catchment). Also, application in the catchment was distributed over the application period

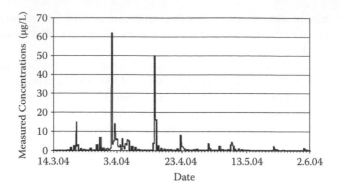

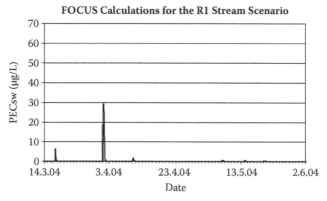

FIGURE 4.3 Comparison between measured monitoring data for isoproturon in the upper Nil catchment in Belgium (top) and simulations with the FOCUS R1 scenario (bottom).

(mid-March to the end of May) rather than simultaneous as in a FOCUS calculation. However, some peaks outside the application period can clearly be attributed to run-off events. Monitoring data for isoproturon collected at an 8-hour interval during events and daily at other times (Holvoet et al. 2007) are compared with a FOCUS simulation for the R1 runoff scenario in Figure 4.3. Despite the limitations to direct comparability, it is clear that the FOCUS SWS give a reasonable representation of the pulsed exposure situation in the field.

Thus, available data suggest that FOCUS simulations reproduce the general characteristics for measured profiles of pesticide concentrations at the edge-of-field scale. This indicates that implications of the complex patterns of exposure that have been measured need to be considered within the risk assessment. At lower tiers, this may simply require a demonstration that existing procedures are sufficiently protective, whereas more sophisticated treatment may be warranted at the higher tiers. Further work is recommended on a range of compounds and across a range of conditions to underpin any use of long-term simulated concentration profiles in risk assessment.

TABLE 4.1
Proposed ELINK Parameters for Characterizing Exposure for Each Exposure Scenario

Parameters	What Do They Show?	Statistical Measure Used for Exposure Parameter
Parameters of Key Exposure Characteristic[a]		
Peak concentration	The parameters "peak concentration," the "duration of peak(s)," the "interval between peaks," the "number of peaks in period between first peak and last peak [$t_{peak(1)} - t_{peak(n)}$]," and the "height of a background concentration" allow allocation of the exposure pattern to one of the generalized exposure profiles illustrated in Figure 4.6. They provide an initial quantitative analysis.	Max (of exposure curve)
Duration of peak		Median
Interval between peaks		Median
Number of peaks > RAC		Total number
Existence and height of background concentration[b]		Yes or no/TWA over time in which background concentration is present
Area under the curve (AUC)	The "area under the curve" is used when comparing longer-term exposures with effects threshold based on the time-weighted averages approach.	Max
Further Parameters to Characterize Exposure Profile		
Single first-order DT50 or other kinetic to describe peak decline	Further detailed information is captured by the parameter "single first-order DT50 or other kinetic" to describe the decline of a peak or cluster of peaks.	
Parameters Supportive for Risk Assessment		
Entry route of peak	The "entry route" of peaks (link to mitigation measures) and the "time of exposure" (link to risk assessment: life cycle and seasonality of exposure) will aid in the final risk assessment.	
Time of exposure (months)		

[a] Preset: Threshold concentration level for data analysis (as default the lower-tier RAC is used).
[b] The background concentration is a longer-lasting elevated concentration that can be found in some drainage scenarios. It may be greater than the threshold level (default as the lower-tier RAC). (Background concentration and threshold level (or RAC) shall not be confounded.)

4.2.2.1 Deterministic and Probabilistic Simulations

The FOCUS SWS currently give a deterministic, single-year output. Clearly, if the interval between successive peaks in exposure or other exposure characteristic parameters are central to the outcome of the risk assessment, then it may be pertinent to determine how frequently a particular outcome will occur and what range in outcomes is possible. Such questions lead to the requirement for long-term or probabilistic estimates of exposure that can account, for example, for variability in weather conditions or usage between years. Tools such as RADAR (risk assessment tool to evaluate duration and recovery) for analysis of exposure profiles are readily applied to multiyear profiles (Williams WM et al. 1999). Any decision on whether multiple-year or probabilistic simulations are required should be led by the risk assessment.

4.2.2.2 Influence of Temporal Resolution

Analysis of monitoring data and model simulations shows that aquatic exposure often varies significantly over subdaily timescales. Exposure inputs to the risk assessment are generally required at daily or multiday resolutions. Temporal aggregation will influence data reported, so exposure endpoints will vary with the resolution of monitoring or simulations (e.g., hourly, daily). It is important to note here that sampling strategies (i.e., the temporal resolutions) and their analysis have to be targeted depending on substance properties, use pattern, and resulting entry patterns. Regular sampling for compounds with highly variable exposure patterns may lead to a large number of results below the detection limit (Crawford 2004); therefore, care needs to be taken when constructing percentile distributions from such situations so that small concentrations below any effect concentration of ecotoxicological or ecological relevance are appropriately excluded when carrying out an exposure frequency distribution analysis. Figure 4.4 shows measured and simulated concentrations of sulfosulfuron for a field drainage study (Brown et al. 2004) at different temporal resolutions. Samples were collected for analysis every 4 hours, and there is a clear difference in exposure pattern at this temporal resolution relative to daily samples collected either as a daily composite sample (daily averaged) or as a single sample collected at a fixed time each day (daily point). There is little difference in simulated data when reported on an hourly or 4-hourly basis. Decreasing the resolution of simulation to daily values reduces the peak concentrations in the sulfosulfuron example but gives a closer representation of hourly values because the simulated profile is smoother than the measured profile.

4.3   STEPS IN THE EXPOSURE ASSESSMENT

4.3.1   Exposure in the ELINK Assessment Scheme

4.3.1.1   Tiered Approach

The concept of tiered approaches is to start with simple conservative tiers and to undertake more work only if necessary (so providing an economic basis for both industry and regulatory agencies). According to Boesten et al. (2007), the general principles of such tiered approaches are as follows:

- Earlier tiers are more conservative than later tiers.
- Later tiers are more realistic than earlier tiers.
- Earlier tiers usually require less effort than later tiers.
- Jumping to later tiers (without considering all earlier tiers) is acceptable. Note, however, that a full presentation of all tiers in the risk assessment may help risk assessors and managers to evaluate individual refinement steps.
- There has to be some balance between the effort and the filtering capacity of the tier.
- The risk assessors have to consider all relevant scientific information provided.

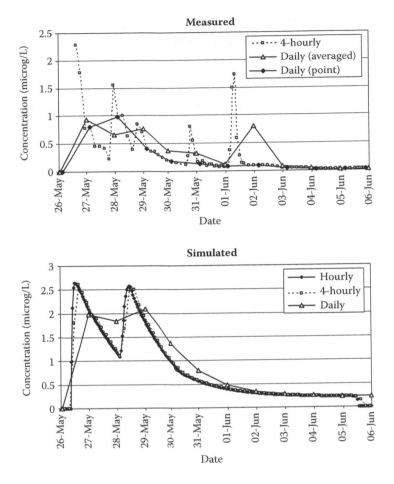

FIGURE 4.4 Influence of temporal resolution of reporting on concentration profiles generated by monitoring and simulation. (Data are based on sulfosulfuron concentrations in drain flow; Brown et al. 2004.)

Although the different steps for modeling aquatic exposure to pesticides as implemented by FOCUS (2001) are in line with the general principles mentioned, a balanced tiered exposure assessment approach is not yet complete. Particularly, more guidance is needed on experimental and monitoring studies that may be used to improve the parameters used for modeling the field exposures (e.g., use of concentration patterns measured in mesocosms and field dissipation and degradation data). This guidance is a prerequisite to estimate more realistic PECs that can be fed into the different boxes of the ELINK decision scheme (see Figure 2.1 in Chapter 2 or Appendix 4). Some building blocks for an improved tiered exposure assessment scheme are provided next.

4.3.1.2   Information Flow from Exposure Assessment to Individual ELINK Risk Assessment Boxes

The specific options are tailored to individual steps of the effect or risk assessment scheme proposed by ELINK and describe how the relevant information can be obtained for this step.

The ELINK decision scheme (Figure 2.1 in Chapter 2 or Appendix 4) consists of 7 boxes, of which Boxes 1 to 6 were discussed during the ELINK workshops.

- For Box 1 in Figure 2.1, the $PEC_{global\,max}$ values of FOCUS SWS Steps 1 to 4 are required for comparison with Tier 1 acute or chronic RACs. The $PEC_{global\,max}$ values are standard outputs from FOCUS.
- For Box 2 in Figure 2.1, the chronic RAC (or RAC curve) is plotted against the predicted exposure profile (PEC curve) of FOCUS SWS Step 1 to 4 calculations. Criteria to determine whether the use of a TWA PEC is appropriate are described in Section 3.4 of Chapter 3.
- For Box 3 in Figure 2.1, the TWA PECs of FOCUS SWS Steps 1 to 4 are required for comparison with chronic Tier 1 (or higher-tier) TWA RACs. Criteria for the duration of the TWA window are given in Box 3 and Section 3.4 of Chapter 3. The TWA PECs are standard outputs from FOCUS.
- For Box 4 in Figure 2.1, laboratory effects studies with additional test species (species sensitivity distribution [SSD] approach), laboratory tests with realistic to worst-case exposure regimes guided by field exposure predictions (refined exposure tests) or ecotoxicological modeling approaches (e.g., use of toxicokinetic and toxicodynamic [TK/TD] models) are foreseen. The key is that the exposure regime adopted in any additional ecotoxicological study or modeling approach allows the derivation of exposure–response relationships that can be used (with or without appropriate extrapolation techniques) to assess the risks of the field exposure prediction. Specific exposure characteristics that would allow proposing refined exposure tests are not standard outputs from FOCUS and have to be generated separately (as described in Section 4.4).
- For Box 5 in Figure 2.1, microcosm/mesocosm experiments or population/ ecosystem model studies (guided by ecological field data) with realistic to worst-case exposure regimes (guided by exposure prediction) are foreseen to derive a RAC representative for the threshold level of toxic effects. Again, the key is that the exposure regimes adopted allow the derivation of exposure–response relationships that can be used (with or without extrapolation techniques) to assess the risks of the predicted field exposure. The generalized exposure characteristics and exposure parameters described in Section 4.4.2 may be essential to allow a proper extrapolation of concentration–response relationships between time-variable exposures. Specific exposure characteristics are not standard outputs from FOCUS and have to be generated separately. Often, $PEC_{global\,max}$ or PECtwa are relevant.
- For Box 6 in Figure 2.1, micro/mesocosm experiments or population or ecosystem model studies (guided by ecological field data) with realistic to worst-case exposure regimes (guided by exposure prediction) are foreseen to derive an RAC based on ecological recovery of sensitive populations. Both the

appropriate RAC based on the threshold level of toxic effects (Boxes 1 to 5) and the RAC based on ecological recovery are plotted against the predicted exposure pattern (PEC curve) calculated for the FOCUSsw scenarios from FOCUS SWS Steps 1 to 4 calculations. The key is that the exposure regimes adopted in the higher-tier tests allow the derivation of exposure–response relationships that can be used (with or without extrapolation techniques) to assess the risks, including recovery times, of the predicted field exposure. Often, $PEC_{global\ max}$ or PECtwa are to be used, and the ERC concept is especially relevant.

4.3.2 Steps in Exposure Calculations

As described in Chapter 3, all exposure calculations that use 1) officially[1] defined scenarios without modification, 2) standard agreed or defined calculation methods, and 3) the base set of physicochemical properties of the active substances result in PECs that can be termed as Tier 1 PECs. At the EU level, this means, for example, that exposure assessments by means of all FOCUS scenarios defined at Steps 1 to 3 and Step 4 calculations that maintain step 3 scenario definitions but introduce mitigation in standard agreed defined ways may generate Tier 1 PECs. The FOCUS SWS exposure flow chart (FOCUS 2001, 2007a, 2007b) has 4 steps that are independent of the risk assessment flow chart of ELINK. FOCUS Step 4 calculations intend to reduce exposure (e.g., by mitigation) or provide a more tailored exposure assessment for the intended uses as it is laid down (e.g., in the report of the FOCUS landscape and mitigation group; FOCUS 2007a, 2007b). The PEC can be used at any point in the aquatic risk assessment, as illustrated in Figure 4.5 with the "crisscross model" adapted from Boesten et al. (2007). Jumping from lower to higher tiers is in principle possible (leaving out intermediate tiers).

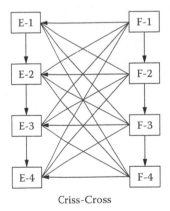

Criss-Cross

FIGURE 4.5 Diagram of the conceptual "crisscross" model of possible routes through combined effects and exposure flowcharts (Boesten et al. 2007). Boxes E-1 to E-4 are 4 effect tiers and F-1 to F-4 are 4 tiers for exposure in the field. (Brock, Forbes. *Conceptual model for improving the link between exposure and effects in the aquatic risk assessment of pesticides.* Ecotoxicol Environ Saf 66:291–308. Copyright Elsevier 2007. Reprinted with permission.)

4.4   APPROACHES TO IDENTIFY THE KEY FEATURES OF THE EXPOSURE PROFILE

There are different objectives for which it might be desirable to abstract the exposure profiles calculated either with the FOCUS surface water scenarios or any alternative approach. In the context of this guidance, 3 objectives are described here (first in brief and then in detail).

> *Objective 1.* At the stage of problem formulation, it may be useful to get a very general view of the exposure profile. This would be achieved through a set of generalized exposure profiles into which individual simulated exposure patterns can be grouped, which may help identify issues for the risk assessment. These generalized profiles aim to provide a theoretical and schematic view of the exposure pattern, giving an indication of peak height and width and an approximate time interval between contamination events. They do not aim to reproduce real exposure profiles.
>
> *Objective 2.* At the next stage of risk assessment, a more detailed characterization of exposure patterns is needed to determine whether particular exposure characteristics are likely to represent a risk. Thus, a detailed description of peak shape, peak/background level ratio, interval between events, and so on would help to identify likely interactions of exposure and effects at different tiers of the risk assessment.
>
> *Objective 3.* A more realistic but, due to the variety of complex simulated exposure patterns, necessarily more generalized exposure profile is introduced that can be used to design the exposure pattern in higher-tier effect studies.

4.4.1   Objective 1: Series of Generalized Exposure Profiles to Aid Problem Formulation

Exposure profiles generated by the FOCUS SWS are very complex, and this may hinder discussions between exposure and effects specialists about how best to capture and accommodate exposure within the risk assessment. The problem formulation step can be aided by generating a series of idealized exposure profiles for streams, ditches, and ponds that capture the overall behaviors. A set of possible generalized exposure profiles is presented in Figure 4.6. The aim of these profiles is to aid discussion about the most appropriate way to develop the risk assessment (i.e., to support problem formulation).

Profiles to the left and right of Figure 4.6 are distinguished by the absence or presence, respectively, of a background concentration that is for prolonged periods above or around the threshold level of concern triggered by the first-tier risk assessment (lower-tier RAC). Short- and long-term pulses can be distinguished by whether pulse duration exceeds the length of acute toxicity tests on the most sensitive group of organisms. Exposure in flowing waters dominated by drainflow and, to a lesser extent, runoff frequently shows highly complex patterns of successive peaks (4th row of Figure 4.6). Exposure in static waters is less dynamic and frequently shows the patterns in the bottom row of Figure 4.6 (in this case, the concept of background concentration has little relevance, and the differentiation is between exposure with a single input [bottom row, left] or multiple inputs [bottom row, right]).

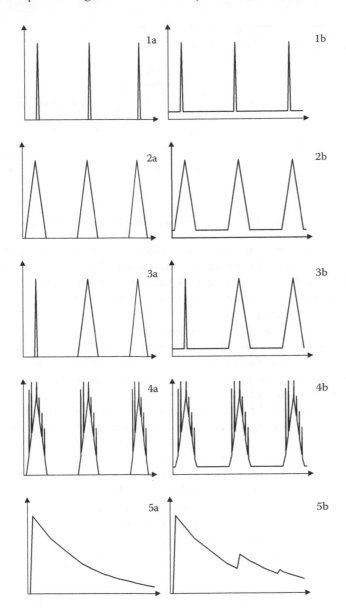

FIGURE 4.6   Ten possible generalized exposure profiles to aid problem formulation.

4.4.2   OBJECTIVE 2: IDENTIFICATION OF KEY CHARACTERISTICS OF THE EXPOSURE PATTERN

The second level at which abstraction of key characteristics of the exposure pattern becomes relevant is during combination of exposure and effects measures to generate the risk assessment. The key topic for objective 2 is to provide the "metrics" to characterize exposure (e.g., "peak height," "peak duration," "interval between peaks,"

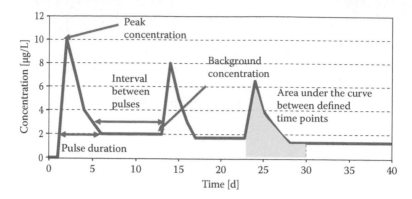

FIGURE 4.7   Key characteristics of the exposure profile that may inform the risk assessment.

etc.), whereas objective 1 provides a description of the general pattern of exposure. The questions that are asked during risk assessment will vary on a case-by-case basis according to patterns of exposure, ecotoxicological properties of the compound, and use characteristics. Generalized characteristics of the exposure profile that may be of relevance are shown in Figure 4.7.

A set of exposure characteristic parameters is proposed to capture the key features of the exposure patterns. The set of parameters should be generally applicable to characterize exposure patterns in monitoring and simulation studies as well as in effect studies. It should be possible to reproduce the exposure pattern from a given set of characteristic parameters.

Not all parameters will be necessary at all tiers of the risk assessment, and some may be found to be more appropriate than others, depending on the individual problem formulation. Essential parameters should however be identified as early as possible to avoid over- or underestimation of risks.

As a prerequisite for the data analysis, a preset threshold concentration (lower-tier RAC) has to be defined in the problem formulation and is used to screen out exposures without concern. Different threshold concentrations may be distinguished for ecological risks of short-term and long-term exposure. Note that the height of the RAC will influence the analysis of the pulse duration.[2]

In the domain of exposure assessment, we propose that special emphasis should be given in characterizing the peak exposure to substances with an acute-to-chronic (A/C) ratio less than 10 since a short-term exposure may already cause the environmental risks. It should be noted that the A/C ratio should be derived for the potentially sensitive taxonomic groups and comparable endpoints. For substances with A/C ratios above 10, expert judgment should be used to decide whether emphasis should be given to characterize peak exposures or TWA exposures (considering the mode of action, the respective endpoint driving the NOEC [no observed effect concentration] and level of effect observed at the LOEC [lowest observed effect concentration]). Other criteria that are of importance to consider if the use of a TWA PEC is appropriate are provided in Section 3.4 (Chapter 3).

The parameters proposed by ELINK as an aid to identifying the key features of an exposure profile are listed in Table 4.1.

The relevant time periods for the exposure characteristic parameters need to be evaluated using a moving time window approach for study durations over the simulated exposure scenario to capture the exposure characteristics. For ease of communication, the exposure characteristic parameters may be described in a first instance as presented in Table 4.1. For detailed analysis, especially of multiyear runs, it will be necessary to obtain statistical parameters to describe the frequency distributions.[3] A clustering of exposure characteristic parameters may be necessary with regard to multipeak events, scenario types, and types of water body. Exposure characteristic parameters will be dependent on fate properties and effect type of the substance, and a multivariate analysis may be necessary to identify the principal components.

4.4.2.1 Tools to Characterize the Exposure Profile

The RADAR tool (Williams WM et al. 1999) developed for the ECOFRAM (Ecological committee on FIFRA Risk Assessment) project (Hendley and Giddings 1999) has been a good starting point for discussion in ELINK, but a more flexible tool that is tailored to the intended use of deriving the statistics and measures of the exposure characteristic parameters listed will be necessary. A specific stand-alone software tool, which meets current software requirements, will be developed by ECPA (European Crop Protection Association) (in line with the ELINK recommendations) and will cover the following aspects:

- ability to calculate exposure characteristic parameters proposed by ELINK and statistics of the parameters in case of multiple seasons (e.g., multiyear runs)
- ability to consider different temporal resolutions (hourly, daily)
- ability to consider any exposure pattern that is provided in a tabular form

4.4.3 OBJECTIVE 3: GUIDANCE CONCERNING EXPOSURE REGIMES IN EFFECTS STUDIES

The description of pesticide usage in the field and the generalized characteristics of the consequent exposure to surface waters can be used to inform the type of exposure regime that may be appropriate to use in effects studies (improvement of study design; Boxes 4 to 6 of scheme). The general procedure is described here, followed by a worked example.

4.4.3.1 How the Exposure Pattern in the Field May Guide the Exposure Regime in Effects Studies

The exposure regime to be tested should be mainly determined by ecotoxicological considerations with respect to mode of action (MoA), A/C ratios, type of endpoint affected (e.g., mortality vs. growth), and sensitive life stages of the test organisms likely to be exposed. An exposure regime in a fish full life-cycle study, for example, is designed to ensure that sensitive life stages of the organism are exposed and should not be driven solely by typical exposure event intervals calculated by FOCUS surface waters tools. However, key exposure parameters should be captured during the exposure regime, especially "peak concentration," "area under the curve (AUC)," "peak duration," and "interval between peaks." Also, the level of the background concentration (if existing) can be important in determining an appropriate concentration regime.

The AUC can be a test variable in determining the exposure regime, and the AUC of the exposure in the effects study should at least cover the AUC of the exposure in the field (simulated by FOCUS or measured), or it should be possible to extrapolate the concentration–response relationships between different exposure regimes.

Peak duration relates to the fact that a certain duration is needed before a toxicological effect is established (see also Chapter 7 on TK/TD models). Time to onset of effect (TOE) studies will allow clear establishment of such a relationship; this may also be relevant in special cases for acute studies with substances showing a low A/C ratio. The design of TOE studies can be guided by information on peak duration.

4.4.3.2 Toxicological and Ecological Independence of Peaks

When using generalized exposure profiles to design the exposure of higher-tier effect studies and the focus of the higher-tier test is to derive an RAC representative for the threshold level of toxic effects (e.g., Box 5 in the decision scheme), it is important to evaluate whether the repeated pulse exposures are toxicologically independent from each other. Note that semistatic and flow-through tests mimic this type of exposure, especially in the case of unstable compounds. Toxicological dependence of repeated pulses may occur if the life span of individuals of the sensitive species is long enough also to experience repeated pulse exposures. If, for example, the generalized exposure profile consists of 2 pulse exposures, the second pulse can be considered toxicologically independent from the first pulse if between the 2 pulses 1) the internal exposure concentrations in the individuals of the sensitive species drop below critical threshold levels and 2) complete repair of damage occurs. The toxicological independence of peaks can be investigated with refined exposure tests (see Chapter 6) or TK/TD modeling approaches (Chapter 7). If the toxicological independence of successive peaks in the generalized exposure profile can be demonstrated for the species of concern, it may be valid to adopt a single pulse exposure regime in a micro/mesocosm study that aims to derive an RAC representative for the threshold level of toxic effects.

Evaluating the ecological dependence/independence of successive pulse exposures will be important when higher-tier tests are used in the risk assessment (e.g., micro/ mesocosm studies) that aim to derive an RAC based on an NOEAEC (no observed ecologically adverse effect concentration; e.g., taking into account ecological recovery of sensitive species). If the interval between peaks is smaller than the relevant recovery time of the sensitive populations of concern, these peaks should be considered as ecologically dependent. For further information on linking exposure and effects when ecological recovery is taken into account, see Section 8.7 in Chapter 8. Since only a limited number of ecological recovery scenarios can be investigated in micro/mesocosm tests (see Sections 8.5 and 8.6 in Chapter 8), modeling approaches may provide an alternative tool for spatiotemporal extrapolation (Chapter 9) and to investigate whether successive pulse exposures are ecologically dependent or not.

The possible ecological independence of pulse exposures may also be of importance in the risk assessment if the potentially sensitive species, or specific sensitive life stages of these species, are not present in the periods when certain pulse exposures occur. The risk assessment, and consequently the design of the exposure in higher-tier tests to support this risk assessment, should then focus on those exposures that coincide with the occurrence of the species (life stages) of concern.

4.5 CHARACTERIZING THE EXPOSURE PROFILE IN SURFACE WATER ACCORDING TO ELINK AND DERIVING A DOSE (EXPOSURE) REGIME IN A HIGHER-TIER EFFECTS STUDY AS ILLUSTRATED FOR ELINKSTROBIN

4.5.1 BACKGROUND

Example Case: ELINKstrobin Use in Cereals

ELINKstrobin — like several other strobilurin fungicides — shows broad aquatic toxicity, mainly via acute effects, and it generally has a very small A/C ratio. Thus, in most studies the dominant endpoint is mortality, and there are no other sublethal effects at concentrations significantly (i.e., < factor 5) below those causing mortality. This was shown for ELINKstrobin in several chronic fish studies (28-day juvenile growth test, a standard [35-day flow-through] and higher-tier (static) ELS [early life-stage toxicity] study). These studies showed that very young fish — at the time shortly after hatching and time of swim up — are the most sensitive life stages.

However, from a single chronic fish study — a 97-day flow-through ELS study with trout — a significantly lower endpoint was derived that was based on a very small, but in this case statistically significant, reduction in fish biomass at the end of the study. It may thus not be excluded that long exposure durations may exert an impact (although being small and likely of little ecological relevance) beyond the general very narrow A/C ratio.

This background (i.e., awareness that the main impact is due to acute toxicity but also trying to cover a potential additional negative impact from long-term chronic exposure) provides the basis for the risk assessment approach shown next to try to cover peak exposure periods as well as potential chronic exposure.

4.5.2 APPROACH

4.5.2.1 Characterization of Exposure Profile

Exposure profiles of ELINKstrobin calculated for the different relevant FOCUS surface water scenarios are characterized using the key parameters proposed by ELINK.

PROPOSED PARAMETERS TO DESCRIBE EXPOSURE CHARACTERISTICS

- Preset: threshold concentration level for data analysis (may be zero)
- Key parameters
  - peak concentration
  - area under the curve
  - duration of peak
  - interval between peaks
  - value of background concentration
  - number of peaks
- Further parameters to characterize exposure profile
- Single first-order DT50 (period required for 50% disappearance) or other kinetic to describe peak decline

TABLE 4.2

Characterization of Scenarios: Statistical Parameter for Each Individual Scenario (Short Version of Table 4.1)

Metric	Number of Pulses above Threshold	Peak Concentration (µg/L)	Pulse Duration (days)	Interval between Pulses (days)	AUC 1-Year Time Window (µg/L * days)	Existence and Height of Background Concentration
Statistical parameter	All	Max	Median	Median	Max	If yes (max, median)

4.5.2.2 Characterization of the Exposure in the FOCUS SWS (Objective 2)

To characterize the exposure patterns in the different FOCUSsw scenarios (objective 2), the exposure characteristic parameters are calculated for each individual scenario as described in Table 4.2. The variability of the metrics of exposure characteristics will reflect the variability in the exposure pattern. However, it is difficult to derive appropriate parameters from highly contrasting scenarios to come up with a proposal for a generalized exposure pattern that may allow defining a dose regime in an effect study (objective 3). Difficulties are associated with the attempt to derive a "single" pulsed-dose regime that should cover "all" FOCUS exposure patterns at the same time, and the individual problem formulation will have to guide the derivation of exposure characteristics (= dose regime). The following section outlines an attempt to capture this variability in exposure patterns into a single, representative dose regime.

From the set of parameter values (e.g., the number of pulses above the threshold level [RAC] or the maximum peak concentration for each of the FOCUS scenarios), the overall statistics for each parameter across all scenarios can be obtained. A proposal for a pulsed-dose regime can be derived from these statistical distributions that may cover generically the exposure pattern in the FOCUS scenarios (see Table 4.3).

The choice of statistical parameter to define the pulse regime is important and needs clear justification.

For the case of ELINKstrobin, the greatest weight has been given to capturing the maximum peak concentration and a realistic worst-case (90th percentile) area AUC (TWA derived from AUC). These joint output values capture the worst-case character. The numbers of peaks and the duration of peaks are seen as comparably less important for this example as the peaks should coincide with sensitive life stages of the test organism. The maximum peak concentration (largest $PEC_{global\ max}$ of all scenarios) and the largest AUC did not occur in the same FOCUS scenario. Worst-case values of important parameters should be chosen, but an aggregation of worst-case parameter values to unrealistic worst-case combinations should be avoided.

A threshold of 0.1 µg/L as the RAC based on Tier 1 chronic assessment (ELS, 97-day constant [flow-through] concentrations: NOEC/AF = 1/10 µg/L) is used in the characterization of the exposure pattern. This threshold is utilized by the RADAR

TABLE 4.3

Proposal for Pulsed-Dose Regime: Overall Statistical Parameters across All Scenarios (Based on the ELINKstrobin Case Study, Explanations of the Basis for the Proposal Are in the Text)

Metric	Number of Pulses above Threshold	Peak Concentration (μg/L)	Pulse Duration (days)	Interval between Pulses (days)	AUC Time Window of Effect Study (μg/L * days)	Height of Background Concentration
Statistical parameter	Median	Maximum of the $PEC_{global\ max}$ across scenarios[a]	Median	10th percentile[b]	Calculate 90th percentile AUC for each scenario, then take the maximum value across scenarios	Calculated to achieve 90th percentile AUC

[a] The maximum of the $PEC_{global\ max}$ is the largest peak concentration across all FOCUS scenarios (= largest value of all $PEC_{global\ max}$). The $PEC_{global\ max}$ is a standard term from FOCUSsw and characterizes the maximum PEC calculated for each single combination of a FOCUS scenario and type of water body (see also this book's Glossary). The exposure characteristic parameter "maximum $PEC_{global\ max}$" used in this context is the highest of all $PEC_{global\ max}$ values across all scenarios (of which each has its own individual $PEC_{global\ max}$).

[b] Zero values are not considered to calculate the 10th percentile (a zero interval would mean either no peak above RAC level or a constant exposure > RAC level during the whole evaluated period).

tool to identify peaks and separate them from very small concentrations of no ecological relevance.

Based on hourly output of concentrations in surface water provided by the TOXSWA[4] (toxic substances in surface waters) model, the maximum values are used to provide a concentration pattern at a daily resolution. The peaks as identified by the RADAR tool are defined as pulses that start with the day when the defined threshold level is exceeded until the day when the concentrations decrease again below this threshold level. In fact, the pulses could have some "peaks," but as long as these peaks do not fall below the threshold level (RAC), the whole (peaky) pulse is counted as a single pulse.

The largest concentrations occurred in runoff stream scenarios (Figure 4.8) with a maximum peak concentration of 4.05 μg/L in the R4 stream scenario (Figure 4.9). The peaks in the runoff scenarios are in general only of a short duration (1 to 2 days) compared to the drainage scenarios with long duration of peaks (Figure 4.10a). The median duration of peaks is 4 days. The exposure in the D1 ditch and D2 ditch scenarios are characterized by significant concentration levels for extended periods with maximum peak concentrations of 2 to 3 μg/L (Figure 4.11 and Figure 4.12, respectively).

The number of peaks is displayed in Figure 4.10b. The greatest number of peaks is in scenario R1 stream ($n = 13$), of which most peaks are only slightly above the

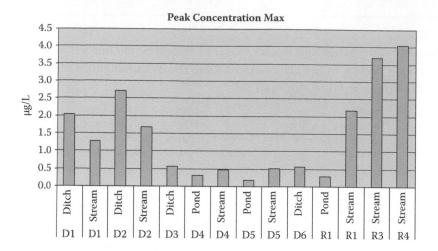

FIGURE 4.8   Maximum concentrations of ELINKstrobin in the FOCUSsw bodies.

threshold level (Figure 4.13, dotted line). The median number of peaks of all scenarios is 3 and, if scenarios D1 and D2 ditch are not considered, the median number is 4.

The AUCs (averaged to 12 months) of the exposure in the FOCUS scenarios are shown in Figure 4.14. As the exposure periods reported by the FOCUSsw tools for the drainage and the runoff scenarios were different (16 and 12 months, respectively) the AUCs were averaged to 12 months. The AUCs (12 months) of the drainage scenarios D1 and D2 are significantly larger (136 to 289 µg/L * day) than those of the other drainage scenarios D3, D4, D5, and D6 and of all runoff scenarios.

The intervals between peaks are shown in Figure 4.15. If the scenarios D2 ditch and D4 pond are excluded (no second peak to calculate an interval), the shortest peak interval of the remaining scenarios (10th percentile equal to the 90th percentile worst case) is 7 days.

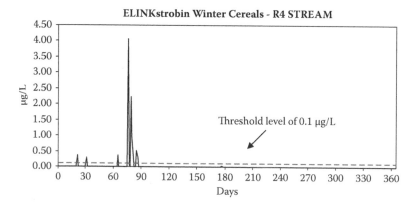

FIGURE 4.9   Exposure profile for ELINKstrobin in the runoff scenario R4 stream.

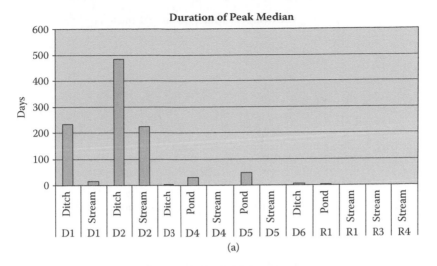

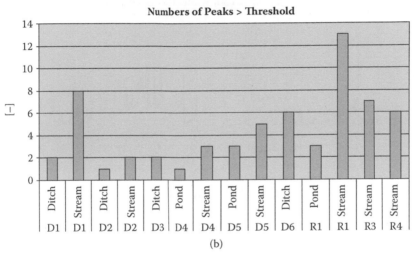

FIGURE 4.10 Duration (a) and number (b) of exposure peaks for ELINKstrobin in the different FOCUSsw scenarios (>threshold of 0.1 μg/L).

The statistical measures for the exposure characteristic parameters are summarized in Table 4.4.

4.5.2.3 Derivation of a Proposal for a Dose Regime in an Effect Study (Objective 3)

A higher-tier effects study is specifically designed to cover the exposure profile in the FOCUS scenarios (objective 3).

For this specific example, 2 cases are outlined for illustrative purposes (because a preliminary assessment showed that passing all scenarios will be difficult, and

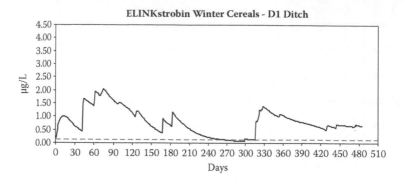

FIGURE 4.11   Exposure profile for ELINKstrobin in the drainage scenario D1 ditch.

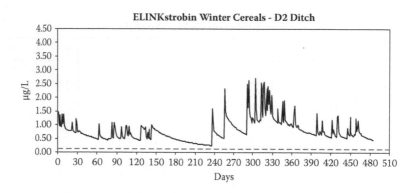

FIGURE 4.12   Exposure profile for ELINKstrobin in the drainage scenario D2 ditch.

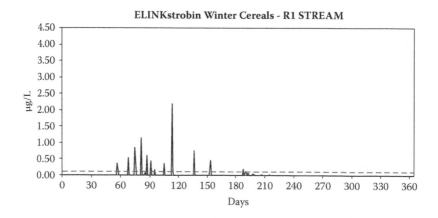

FIGURE 4.13   Exposure profile for ELINKstrobin in the runoff scenario R1 stream.

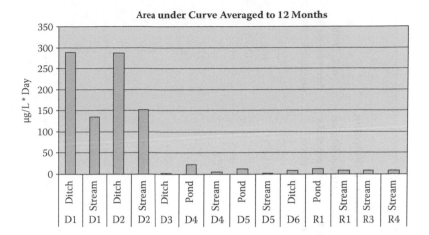

FIGURE 4.14 Area under the curve of exposure peaks for ELINKstrobin averaged to 12 months to make the runoff and drainage scenarios comparable (whole period: drainage scenarios 16 months, runoff scenarios 12 months).

the notifier may not be interested in defending all scenarios with the refined effects study):

- Case 1: The risk assessment for all scenarios and water bodies should be passed.
- Case 2: The notifier considers suitable mitigation measures and restricts the use to situations in which the conditions of the D1 and D2 scenarios do not play a role (here, simplified assumption to restrict the area of use to D3, D4, D5, D6, and the runoff scenarios).[5]

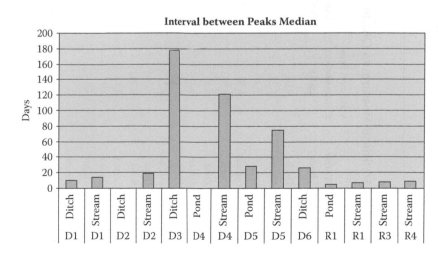

FIGURE 4.15 Interval between exposure peaks for ELINKstrobin in the FOCUS SWS (>threshold of 0.1 µg/L).

TABLE 4.4
Summary of the Statistical Measures to Quantify the Exposure Characteristic Parameters for ELINKstrobin

FOCUS Scenario	Water Body	Maximum Peak Concentration (µg/L)	Median Duration of Peak (days)	Numbers of Peaks (>Threshold of 0.1 µg/L)	AUC Averaged 12 Months (µg/L * day)	Median Interval between Peaks (days)	AUC (90th Percentile for 90 days) (µg/L * day)
D1	Ditch	2.04	232	2	289.07	10.5	120.9
D1	Stream	1.27	15	8	135.50	14	65.0
D2	Ditch	2.71	485	1	287.00	—	112.2
D2	Stream	1.69	223	2	152.95	19.5	61.2
D3	Ditch	0.555	3	2	1.65	178	2.3
D4	Pond	0.31	31	1	22.03	—	17.7
D4	Stream	0.474	1	3	5.19	121	6.7
D5	Pond	0.188	47	3	13.03	28	7.7
D5	Stream	0.515	1	5	1.93	75	3.8
D6	Ditch	0.565	5	6	9.19	26.5	4.9
R1	Pond	0.292	3	3	12.90	5	12.2
R1	Stream	2.18	1	13	8.32	7	6.4
R3	Stream	3.7	1	7	8.92	8	3.2
R4	Stream	4.05	1	6	9.43	9.5	8.8
Parameters for Dose Regime (Statistic across Scenarios)		Max	Median	Median	Not Used for Dose Regime	10th Percentile	Max
		4.05	4	3	—	7	120.9

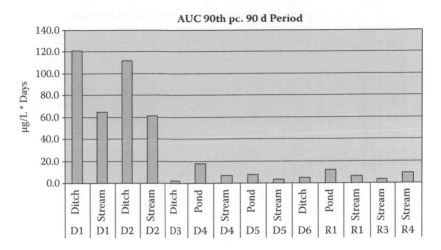

FIGURE 4.16 Area under the curve (90th percentile) for a time period of 90 days (= duration of the effect study) obtained by a moving time frame from the exposure profile for ELINKstrobin in the FOCUSsw scenarios.

The 2 cases are used to reflect 2 major types of exposure scenarios; this may be applied to reduce uncertainties for risk assessment for different conditions and thus allow optimized use of assessment factors. However, generally strive to reduce the number of testing scenarios and instead perhaps reduce uncertainty factors for apparently overconservative conditions in comparison to less-severe scenarios.

- The exposure regime in the effects study shall be based on the exposure characteristics in close interaction (linkage) with the ecotoxicological knowledge about the mode of action of the compound and the biology (sensitive life stages) of the most sensitive test organism.
- The duration of the higher-tier effects study is about 90 days. Therefore, the 90th percentile worst-case AUCs for this 90-day period are derived from the simulated exposure profiles in the different FOCUSsw scenarios[6] (Figure 4.16).
- The largest 90th percentile worst-case AUCs in all FOCUS scenarios (= case 1) for a 90-day period are found in the D1 and D2 scenarios ranging from 61.2 µg/L * days (D2 stream) to 120.9 µg/L * days (D1 ditch). This is equivalent to a TWA chronic exposure concentration of 0.68 to 1.34 µg/L over 90 days (Figure 4.17).
- The 90th percentile worst-case AUCs in the FOCUS D3 to D6 and the runoff scenarios (= case 2) for a 90-day period range from 3.2 to 17.7 µg/L * days (D4 pond). This is equivalent to a TWA chronic exposure concentration of 0.04 to 0.20 µg/L over 90 days (Table 4.5).

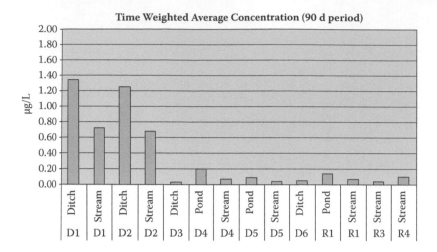

FIGURE 4.17   Time-weighted average concentration for ELINKstrobin obtained for a time period of 90 days (based on 90th percentile worst-case AUC).

TABLE 4.5

Area under the Curve (90th Percentile) for a Time Period of 90 Days (Duration of Effect Study) Obtained by a Moving Time Frame from the Exposure Profile of the FOCUSsw Scenarios and Time-Weighted Average Concentration of ELINKstrobin

FOCUS Scenario	Water Body	AUC (90th Percentile) for Time Period of 90 Days (µg/L × day)	Time-Weighted Average Concentration for Time Period of 90 Days (µg/L)
D1	Ditch	120.9	1.34
D1	Stream	65.0	0.72
D2	Ditch	112.2	1.25
D2	Stream	61.2	0.68
D3	Ditch	2.3	0.03
D4	Pond	17.7	0.20
D4	Stream	6.7	0.07
D5	Pond	7.7	0.09
D5	Stream	3.8	0.04
D6	Ditch	4.9	0.05
R1	Pond	12.2	0.14
R1	Stream	6.4	0.07
R3	Stream	3.2	0.04
R4	Stream	8.8	0.10

4.5.2.4 Dose Regimes for Effects Studies (Objective 3)

Based on the exposure characteristics as described, the proposed exemplified dose regimes for the higher-tier effects study (~90 days) for the 2 cases are

- Case 1 (all FOCUS scenarios passed):
  - Maximum peaks of 4.05 µg/L of short duration (4 days, exceeding a significant background concentration level) in the sensitive life stage of the test organism starting at day 30.
  - Repeated maximum short-term peaks with an interval of 7 days after the first peak. In total, 3 peaks are envisaged (= median number of peaks for all scenarios).
  - The AUC in the effect study should capture the largest value of the 90th percentile worst-case AUC found for 90 days (121 µg/L × day).
  - The background concentration in the effect study should comprise the TWA for the 90-day period with the highest AUC including 3 peaks of 4.05 µg/L. Given a period of 90 days, a chronic exposure of 1.15 µg/L is proposed plus 3 peaks of 4 days duration each generated with doses of 2.9 µg/L (1.15 µg/L + 2.9 µg/L = 4.05 µg/L). The maximum PEC of 4.05 µg/L dissipates rapidly in the days following the peak, as in the runoff scenarios that generated the maximum peaks. The peaks and the background concentration sum to an AUC in the exposure study of 121 µg/L × day, which matches the target AUC of 121 µg/L × day.
  - A dose regime in a 90-day study that would encompass the characteristics for case 1 is shown in Figure 4.18.
- Case 2 (subset of FOCUS scenarios "all except D1, D2"):

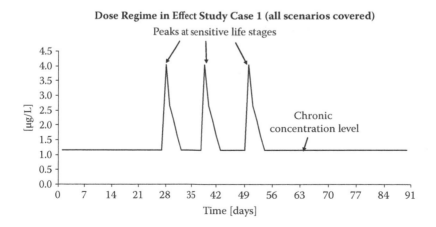

FIGURE 4.18 Proposed dose regime for ELINKstrobin in the effects study for case 1 (all scenarios) (chronic concentration level of 1.15 µg/L and 3 peaks with 4 days duration with PEC$_{max}$ of 4.05 µg/L followed by fast dissipation).

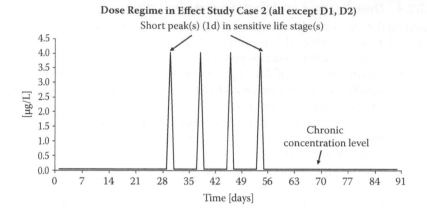

FIGURE 4.19 Proposed dose regime for ELINKstrobin in the effects study for case 2 (all scenarios except D1, D2) (chronic concentration level of 0.05 µg/L and 4 peaks of 4.05 µg/L with 1-day duration).

- Maximum peaks of 4.05 µg/L of short duration (1 day) at a sensitive life stage of the test organism starting at day 30.
- Repeated short-term peaks at the sensitive life stages of the test organism. A worst-case combination of numbers and interval of peaks would be 4 pulses in total (= median peak number of all scenarios except D1 and D2) with a duration of 1 day in intervals of 7 days.
- The AUC in the effects study should capture the largest value of the 90th percentile worst-case AUC found for 90 days in the scenario sub-set (17.7 µg/L × day).
- The background concentration in the effects study should comprise the TWA for the 90-day period with the largest AUC in the scenario subset D3 to R4), including the 4 peaks of 4.05 µg/L with 1 day duration. A chronic exposure of 0.05 µg/L plus 4 peaks of 1 day duration each generated with doses of 4.0 µg/L (0.05 µg/L + 4 µg/L = 4.05 µg/L) sums to the AUC in the exposure study of 21 µg/L × day, which would cover the target AUC of 17.7 µg/L × day.
- A dose regime in a 90-day study that would encompass these characteristics for case 2 is shown in Figure 4.19.

Dose regimes with different concentration levels to establish a dose–response curve and to cover the required assessment factors can be derived from these basic dose regimes.

NOTES

1. Defined by regulatory authorities of governments or in agreed EU guidance documents.
2. For example, the pulse durations in Figure 4.7 would be different using a threshold level (RAC), for example, of 2 µg/L compared, for example, to 6 µg/L (longer time periods for exceeding the threshold are obtained with the lower RAC).

3. Multiyear runs with the FOCUSsw models can be made manually by transferring multiyear output provided by MACRO (model of water movement and soluable transport in macroponous soils) and PRZM (pesticide root zone model) into the appropriate format used by TOXSWA. Modeling and evaluation of results have to be done outside the FOCUS SWASH (Surface Water Scenarios Help software) shell.

4. From the hourly output of the concentration in surface water provided by the TOXSWA model, the maximum values are used to provide a concentration pattern at a daily resolution with the help of a software script (the maximum hourly value from each 24-hour period is used as the value for the whole day). For the purpose of this exercise, the daily resolution is used. This is done to handle the concentration patterns in Excel (limited number of possible lines). To be consistent, the values at a daily resolution are also used in RADAR. The transfer from maximum hourly to maximum daily values is a simplification (however, a conservative approximation) used for the purpose of this case study. This means that (for this case example only) a peak of the duration of 1 hour is transferred to a peak of the same concentration but with the duration of 1 day. In the future, the hourly resolution is to be used as it actually reflects the dynamic behavior simulated with the FOCUSsw tools.

5. For an assessment performed in the context of Annex I listing, it may only be necessary to identify a single safe use (which might be demonstrated by acceptable risk assessments for a subset of the standard FOCUS scenarios). The exemplified case shall not preclude regulatory decisions on member-state level as acceptance of soil-type restrictions across MS (Member State of Europe) may differ.

6. The AUC is obtained with a moving time frame (like in the FOCUSsw tools). This can be done with specific software or in an Excel spreadsheet, where the sum of concentrations in the respective time window is calculated, and the 90th percentile is used.

5 Species Sensitivity Distribution Approach in the Risk Assessment of Plant Protection Products[*]

CONTENTS

5.1 GENERAL INTRODUCTION TO THE SPECIES SENSITIVITY DISTRIBUTION CONCEPT

Species vary markedly in their sensitivity to pesticides, and this variation as a result of direct toxicity can be described by constructing a species sensitivity distribution (SSD). The SSD is a statistical distribution estimated from a sample of laboratory toxicity data and visualized as a cumulative distribution function (see Figure 5.1).

The SSDs are used to calculate the concentration at which a specified proportion of species are expected to suffer direct toxic effects. When compared with the first-tier effects assessment on the basis of standard test species, SSDs have the advantage of making more use of the available laboratory toxicity data for a larger array of species. They describe the range of sensitivity rather than focusing on a single value, they enable estimates to be made of the proportion of the species affected at different concentrations, and they can be shown together with confidence limits indicating the sampling uncertainty due to the limited number of species tested. They can be used

---

[*] Box 4 in the decision scheme; see Figure 2.1 in Chapter 2 or Appendix 4.

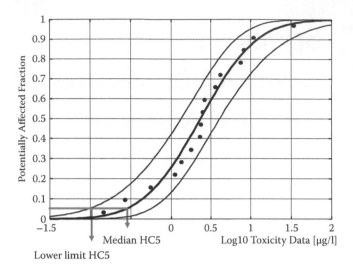

FIGURE 5.1 Graphical presentation of the species sensitivity distribution curve, its 90% confidence interval, and the derivation of the lower limit and median hazardous concentration to 5% of the species (HC5).

in a deterministic risk assessment by taking an appropriate percentile from the SSD or in a probabilistic risk assessment by using the whole SSD (European Food Safety Authority [EFSA] 2006a). The use of the SSD concept in ecological risk assessment is based on several assumptions (Versteeg et al. 1999; Forbes and Calow 2002; Posthuma et al. 2002; Van den Brink et al. 2008), namely:

1) The sample of the species on which the SSD is based is a random selection of the community of concern and is herewith representative for this community.

    The toxicity data used to construct an SSD, however, normally are not derived solely from species in a specific community of concern. Often, the available toxicity data of standard test species and of other species published in the literature are used as well. So, the SSD usually does not represent a known community but is often interpreted as if it does (Forbes and Calow 2002). For further discussion on this topic and the implications for pesticide effect assessment, see Section 5.2 on SSDs in pesticide risk assessment.

2) Interactions among species do not influence the sensitivity distribution.

    Indirect effects, which occur in stressed communities as a result of shifts in species interactions due to direct toxic effects, are not necessarily linked to a certain percentile of the SSD curve. Consequently, values higher than the threshold level for direct toxic effects (e.g., values higher than the [lower limit] HC5 or HC1 [hazardous concentration to 5% or 1% of the tested taxa, respectively]; see Section 5.2) may not correspond to the overall impact observed because of the combination of direct and indirect effects at higher exposure levels (Van den Brink, Brock, et al. 2002).

3) Since functional endpoints are normally not incorporated in the SSD, community structure is the target of concern.

   This assumption suggests that by protecting the structure of a community, its functions, including energy flows within food webs, are also protected. The validity of this assumption probably is generally applicable for most pesticides except possibly photosynthesis-inhibiting herbicides (see, e.g., Brock, Lahr, et al. 2000).

4) The laboratory sensitivity of a species approximates its field sensitivity.

   There is an extensive literature on the comparison between laboratory and (semi)field sensitivity of species as a result of direct toxic effects. The general conclusion of these comparisons is that, when exposure and developmental stages of the organisms are similar, the laboratory sensitivity of a species to certain pesticides is representative for its field sensitivity. In comparing SSDs for the insecticide chlorpyrifos based on mesocosm-derived toxicity values (mesocosm SSD) with an SSD based on results from laboratory toxicity experiments (lab SSD), Van den Brink, Brock, et al. (2002) found an exact match between the 2 when both SSDs were based on sensitive groups (in these cases, arthropods), and the lab SSD was based on the same concentration range used in the mesocosm experiment. A similar result was obtained for the insecticide lambda-cyhalothrin by Schroer et al. (2004). For the insecticide endosulfan, Hose and Van den Brink (2004) found a difference of a factor of 3 between lab- and mesocosm-SSD-based HC5 values, although confidence intervals overlapped. The HC5 value derived from laboratory data was less than that derived from the mesocosm data and was thus protective of those populations. Results of experiments performed by Roessink et al. (2006) with the fungicide or biocide triphenyltin-acetate, however, deviated from these findings for insecticides. They studied the responses of aquatic communities in microcosms to a single application of triphenyltin-acetate, a persistent substance that rapidly dissipated from water and partitioned into the upper sediment layer. In addition, they compared the population responses observed in the microcosms with that of acute single-species toxicity tests. They found that the SSD constructed with microcosm-derived EC50s (median effective concentrations that affect 50% of the test organisms) resulted in a significantly lower HC5 value than that derived from an SSD constructed with acute laboratory EC50 data. This was explained by latency of effects and the fact that triphenyltin-acetate accumulated in the food chain of the microcosms, causing effects due to chronic exposure (Roessink et al. 2006).

5) The endpoints measured in the toxicity tests on which the SSD is based are ecotoxicologically relevant.

   This assumption implies that the endpoints measured in the toxicity tests on which the SSD is based must be toxicologically and ecologically relevant. Mortality and immobility are the most frequently studied endpoints in acute laboratory tests. In chronic tests, endpoints such as reproduction and inhibition of growth are also studied. Forbes et al. (2001) argued that individual-level endpoints like survival, fecundity, and growth may not reflect effects at the population level. They recommended that additional consideration be given to the relative frequency of different life-cycle types, to the proportion of sensitive and insensitive taxonomic groups in communities, and to the role of density-dependent influences on population dynamics.

6) Since in SSDs all species have equal weight, it is assumed that all species are equally important for the structure and functioning of the ecosystem of concern.

For the SSD concept, all species are considered to be equal, although we know that some species are more important for the functioning of eco-systems than others. Also, the concept of keystone species (i.e., species that have effects on ecosystems that are disproportionate to their relative abundance) acknowledges that not all species are equal. Although it is possible to weight species in an SSD based on their importance for the ecosystem, there is no consensus in the field of ecology on what to measure as an indicator of the importance of a species to an ecosystem, and such a consensus does not seem likely (Van den Brink et al. 2008).

7) The real distribution of the sensitivity of the community is well modeled by the selected statistical distribution.

When choosing a model to fit the distribution, the choice can be between distribution-free and distribution-based methods and between methods based on classical or Bayesian statistics. The most popular are those based on the logistic and lognormal distributions (e.g., Aldenberg et al. 2002) because they require fewer data than distribution-free methods and are relatively easy to fit with standard statistical software (Van den Brink et al. 2008).

8) The number of species data used to fit the distribution is adequate from a statistical, ecological, and animal welfare point of view to describe the real distribution of the sensitivity of the community.

The adequate number of data points needed to construct an SSD depends on the method used. Suter et al. (2002) reviewed the literature on this matter and found numbers that were considered adequate between 3 and 30.

9) The protection of the prescribed fraction of species (e.g., HC5 or HC1) ensures an "appropriate" protection of the structure of ecosystems.

More and more pesticide studies evaluating whether the HC5 values are protective for the structure and function of microcosms and mesocosms have become available (see, e.g., the reviews of Brock et al. 2006; EFSA 2006a; Maltby et al. submitted). In general, it can be concluded from these studies that the (lower-limit) HC5 or HC1 values derived from SSDs can provide a cost-effective risk evaluation to establish threshold concentrations for toxic effects of pesticides in the aquatic environment.

10) The validity of the toxicity data used to construct the SSD is ensured.

When use is made of open literature or Internet databases from which it is difficult to check the validity of the data, it is not known what is modeled.

In practice, a combination of differences between laboratories, between endpoints, between test duration, between test conditions, between genotypes, between phenotypes, and eventually between species is modeled (Van den Brink et al. 2008). However, the differences in quality of the toxicity data used to construct the SSD may have a smaller impact on the overall position and shape of the SSD curve if a more toxicity data are available.

5.2 SSDs IN PESTICIDE RISK ASSESSMENT

The SSD can be based on either acute or chronic toxicity data. According to the HARAP (Higher-Tier Aquatic Risk Assessment for Pesticides) guidance document (Campbell et al. 1999), the toxic mode of action of the pesticide should be taken into account when constructing SSDs to derive acceptable concentrations. If the first tier indicates that 1 standard test species of the basic set is considerably more sensitive (e.g., differing by a factor greater than 10), an SSD should be constructed that is representative for the sensitive taxonomic group. According to Campbell et al., toxicity data for at least 8 different species from the sensitive taxonomic group are recommended to construct SSDs. In the case of herbicides, vascular plants or algae usually comprise the most sensitive groups (see Van den Brink et al. 2006). In the case of insecticides, arthropods (crustaceans and insects) usually are most sensitive (Maltby et al. 2005). For fish, the HARAP guidance document recommends using a minimum of 5 toxicity data to construct SSDs specific for fish. This lower number of toxicity data is chosen, among other reasons, for animal welfare considerations and because of the overall lower variability in toxicity data when, for example, compared with that of invertebrates. In addition, constructing a separate SSD for fish may be necessary if, for example, the risks of a plant protection product to populations of invertebrates, and primary producers have been assessed by means of an appropriate microcosm or mesocosm experiment. Fish were not present in these test systems, but potential risks to fish cannot be excluded.

For pesticides with a narcotic mode of action, such as several fungicides for which the basic set of standard test species shows a more or less equal sensitivity, at least 8 toxicity data representing different taxonomic groups should be used according to Campbell et al. (1999). The HARAP guidance document, however, does not specify the taxonomic groups and level of taxonomic resolution when selecting toxicity data for this generic SSD. Although the guidance provided by the HARAP guidance document is more or less accepted within the context of Directive 91/414/EEC (see EC 2002), other criteria may be used in other jurisdictions. For example, in setting environmental quality standards within the context of the Water Framework Directive (WFD), the construction of an SSD requires preferably more than 15 but at least 10 toxicity data for different species covering at least 8 taxonomic groups (Lepper 2002; EC 2003). The SSD procedure currently used under the umbrella of the WFD initially seems to ignore knowledge on the specific toxic mode of action of pesticides, which may lead to unnecessary (animal) testing and a less cost-effective risk assessment approach.

According to the *Guidance Document on Aquatic Ecotoxicology* (European Union 2002), the lower-tier uncertainty factors (UFs) may be reduced if additional sensitive species are tested. For example, if in the first tier *Daphnia magna* is the most sensitive standard test species for a specific insecticide (EC50 more than a factor of 10 lower than the acute toxicity data for other standard test species) and acute laboratory toxicity tests with additional arthropod species reveal that *Daphnia magna* is among the most sensitive arthropod species tested, the UF of 100 may be lowered to 10. The extent to which the first-tier UF is reduced, however, largely depends on expert judgment.

A statistical extrapolation technique, such as the method described by Aldenberg and Jaworska (2000), can also be used to calculate the concentration at which a specified proportion of species p is expected to suffer direct toxic effects, referred to as the hazardous concentration (HC) to $p\%$ of the species (HCp). The SSD from which the HCp is derived can be based on either acute or chronic toxicity data. However, the smaller the number of data available for the calculation, the larger the confidence interval around the SSD (and the HCp) will be. It is common to take the 5th percentile of the SSD (median HC5) or the lower 90% confidence bound for it (lower-limit HC5) (see Figure 5.1). When based on chronic toxicity data, the median HC5 is the concentration that protects 95% of the tested species with 50% certainty, while the lower-limit HC5 protects 95% of the species with 95% certainty. The HARAP guidance document (Campbell et al. 1999) mentions HC5 and HC10 values as possible assessment endpoints. However, in the *Guidance Document on Aquatic Ecotoxicology* (European Union 2002) currently no established guidance is provided on which HCp is appropriate for assessments under Directive 91/414/EEC. The technical guidance document underlying the WFD mentions the median HC5 divided by an UF as a possible assessment endpoint to derive a water quality standard (EC 2003).

Calibration of the relationship between HC5 and HC1 values and results of semi-field experiments for insecticides, herbicides, and fungicides was funded by DEFRA (UK Department of Environment, Food, and Rural Affairs) and the Dutch ministry of LNV (Dutch ministry of Agriculture, Nature, and Food Quality) and published by Maltby et al. (2005), Brock et al. (2006), EFSA (2006a), Van den Brink et al. (2006), and Maltby et al. (submitted). In their evaluations, the median HC5 value, the lower limit HC5, and the median HC1 estimate were used. They demonstrated that the key point is to focus on a specific taxonomic group for which the assesment concerns a pesticide with a specific toxic mode of action.

Maltby et al. (2005) showed for insecticides and aquatic arthropods that the lower-limit HC5 estimate on the basis of acute toxicity data provides a lower concentration than the ecological threshold concentration in micro/mesocosms (classified as Effect class 1 to 2 and measured as peak concentration; for explanation of Effect classes, see Section 8.2 in Chapter 8). This was not only valid for single but also for multiple applications of the insecticides evaluated (azinphos-methyl, carbaryl, carbofuran, chlorpyrifos, diflubenzuron, fenvalerate, fenitrothion, lambda-cyhalothrin, lindane, methoxychlor, parathion-ethyl). The median HC5 estimate based on acute toxicity for freshwater arthropods was for 6 of the 7 insecticides tested protective for single applications. This median acute HC5 appears also protective of longer-term exposure (due to multiple applications) to the insecticides tested in micro/mesocosms when at least a safety factor of 5 is applied (Maltby et al. 2005). Only a few published data for insecticides are available to evaluate whether the HC5 value derived from chronic laboratory no observed effect concentrations (NOECs) is protective when compared with threshold levels of long-term exposure in micro/mesocosms. The available data for the insecticide azinphos-methyl, however, showed that the median HC5 based on chronic laboratory toxicity data for arthropods is lower than the threshold level for direct toxic effects due to long-term exposure (expressed as time-weighted average [TWA] concentration) in freshwater mesocosms (Giddings et al. 1994; Maltby et al. 2002).

Van den Brink et al. (2006) showed that for herbicides and primary producers the lower limit of the acute HC_5 and the median value of the chronic HC5 were protective of adverse effects in aquatic micro/mesocosms (classified as Effect class 1 to 2) even under a long-term exposure regime. The evaluated herbicides were 2,4-D, atrazine, diuron, linuron, metamitron, metribuzin, pendimethalin, and simazine. The median HC5 estimate based on acute toxicity data of herbicides was protective of adverse effects in aquatic micro/mesocosms when a short-term exposure regime (pulse application in flow-through system; single application of a nonpersistent [$DT50_{water}$ < 10 days] herbicide in stagnant test system) was studied (Van den Brink et al. 2006).

Maltby et al. (submitted) reanalyzed the relationships between SSDs constructed with acute toxicity data and threshold concentration derived from micro/mesocosm experiments for insecticides (as published by Maltby et al. 2005) and herbicides (as published by Van den Brink et al. 2006) and demonstrated that for these groups of pesticides also the median HC1 can be used to derive an appropriate regulatory acceptable concentration (RAC), even for repeated-exposure regimes.

Results of the comparison of the SSD approach and the model ecosystem approach in the effect assessment for fungicides can be found in the work of Brock et al. (2006), EFSA (2006a), and Maltby et al. (submitted). Both the median HC1 and the lower limit of the HC5, based on acute toxicity data for the relevant taxonomic group (which may be a wider array of species tested due to the general biocidal properties of certain fungicides), result in concentrations that are lower than the threshold level for toxic effects (classified as Effect class 1 to 2 and expressed in terms of peak concentration) in a micro/mesocosm test treated repeatedly with the same fungicide (carbendazim, chlorothalonil, fluazinam, kresoxim-methyl, mancozeb, picoxystrobin, tolylfluanid, trifloxistrobin). The median HC5 estimate based on acute toxicity data was also protective of a single azoxystrobin application. The median HC5 (based on acute toxicity data) for carbendazim and triphenyltin-acetate, however, resulted in a higher concentration than the Effect class 1 to 2 threshold concentration in model ecosystems treated once with these compounds. For carbendazim, this can be explained by its persistence in the water column, so that a single application already results in a chronic exposure regime. Although triphenyltin-acetate dissipates quickly from the water column, this compound is persistent in the sediment and bioaccumulates in the food chain, also leading to long-term exposure (see Roessink et al. 2006).

5.3 SSDs AND QUALITY OR VALIDITY CRITERIA FOR EXPOSURE REGIMES IN THE TOXICITY TESTS

In discussing the linking of exposure and effects when using the SSD concept in the risk assessment for pesticides, a distinction should be made between the concentration–response relationships in the laboratory toxicity tests of which the results are used to construct the SSDs and the linking of the assessment endpoints derived from these SSDs (e.g., HC5 values) to the exposure concentrations predicted for the field. When performing laboratory toxicity tests with pesticides and freshwater organisms, the EC50 and NOEC values usually are expressed in terms of exposure concentrations in water.

The following quality or validity criteria with respect to exposure regime may be considered in deriving toxicity data and SSDs, namely:

1. Preferably, gather the toxicity data used for additional test species as much as possible in line with internationally accepted testing guidelines developed for standard test species (e.g., of the Organisation for Economic Cooperation and Development, OECD) and performed according to good laboratory practice. For example, the exposure regime in acute laboratory tests with invertebrates is usually static or semistatic, while that in chronic tests with invertebrates is usually semistatic or flow through. It is anticipated that for additional test species the test conditions do not totally comply with the specific testing guidelines for standard test species. The deviating test conditions and the properties of the test organisms, however, need to be documented.

2. Make a proper distinction between acute and chronic toxicity data. "Acute toxicity" is the ability of a substance to cause adverse effects within a short period following dosing or exposure. Acute tests with freshwater organisms usually have a duration of 48 to 96 hours. Generally, a chronic toxicity test is defined as a study in which the species is exposed to the pesticide for at least 1 full life cycle or the species is exposed to the pesticide during 1 or more critical and sensitive life stages (see, e.g., Holland 1996). Consequently, what is considered chronic or acute is very much dependent on the species and studied endpoint considered. Separate SSD curves should be constructed for acute and chronic toxicity data. Usually, acute EC50–LC50 values are used to construct SSDs for short-term exposure. However, in some European member states, acute LC10 and acute NOEC values for fish are used to construct the SSD and to calculate the HC5 for fish since a higher protection level is desired for vertebrates than for invertebrates and plants. Chronic EC10 and chronic NOEC values usually are used to construct SSDs for long-term exposure.

3. The exposure concentrations in the laboratory toxicity test need to be well characterized and preferably based on measurements of the active substance in the application solution or test medium.

4. Preferably, acute toxicity values (e.g., EC50s and LC50s) are expressed in terms of the mean exposure concentration as measured or calculated in the medium of the short-term test. The nominal concentration to calculate the acute toxicity values may be used if the measured concentration are in the range of 80% to 120% of nominal.

5. In the case of substances that are labile in water and the predicted or measured field exposure regimes comprise very short-term pulses, it may be valid to use the measured initial concentration (or the checked nominal

concentration on the basis of measurements in the dosing solution) to calculate acute toxicity values, at least if the water dissipation rate of the active substance is not faster in the laboratory test vessels than predicted for (or measured in) the field. In that case a PEC_{max} should be used in the risk assessment.

6. Chronic toxicity values (e.g., chronic EC_{10}s and chronic NOECs) usually are expressed in terms of the mean exposure concentration (e.g., TWA concentrations) as measured or calculated in the medium of the long-term test.

7. The relevant toxicity values should not be above the water solubility of the active substance, and the derived toxicity values should fall within the range of concentrations tested. Toxicity values that are much higher or lower than the highest or lowest concentration tested should be expressed as "higher-than" or "lower-than" values, respectively. These values are not used to construct the SSD, but they may be useful in the evaluation of the assessment endpoint (e.g., HC5) or in the final risk assessment. For example, the higher-than values may be used to determine rank for plotting purposes.

5.4   SSDs AND LINKING EXPOSURE AND EFFECTS

5.4.1   ACUTE RISK ASSESSMENT

According to the decision scheme (see Chapter 2), in the acute risk assessment the peak concentrations of the pesticide in the relevant exposure scenarios are usually the appropriate predicted environmental concentrations (PECs) to evaluate risks due to short-term exposure (see Box 1 in Figure 2.1). When using the SSD approach, these PEC_{max} values should be lower than the effect assessment endpoint (RAC) derived from the acute SSD. The median acute HC5 value divided by an appropriate UF (which may be 1 for single short-term pulse exposures) may be used as the RAC. In the case of repeated pulsed exposures that are toxicologically not independent, also the lower-limit acute HC5 value or the median HC1 value might be used as the appropriate RAC (see Maltby et al. 2005; Brock et al. 2006; Van den Brink et al. 2006; Maltby et al. submitted). However, for substances differing in toxic mode of action from the plant protection products evaluated by these authors, a prudent approach is advocated. Alternatively, to estimate the RAC for short-term exposure, the UF of 100 that is used in the first-tier risk assessment may be lowered (see EC 2002) if the test species that triggered the risk belongs to the most sensitive species tested.

5.4.2   CHRONIC RISK ASSESSMENT

According to the decision scheme presented in Chapter 2, in the chronic risk assessment either the predicted peak concentration (PEC_{max}) (see Box 1 in Figure 2.1) or TWA concentration of the pesticide in the relevant exposure scenario (see Box 2 and Box 3 in Figure 2.1) may be the appropriate PEC to evaluate risks due to long-term exposure. Criteria to decide whether the TWA approach is appropriate are given in Section 3.4 of Chapter 3. The selection of the length of the TWA time window should be based on ecotoxicological information related to organisms that triggered the risk (e.g., acute-to-chronic [A/C] ratio; time to onset of effect [TOE] information in the toxicity tests; information on duration of possible sensitive life stages). The majority of the ELINK workshop participants considered a default time window of 7 days for the TWA estimate of the long-term PEC a good option (see Box 3 in Figure 2.1) in case no scientifically underpinned arguments are provided to shorten or lengthen this default time window. When using the SSD approach, either the PEC_{max} or the appropriate TWA PECs should be lower than the RAC derived from the chronic SSD. On the basis of the data presented here, the median chronic HC5 value divided by an appropriate UF (which may be 1) or the lower-limit chronic HC5 value can be used as the RAC for long-term exposure, although to date only a limited number of chronic HC_5 values could be compared with threshold levels derived from micro/mesocosm tests simulating a chronic exposure regime (see Brock et al. 2006). Alternatively, to estimate the RAC for long-term exposure, the UF of 10 that is used in the first-tier risk assessment may be lowered (see EC 2002) if the test species that triggered the risk belongs to the most sensitive species tested.

Case Study 5.1   ELINKesterase (insecticide; Cholinesterase Inhibitor): Refining the First-Tier RAC for Acute Exposure by the SSD Approach

The calculated PEC_{max} from the relevant FOCUS scenarios was 0.17 µg as/L, and the relevant generalized exposure profile selected consisted of 3 pulse exposures with one high pulse (0.17 µg as/L) and two smaller ones (pulse height, respectively, 0.12 and 0.07 µg as/L), while the interval between pulses was 5 to 9 days. Pulse duration varied between 2 and 3 days.

The most sensitive standard test species was *Daphnia magna* (48-hour flow-through test; EC50 = 1.8 µg as/L), followed by fish (*Oncorhynchus mykiss*; 96-hour LC50 = 32.0 µg as/L). The reported bioconcentration factor (BCF) for fish was less than 100. The reported EC50 value for the green alga *Chlorella vulgaris* was greater than 1000 µg as/L. Applying an UF of 100 resulted in a first-tier RAC of 0.018 µg as/L. This first-tier RAC was refined by means of the SSD approach.

Additional semistatic (daily renewal) acute toxicity tests were performed with 9 freshwater arthropods. These acute tests lasted 2 days, and 48-hour EC_{50} values could be calculated. The measured concentrations in the test vessels varied between 85 and 106% of the nominal concentration approximately 1 hour after application. On average, 60% of the test compound was still available in the test medium after 24 hours (just before renewal of the medium). The EC_{50} values were based on nominal test concentrations. The SSD approach (applying the methods described in Aldenberg and Jaworska 2000) resulted in the following acute HC5 values and 95% confidence intervals: 0.34 (0.21 to 2.70) µg as/L.

Since a semistatic exposure regime was adopted for the additional test species and dissipation in the test vessels was slower than predicted for the field, it was decided that the EC50 values provided could be used to calculate an HC5. The lower limit of the HC5 (0.21 µg as/L) was used for the higher-tier RAC because of 1) the repeated pulse exposure regime in the relevant FOCUS scenarios and 2) no additional evidence was provided that the different pulses can be considered as independent from a toxicological point of view.

This SSD-based RAC value of 0.21 µg as/L for acute exposure was higher than the PEC$_{max}$; consequently, risks were considered acceptable for aquatic invertebrates and the pulsed exposure regime under evaluation. Since the lower-limit HC5 was based on toxicity data for invertebrates, it was checked if the revised RAC was also protective for fish and primary producers. The first-tier risk assessment procedure for short-term exposure, however, indicated that fish and primary producers were not at risk at concentrations below $31.0/100 = 0.31$ µg as/L and $>1000/10 = >100$ µg as/L, respectively. In addition, the potential of this substance to bioaccumulate was considered small (BCF < 100).

Case Study 5.2   ELINKthiol (Fungicide; Nonspecific Thiol Reactant): Refining the First-Tier RAC for Acute Exposure by the SSD Approach

A risk assessment was performed for a fungicide that was observed to be very labile in water with a DegDT50 (20 °C) of 2.4 hours. Spray drift was the main entry route to surface waters, and the fungicide was applied 4 to 6 times at weekly intervals. The calculated PEC$_{max}$ for this fungicide was 17.3 µg as/L. This PEC$_{max}$ was derived from a single application, which gave higher concentrations than the multiple application good agricultural practice (GAP). A typical predicted exposure pattern for this fungicide in surface waters is presented next.

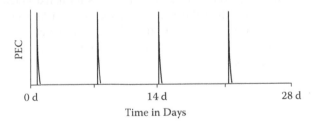

The most sensitive standard test species was *Oncorhynchus mykiss* (96-hour flow-through test; LC50 = 102 µg as/L). Of the basic set of standard test species, *Daphnia magna* (96-hour LC50 = 4286 µg as/L) was relatively insensitive. The reported EC50 value for the green alga *Pseudokirchneriella subcapitata* was greater than 1000 µg as/L. Applying a UF of 100 on the endpoint for fish resulted in a first-tier RAC of 1.02 µg as/L. This first-tier RAC was refined by means of the SSD approach.

Additional static acute toxicity tests were performed with 6 fish species, including *Oncorhynchus mykiss*. These acute tests lasted 4 days and 96-hour LC50 and LC10 values could be calculated. The measured concentration in the test vessels 15 minutes after application was on average 85% of the nominal concentration, and after 2 hours approximately 50% of the nominal concentration still could be detected. At the end of the acute test, the fungicide could not be detected anymore in the water of the test vessels. The LC50 and LC10 values of the 6 fish species were based on measured initial concentrations in the test vessels. The 96-hour LC50 value for *Oncorhynchus mykiss* in the static test was 199 µg as/L (so approximately 2 times that of the flow-through study).

The SSD approach (applying the methods described in Aldenberg and Jaworska 2000) resulted in the following acute HC5 values and 95% confidence intervals:

- 40.3 (8.9 to 97.3) µg as/L on basis of 96-hour LC50 values
- 36.4 (6.9 to 89.0) µg as/L on basis of 96-hour LC10 values

In addition, a 28-day pulsed exposure study was performed with *Oncorhynchus mykiss* in the laboratory. In this test, the fungicide was applied 4 times at weekly intervals. From this repeated pulse exposure study, at the end of the experiment an LC50 could be calculated of 193 µg as/L. This suggests that the effects observed in the repeated pulse exposure study can be largely explained by the first pulse exposure (toxicological independence of peaks).

In the acute risk assessment, it was decided to use the median HC5 value based on LC10 values divided by a UF of 2. LC10 values were used to construct the SSD because the sensitive test species concerned vertebrates. The median HC5 value was used since the different pulse exposures seem toxicologically independent. An AF of 2 was used to address remaining uncertainty (refined repeated pulse exposure test was performed with *Oncorhynchus mykiss* only). Consequently, the SSD-based acute RAC for this fungicide was assessed to be 18.2 µg as/L (median HC5 on basis of 6 LC10 values for fish divided by 2). This value is somewhat higher than the PEC_{max}, so the acute risks were considered acceptable. Note that in this case study the AF of 2 and the choice to use LC10 values to construct the acute SSD should be considered examples, and that regulators may adopt other choices.

6 Use of Refined Exposure Single-Species and Population Studies in the Risk Assessment of Plant Protection Products*

When predicted (modeled) field exposure profiles differ considerably from exposure regimes in standard aquatic toxicity studies, it may be appropriate to design higher-tier laboratory toxicity tests that more closely resemble realistic exposure profiles. However, before embarking on a new study design with a refined exposure profile, it is necessary to consider the effects (acute or chronic) and the nature of the ecotoxicologically relevant concentration (ERC) (e.g., peak or time-weighted average [TWA]) and whether the focus is the concentrations of the pesticide in water or sediment.

If the use of the TWA approach in the chronic risk assessment is appropriate (see Boxes 2 and 3 in Figure 2.1 and the criteria in Section 3.4 of Chapter 3), refined exposure studies are not necessary. In that case, the concentration–response relationships observed in chronic toxicity tests, as well as the derived regulatory acceptable concentration (RAC), can be expressed in terms of TWA concentrations. This RAC value can be compared in the risk assessment with the appropriate TWA predicted environmental concentration (PEC). If, however, the TWA approach cannot be applied in the chronic risk assessment, refined ecotoxicological exposure studies may be a higher-tier option.

In designing refined exposure studies with standard or additional species, parameters like "mode of action," time to onset of effect, and the appropriateness of using sensitive life stages should be considered. To adopt a realistic worst-case scenario, the refined exposure regime tested should be guided by relevant exposure predictions for the intended agricultural uses (e.g., as deduced from FOCUS [Forum for the Co-ordination of Pesticide Fate Models and Their Use] surface water scenarios or from national exposure scenarios but based on the most representative generalized

---

* Box 4 in the decision scheme; see Figure 2.1 in Chapter 2 or Appendix 4.

exposure profile as described in Chapter 4). The recommendation to use generalized exposure profiles prevents the use of specific scenarios that may change over time or location of product use, which could lead to additional and possibly unnecessary work.

The likely output of a refined exposure study should be carefully considered before the start of the study (i.e., whether it may change the outcome of the overall risk assessment). For example, if the time to onset of effects is very short, a narrow pulsed exposure may already yield the effects irrespective of the dissipation rate of the compound. If the concerns are chronic sublethal effects (e.g., to *Daphnia*), then the duration of the study, even if the system is exposed to a single narrow pulse, must extend beyond a week to ensure that potential effects on reproduction and survivorship of neonates are monitored.

According to the ELINK assessment scheme (Box 4 in Figure 2.1), the option for refined exposure studies may apply to acute and chronic risk assessments and to standard or additional test species, and it may apply to individual-level or population-level studies (see also Campbell et al. 1999). In the event that standard test species are assessed in refined exposure studies, a reduction of the uncertainty factor (UF) is not justified when deriving the RAC. However, a higher toxicity value (e.g., acute EC50 or chronic NOEC) from the refined exposure study with standard test species and the application of the appropriate UF (e.g., 100 to derive the acute RAC and 10 to derive the chronic RAC) may change the overall risk assessment. Refined exposure studies may also be used to put other higher-tier data into perspective. If, for example, the sensitive species were identified from SSDs or mesocosm experiments but these tests did not fully address the environmentally realistic exposure regime, then refined exposure studies with these sensitive species might provide relevant results for the final risk assessment (see, e.g., case study examples in Chapter 5 [species sensitivity distribution approach] and Chapter 8 [model ecosystem approach]). In principle, refined exposure profiles may also be used for recovery studies; however, the same limitations as mentioned apply.

Population-level experiments with refined exposure regimes are usually performed with populations of individuals that differ in age and developmental state. If, for example, *Daphnia* is the most sensitive standard test species for a specific substance, the NOEC of a long-term refined exposure study with a mixed population of *Daphnia* (addressing an exposure regime guided by relevant field exposure predictions) might be used in the chronic risk assessment by applying the standard UF of 10 to derive a chronic RAC. It should be mentioned, however, that this approach warrants "confirmation or validation," such as by comparing the results with threshold concentrations (or RACs) derived from appropriate semifield experiments in which chronic exposure regimes with the same substances were investigated. Because sensitivities of different life stages are usually different, it is, from a scientific point of view, appropriate also to study the recovery potential of an affected population. However, the relevance of the recovery of the test species in relation to field conditions and to that of other species should be carefully considered. *Daphnia*, for example, does not represent k-strategists.

A case study is presented next that outlines a higher-tier population-level approach.

Case Study 6.1 ELINKpyrimate

ELINKpyrimate is a fungicide used in orchards. One application is intended as a spray on late (fruit) growth stages of trees.

ELINKpyrimate is toxic to fish, invertebrates, and algae. It affects fish and invertebrates mainly on survival. The acute–chronic (A/C) ratio was less than 10 for *Oncorhynchus mykiss* and *Daphnia magna*. *Oncorhynchus mykiss* showed the highest sensitivity, with an acute LC50 of 0.75 µg/L and a chronic NOEC of 0.18 µg/L based on mortality, resulting in a tier 1 acute RAC of 0.0075 µg/L and a tier 1 chronic RAC of 0.018 µg/L (see Table CS6.1A).

As ELINKpyrimate dissipates rapidly from the water column (partitions rapidly to the sediment, where its DT50 [time until 50% decay] is around 30 days) and mainly acts acutely, a refined exposure test was performed to assess the level of effects under more realistic exposure conditions. Since the occurrence of one main peak due to spray drift was calculated for the relevant FOCUS scenarios and exposure due to drainage was estimated to be low, the most relevant exposure profile was identified as a single-pulsed exposure. A single-pulsed exposure regime was employed in a laboratory water–sediment study with rainbow trout, the most sensitive standard test species.

TABLE CS6.1A

Overview of First-Tier Toxicity Data for ELINKpyrimate

Study Type	Test Duration	Test Organism	Findings
Acute test in fish (flow through)	96 hours	*O. mykiss*	LC50 (96 hours) = 0.75 µg/L
Acute test in fish (flow through)	96 hours	*L. macrochirus*	LC50 (96 hours) = 3.5 µg/L
Acute test in fish (flow through)	96 hours	*C. carpio*	LC50 (96 hours) = 7.04 µg/L
Chronic test in fish (flow through)	28 days	*O. mykiss*	NOEC (28 days) = 0.18 µg/L
Acute test in invertebrates (flow through)	48 hours	*D. magna*	EC50 (48 hours) = 15 µg/L
Chronic test in invertebrates (flow through)	21 days	*D. magna*	NOEC (21 days) = 4 µg/L
Chronic test in sediment-dwelling organisms	28 days	*C. riparius*	NOEC (28 days) = 2 mg/L
Growth test in algae (static)	72 hours	*S. subspicatus*	ECb50 = 9.8 µg/L ECr50 = 6.4 µg/L

TABLE CS6.1B

Long-Term Effects of a Single Application of ELINKpyrimate on Mortality of Juvenile Rainbow Trout

Time of Sampling	Mortality (%) at the Following Concentrations (μg/L)						
	Control	0.1	0.32	0.56	1.0	3.2	10
24 hours	0	0	0	0	0	40	95
48 hours	0	0	0	0	0	90	100
72 hours	0	0	0	0	0	95	100
96 hours	0	0	0	0	0	100	100
5 days	5	0	0	0	0	100	100
7 days	5	5	0	0	0	100	100
14 days	5	5	0	5	5	100	100
21 days	5	5	0	5	5	100	100
28 days	5	5	5	5	5	100	100

To investigate the short- and long-term effects of the single pulsed exposure on sensitive life stages of rainbow trout, the test was performed on juveniles (standard 28-day test). In this test, the substance was added in water at $t = 0$ and thereafter followed in the system by regular samplings of the water and the sediment phases. The observed dissipation DT50 in the water column was 7 days and was 28.8 days in the sediment. The observed DT50 in the water phase of the water–sediment test systems was in the higher range of the predicted water dissipation rate for the relevant FOCUS scenarios (DT50 1 to 8 days).

Several concentrations (0.1, 0.32, 0.56, 1.0, 3.2, and 10 μg/L) were tested in the water–sediment systems. Effects on fish were checked every day during the whole test period. Results are summarized in the Table CS6.1B based on nominal initial concentrations.

Some fish displayed temporary behavioral impairments like slowed-down movements during the first 2 days at the 1 μg/L level, but it was not related to any other effect. The NOEC of the study was calculated to be 1 μg/L (nominal).

Mainly lethal effects were observed on juvenile O. mykiss. At the LOEC (3.2 g/L) mortality was observed during the first 96 hours. Since the dissipation of Elinkpyrimate was somewhat faster in the modified exposure test (DT50 = 7 d) than in the worst case FOCUS surface water scenario (DT50 = 8 d) it was decided not to express the NOEC of juvenile rainbow trout in terms of nominal concentration (1 g/L) but in terms of a 96 h TWA concentration (100% mortality within 96 h). The 96 h TWA NOEC in the modified exposure test was 0.61 μg/L. This 96 h TWA NOEC was then divided by 10, resulting in a RAC of 0.061 μg/L (as compared to the Tier 1 chronic RAC of 0.018 μg/l). Since short-term exposure could induce mortality, the risks are assessed based on PEC_{max} values.

7 Toxicokinetic and Toxicodynamic Modeling in the Risk Assessment of Plant Protection Products[*]

CONTENTS

7.1   INTRODUCTION

In the ELINK (EU Workshop on Linking Aquatic Exposure and Effects in the Registration Procedure of Plant Protection Products) decision scheme (Figure 2.1), the higher-tier risk assessment procedures mentioned in Box 4 refer to toxicokinetic and toxicodynamic (TK/TD) modeling approaches. This chapter aims 1) to shed light on the current state of the art of TK/TD modeling and 2) to describe how TK/TD models can be used in the risk assessment for pesticides. Although some

---

[*] Box 4 in the decision scheme; see Figure 2.1 in Chapter 2 or Appendix 4.

of the models discussed have been available for a significant period of time (e.g., DEBtox, dynamic energy budget), the science base continues to develop; use in regulatory risk assessment is not yet fully developed.

Pesticides may reach water bodies via various pathways, and typically aquatic organisms are exposed to sequential pulses with fluctuating concentrations (Reinert et al. 2002). Standard toxicity tests with aquatic organisms are performed either at maintained concentration (flow through) or under static (single initial input of test material) or semistatic conditions. Hence, any extrapolation to more realistic patterns of exposure must rely either on specifically designed experiments for specific exposure scenarios or on modeling (Boxall et al. 2002). Modeling has the advantage that we can extrapolate to a wide range of field exposure scenarios. A variety of approaches has been developed (Reinert et al. 2002; Ashauer et al. 2006a), and the importance of the response by individual organisms to recovery periods between successive pulses has been recognized (Kallander et al. 1997; Ashauer et al. 2007a). After selecting a model that simulates effects in an organism based on the time course of the contaminant concentration, the model parameters are estimated by calibration on experimental data, and the model performance is evaluated against independent experimental data; this establishes a range of validity within which the model can be used to extrapolate to effects from specific patterns of exposure.

7.2   TK/TD MODELS AVAILABLE

Toxicokinetic and toxicodynamic (TK/TD) models describe the processes that link exposure to effects in an organism. Toxicokinetics consider the time course of concentrations within an aquatic organism in relation to concentrations in the external medium. Toxicodynamics describe the time course of damage and repair to the affected organisms based on specific patterns of exposure to the test compound. The models fitting into the TK/TD category have been reviewed by Ashauer et al. (2006a). The models were considered in detail by ELINK working group (WG) 5 and the outputs from their work and discussions are provided as a WG report in Chapter 14. Only the most important points are covered here to avoid repetition. The main TK/TD models available are summarized along with their major characteristics in Table 7.1.

7.2.1   Toxicokinetics

The predicted endpoint of the TK submodel is the concentration at the target site. The simplest description of TK is the 1-compartment first-order kinetics model, which is also the most commonly used in aquatic ecotoxicology. It describes the dynamics of the internal (whole-body) concentration of the toxicant C_{int} depending on the external concentration C_{ext} using uptake and elimination rate constants (k_{in} and k_{out}, respectively) and the external concentration (Equation 7.1).

$$\frac{dC_{int}(t)}{dt} = k_{in} \times C_{ext}(t) - k_{out} \times C_{int}(t) \qquad (7.1)$$

TABLE 7.1

Overview of Toxicokinetic and Toxicodynamic Models (See the Work Group Report on Extrapolation Tools in Chapter 14 for Detailed Comparison)

	Toxicokinetics: Uptake and Elimination	Hazard Proportional to C_{int}	Hazard Proportional to Damage	Threshold Concentration for Effect	Target Interaction/ Reversibility of Binding	Recovery/Repair of Effects
SHM (simple hazard model)	x	x	—	—	—	Instantaneous and complete
DEBtox (dynamic energy budget)	TK related to C_{ext}, only elimination considered	x	—	x	—	Instantaneous and complete
THM (threshold hazard model)	x	x	—	x	—	Instantaneous and complete
DAM (damage assessment model)	x	—	x	—	—	x
DHM (damage hazard model)	x	x	x	—	—	x
TDM (threshold damage model)	x	—	x	x	—	x
RKM (receptor kinetics model)	x	—	x	x	x	x

7.2.2 Toxicodynamics

The models summarized in Table 7.1 are hazard models; that is, they assume that death (or a sublethal endpoint), although depending on the toxicant concentration in the organism or the damage, is at least partly stochastic. Hence, the hazard rate $h(t)$ is the probability of death at a given point in time. The integral of the hazard rate, the cumulative hazard $H(t)$ appears in the equation that defines the survival probability $S(t)$, which is the probability that an organism survives until time t (Equation 7.2):

$$S(t) = \exp[-H(t)] \tag{7.2}$$

The main difference in concepts between the TK/TD models is in the component to which the toxic effect (i.e., hazard rate) is related. Consequently, the TD concepts differ in their TD assumptions and in the range of toxic mechanisms for which they are valid. Explicit statement of the underlying assumptions should help to select an appropriate model for a compound with a given mechanism of action. The models are briefly summarized next; the WG report "Extrapolation Methods in Aquatic Effect Assessment of Time-Variable Exposures to Pesticides" in Chapter 14 gives a more detailed analysis.

- Effect (hazard rate) is proportional to internal concentration (critical body residue models):

  TD equation: $h(t) = k * C_{int}$
  Assumption: Effect is proportional to internal concentration; the time course of effect will mimic the time course of the body residue.
  Examples: Widianarko and van Straalen (1996); Jagers op Akkerhuis, Seidelin, et al. (1999)

- Effect (hazard rate) is proportional to internal concentration above a threshold:

  TD equation: $h(t) = k * \max[C_{int} - threshold, 0]$
  Assumption: Effect is proportional to internal concentration above a threshold for effects; implies instantaneous and complete recovery of the fitness of an organism.
  Examples: DEBtox (Kooijmann and Bedaux 1996; Péry et al. 2001; Pieters et al. 2006).

- Effect (hazard rate) is proportional to damage:

  TD equation: $dD(t)/dt = k_k * C_{int} - k_r * D(t)$,
  where k_k and k_r are killing and recovery rate constants, respectively, and $D(t)$ is the damage at any point in time.
  Assumption: Allows the time course of effects to differ from the time course of the internal concentration. Recovery is a function of the damage incurred. Even small amounts of toxicant will exert an effect.
  Examples: DAM (damage assessment model; Lee et al. 2002) for constant exposures; not yet applied to time-varying exposures.

- Effect (hazard rate) is proportional to damage above a threshold:
  TD equation: $dD(t)/dt = k_k * C_{int} - k_r * D(t)$, and $h(t) = \max[D(t) - \text{threshold}, 0]$
  Assumption: Allows the time course of effects to differ from the time course of the internal concentration plus the introduction of a threshold for effects.
  Examples: TDM (threshold damage model; Ashauer et al. 2007a, 2007b, 2007c); RKM (receptor kinetics model; Jager and Kooijman 2005) differs in simulating saturating of target sites for compounds that occupy specific receptor sites.

7.3 EXPERIMENTATION

7.3.1 UPTAKE–ELIMINATION EXPERIMENTS

The purpose of uptake–elimination experiments is to determine the time course of concentrations within an aquatic organism in relation to concentrations in the external medium. Generally, the work is undertaken to derive model input parameters that will allow prediction of internal concentrations of chemicals for new situations (Figure 7.1). Expression of the TK within the main models available generally relies on 2 competing first-order expressions and assumes that organisms comprise a single compartment with uniform chemical concentration. Equation 7.1 assumes first-order uptake and first-order elimination as well as nongrowing organisms. More complex TK models are available and may be more suitable for organisms for which the single-compartment assumption is inadequate (e.g., Barber 2003; Nichols et al. 2004). TK parameters may vary with life stage of a species, primarily in response to changes in lipid content.

Uptake–elimination experiments comprise the following phases:

1) Equilibrate and characterize the system. Ensure that test organisms are established in experimental media and that test conditions are established (temperature, oxygenation of water, target pH if required, etc.); determine size, weight, and lipid content for test organisms; and begin to monitor conditions in the aqueous medium. Take organisms for determination of background levels of test compound.

2) Introduce test compound into the experiment, normally as radiolabeled material. Monitor concentrations of the test compound in the aqueous medium and (sacrificed) test organisms over time. This phase tracks the assimilation of test compound into the organisms. The concentration added needs to be sufficient that it can be detected in the test organism but sufficiently small that there is no significant ecotoxicological effect. If a compound is known to be stable in both water and the test organism, analysis may be on the basis of total radioactivity; if stability cannot be assumed, radioactivity should also be characterized to determine any formation of metabolites as these may have different uptake and elimination properties from the parent compound.

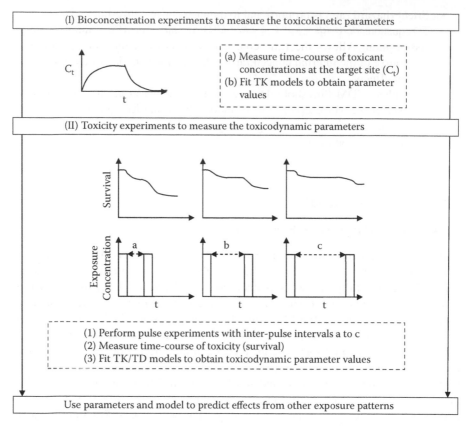

FIGURE 7.1 Experiments required to parameterize a toxicokinetic and toxicodynamic model, for example, the TDM. Interpulse interval a should be long enough for complete (95%) depuration but not much longer to ensure that toxicodynamic recovery has not occurred yet. Interval c should allow for complete organism recovery.

3) Transfer organisms to clean medium. Continue to monitor concentrations in the test organism over time as well as any measurable concentrations in the aqueous medium (through release [depuration] from the organisms). This phase tracks elimination of test compound from the organisms.

4) Use appropriate fitting software[1] to fit Equation 7.1 to the experimental data, thus obtaining estimates for the uptake rate constant and elimination rate constant. Initial parameter estimates may influence the success of the calibration step. Consideration should be given to obtaining initial estimates that are as robust as possible (e.g., through use of quantitative structure–property relationships). Multiple combinations of initial parameter estimates should be used to assess nonsingularity in the fitting procedure (i.e., whether fitting may deliver more than 1 set of [locally] optimal parameters).

Detailed descriptions of uptake–elimination experiments with pesticides were given by Muir et al. (1994), Nuutinen et al. (2003), and Ashauer et al. (2006b).

Variability in experimental results is generally rather large in these experiments, and careful attention should be paid to the number of time intervals sampled and the extent of replication.

7.3.2 TOXICODYNAMIC EXPERIMENTS

The purpose of TD experiments is to infer the time course of damage and repair to the target organisms based on the time course of survival in response to specific patterns of exposure to the test compound. The information can be used to estimate TD parameters within TK/TD models (Figure 7.1). Generally, the TK parameters will already have been estimated from independent experiments and are then held constant. Whereas TK descriptions in the primary models are generally identical or similar, the TD assumptions vary from model to model. This variability will influence test design depending on the modeling tool selected (Section 7.2.2 and WG report in Chapter 14). For example, the main models include either 1, 2, or 3 TD parameters; the greater the number of parameters to be estimated, the greater the amount of experimental data that will be required to obtain robust parameter estimates. Ashauer et al. (2007b, 2007c) suggested that between 2 and 6 TD experiments with contrasting, time-varying exposure patterns were sufficient to obtain robust estimates for the 3 TD parameters in the TDM model.

Toxicodynamic experiments have the following phases:

1) The exposure patterns to be studied in the experiments are defined. This may require preliminary toxicity tests to determine the range of concentrations likely to cause the relevant effect in the test organism (frequently, these data may be available from standard or additional single-species tests at lower tiers). Exposure scenarios should then be defined to generate clear differences in effects. The selected TK/TD model can be parameterized with measured uptake and elimination constants and with other parameters taken from the literature or QSAR (quantitative structure–activity relationship) calculations to aid exposure definition. Exposure scenarios should target significant but not complete effects (e.g., a mortality test might target mortality of between 20 and 80% of individuals over the full test duration). Clear differences in effects from different exposure patterns are also desirable.

2) Experiments with time-varying exposures are undertaken, and frequent (e.g., daily) measurements of exposure concentration and effect endpoint are made. Although uptake and elimination constants will have been determined independently, measurement of internal concentrations within the organism at less-frequent intervals allows checking that TK have not deviated markedly from that expected. The number of experiments to undertake will vary according to the complexity of the TD description within the selected model. Numbers between 1 exposure scenario with several dose levels (e.g., Widianarko and Van Straalen 1996; Jagers op Akkerhuis, Kjæ, et al. 1999; or standard DEBtox) and 2 to 6 exposure scenarios with only 1 dose level (Ashauer et al. 2007b, 2007c) have been recommended.[2] Each exposure scenario stands for a pattern of time-varying concentration.

3) With TK parameters held constant, appropriate fitting software[3] is used to fit the TD element of the model to the experimental data, thus obtaining estimates for the TD parameters. The model should be fitted to all available experimental data concurrently to maximize the rigor of the fitting procedure. Initial parameter estimates may influence the success of the calibration step. Consideration should be given to obtaining initial estimates that are as robust as possible (e.g., by considering independent evidence for the presence or absence of a threshold). Multiple combinations of initial parameter estimates should be used to assess nonsingularity in the fitting procedure.

7.3.3 VALIDATION EXPERIMENTS

None of the TK/TD models has been extensively validated to date. Regulatory use of the models should thus be supported with a validation experiment for the particular combination of compound and organism. Ideally, validation experiments should include an exposure profile that contrasts markedly with those used in model calibration (e.g., more or fewer pulses of shorter or longer duration than previously tested). Longer-term experiments are also useful to demonstrate the ability to extrapolate beyond the precise conditions of the calibration experiments. Consideration of this evaluation phase requires careful definition of validity criteria.

7.4 APPLICABILITY OF THE APPROACH

7.4.1 EXPOSURE PROFILE

A major advantage of the TK/TD modeling approach is that it should not be limited in terms of the kind of exposure profile that can be considered. Most of the effort is required in undertaking experiments to determine and evaluate model input parameters that are robust and broadly applicable. Once this has been done, the models are quick to run, and predictions for effects can be generated for a large number of exposure situations. The FOCUS surface water scenarios already generate output at an hourly resolution, and it would be most appropriate to use the exposure profile at this resolution rather than the daily data calculated prior to standard report production.

7.4.2 EFFECTS ENDPOINTS

The TK/TD modeling is most highly developed for aquatic invertebrates at present. For relatively simple animals, the assumption of uniform internal concentration appears to hold, and the TK can be simulated with a single-compartment, first-order model. A major constraint to current models is that they apply when the duration of exposure is less than the duration of life of the test species and generally assume negligible growth and negligible change in lipid content during the period of exposure (such conditions might reasonably by applied to *Gammarus* and *Asellus*, for example). There is no intrinsic reason that the models cannot be developed to account for species with short generation times for which growth or reproduction are significant within the duration of exposure. However, further work is required on this aspect.

The methodology has generally been applied to effects on survival. However, there is no intrinsic limitation to use of the approach with sublethal endpoints (e.g., growth or reproduction) provided that the model has been appropriately calibrated on the respective endpoint. The TK/TD models may require a limited amount of modification to be applicable to nonlethal endpoints, but the overall conceptualization of the system remains the same. It is noted in the WG report in Chapter 14 that the DEB theory may be particularly suited to simulation of sublethal effects that result from changes in energy allocation within the organism because energy usage and budgets within organisms are explicitly simulated.

Modeling using TK/TD has been applied to simulation of effects in fish (Jager and Kooijman 2005 for AChE [acetylcholinesterase] inhibition). Generally, model complexity increases with respect to the TK because of the need to model partition and depuration processes for individual compartments within the organism. Given the intensive data demands at the parameterization stage, it seems unlikely that these models will be extensively applied to vertebrates because of animal welfare concerns. Ashauer et al. (2007b) noted that application of a time-weighted averages approach that was fitted to pulsed rather than maintained exposure data for macroinvertebrates greatly improved subsequent extrapolation to effects from new exposure profiles. The approach is comprised of running pulsed dose ecotoxicity tests and then relating the time course of the area under the dose curve to the time course of the ecotoxicological effect. Such an approach may offer a compromise between accuracy of extrapolation and extent of additional data requirements for aquatic vertebrates.

No work is available on simulation of effects of time-varying exposures on aquatic plants. Extension to plants with simple structures seems relatively straightforward.

7.4.3 ADDITIONAL OUTPUTS FROM THE MODELING

Since the TK describe the time course of the toxicant concentration in the organism, they can also be used to simulate whether repeated pulsed exposures lead to a buildup of internal concentrations or whether the time between pulses is long enough for complete elimination (depuration) of the compound. The time to eliminate 95% of the compound is approximately $t_{95} = 3 / k_{out}$. This information will have application, for example, in characterizing risks from bioaccumulation. Equally, it might assist in setting acceptable intervals between successive pulses (e.g., where spray drift is the primary route of entry to water).

Similarly, it is also possible to calculate total recovery times for the individual organisms. Since TK (elimination) and TD (recovery) occur in parallel, the calculation of total recovery times can be determined by running the model for the exposure in question. For example, following exposure to a 1-day pulse that kills 50% of the population, the total recovery times of surviving individuals of *Gammarus pulex* are 3, 15, and 25 days for pentachlorophenol, carbaryl, and chlorpyrifos, respectively (Ashauer et al. 2007c). The total recovery time of individual organisms could be used to define the length of the exposure concentration profile that has to be considered as 1 exposure window from a TK point of view, that is, the window within which effects from subsequent exposures are toxicologically not independent of each other.

7.5 REQUIREMENTS FOR RESEARCH AND DEVELOPMENT

Use of TK/TD modeling for effects of pesticides on aquatic organisms has been primarily a research activity to date. Models are available, and their applicability in extrapolating to effects from complex exposure patterns has been demonstrated (e.g., Jager and Kooijman 2005; Ashauer et al. 2007a, 2007b, 2007c). The primary areas for research and development are the following:

1) Broader application and evaluation of the models for aquatic invertebrates with insignificant growth and reproduction over the course of the exposure period (e.g., Asellidae, gammarids, chironomids). Models to describe lethality under such circumstances are fully available and can be applied with care.
2) There may be significant interspecies variability in TK/TD for a particular chemical. The extent of this variability needs to be quantified and decisions on utility in risk assessment, selection of representative organisms, and so on guided accordingly.
3) Development of models for sublethal endpoints in aquatic invertebrates. Such models should be only minor developments from the existing science base, but examples are required to demonstrate application of the approach. There will also be methodological challenges associated with studying sublethal effects from pulsed exposures.
4) Development of the models to account for aquatic invertebrates with short generation times and rapid growth of individuals (e.g., *Daphnia*).
5) Work with fish and aquatic plants is relatively underdeveloped; substantial research would be required to develop generally applicable methods for these groups.

A case study using TK/TD approaches demonstrates toxicological independence of pulsed exposures.

Case Study 7.1   Use of Toxicokinetic/Toxicodynamic (TK/TD) Modeling to Demonstrate Independence of Successive Pulses of Exposure

BACKGROUND

- Insecticide A is an AChE inhibitor. There is an existing registration for a single use on cereals at 250 g as (active substance)/ha.
- The registration is supported by a mesocosm study demonstrating no unacceptable effects (Effect class 1 to 2) following a single dose at the PEC_{max}. Clear effects with no recovery within 8 weeks are shown for

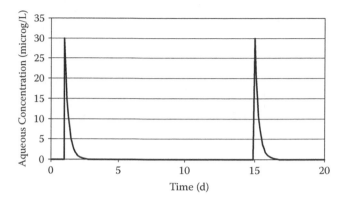

FIGURE CS7.1A Indicative profile for PECsw derived using the FOCUS surface water scenarios.

a dose 4 times larger than the PEC_{max}. The most sensitive species by a little over an order of magnitude is *Gammarus pulex*.[4] Spray drift entry is the dominant route of entry to surface waters.

- The notifier wishes to extend the registration to support 2 applications to cereals, both at 250 g as/ha and with a 14-day interval (Figure CS7.1A shows the PEC profile). As the critical mesocosm study was undertaken with a single dose, the regulatory authority expresses concern over the possibility of additive effects from 2 pulses of exposure. The notifier is requested to undertake a mesocosm experiment with 2 doses spaced 14 days apart to reflect the new GAP (good agriculture practice) or to provide alternative evidence that additive effects will not occur.

APPROACH

- The notifier decides to use TK/TD modeling to determine whether pulsed exposures spaced 2 weeks apart can be considered toxicologically independent. A case can then be prepared in conjunction with the existing mesocosm study. The notifier anticipates anticipate a good prediction of effects for the new pattern of exposure while eliminating the need for a new mesocosm experiment and its associated costs.
- An initial experiment measures uptake and elimination of insecticide A from *Gammarus pulex* (Figure CS7.1B; Table CS7.1A). Test concentrations are set at sublethal values, and the durations of exposure and depuration are set to times that allow significant accumulation and elimination of the test compound, respectively. The time for 95% elimination of internal residues of insecticide A after transfer to clean water is 11 days.
- Subsequent experiments monitor survival of *Gammarus pulex* in response to 2 contrasting patterns of pulsed exposure (Figure CS7.1C). Test concentrations are selected to ensure significant effects but not

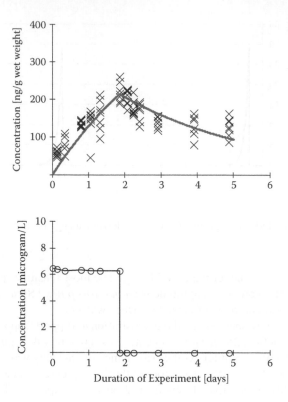

FIGURE CS7.1B   The measurement of uptake and elimination of insecticide A in *Gammarus pulex*. Mean aqueous concentrations of insecticide A (O, *n* = 7, bottom graph) and measured internal concentrations (×, top graph). The solid line (top graph) shows the fitted toxicokinetic model with k_{in} = 23.4 L kg^{-1} day^{-1} and k_{out} = 0.27 day^{-1}.

TABLE CS7.1A

Toxicokinetic/toxicodynamic Parameters Determined with the TDM

Name	Symbol	Units	Value (±SE)
Uptake rate constant	k_{in}	L × kg^{-1} × day^{-1}	23.4 (±0.9)
Elimination rate constant	k_{out}	day^{-1}	0.27 (±0.04)
Killing rate constant	k_k	$g_{wet.w.}$ × day^{-1} × $\mu g_{a.i.}^{-1}$	0.42 (±0.10)
Recovery rate constant	k_r	day^{-1}	0.97 (±0.24)
Threshold	*Threshold*	L × mg^{-1} × day^{-1}	0.067 (±0.010)

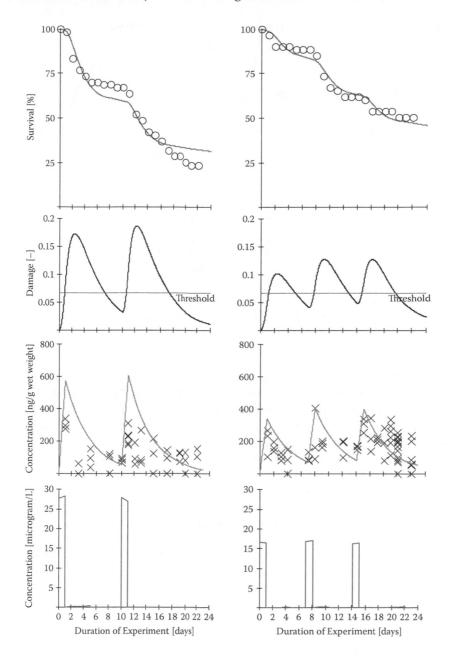

FIGURE CS7.1C   The measurement of toxicodynamic parameters for insecticide A in *Gammarus pulex*. The graphs show aqueous concentrations of insecticide A (bottom) as well as measured (×) and predicted (solid line) concentrations of insecticide A in *Gammarus pulex* (second from bottom). Above that is the simulated time course of damage (damage is operationally defined to link internal concentrations of toxicant with the ecotoxicological effect; further effect [e.g., mortality] is incurred whenever the level of damage exceeds the threshold) in the TDM (third graph from bottom) and the observed (O) survival with the fitted threshold damage model (solid line).

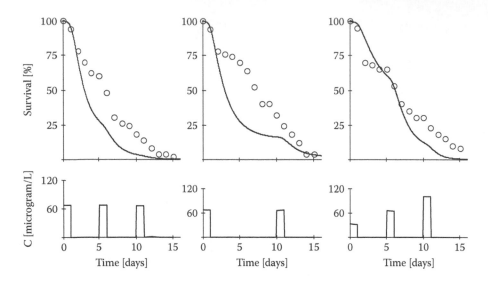

FIGURE CS7.1D   Experiments to evaluate the predictive ability of the TDM using previously calibrated parameters (Table CS7.1). Figures show the aqueous concentrations of insecticide A (bottom) and observed (O) survival with the predictive simulations of the threshold damage model (solid line) in the top graph.

100% mortality over the test period (note that although exposures in Figure CS7.1C are in the same range as PEC_{max} in Figure CS7.1A, the duration of exposure at elevated concentrations is significantly longer in Figure CS7.1C). Intervals between pulses and duration of pulses are set according to the guidance in Section 7.3 and Figure 7.1. The TDM model (Ashauer et al. 2007a) is used in conjunction with the ModelMaker software package to derive TD parameters to describe the effect (Table CS7.1A).

- Mortality of *Gammarus pulex* in response to 3 different pulsed exposures is measured in independent experiments (Figure CS7.1D). Using the model in predictive mode with parameters from Table CS7.1A, a satisfactory fit between observed and simulated mortality is demonstrated.
- The model is used in predictive mode to simulate the mesocosm experiment (Figures CS7.1Ea, CS7.1Eb). The model agrees with experimental results in simulating slight effects for a single dose at the PEC_{max} and clear effects at a dose 4 times larger than the PEC_{max}. The model and parameters are thus considered to be valid to simulate effects of insecticide A on *Gammarus pulex*.
- The TDM model is used to calculate that 95% recovery of damage from a 1-day pulsed exposure will occur within 15 days. This value is taken to define the total recovery time of individual organisms and thus the period beyond which separate pulses of exposure can be considered toxicologically independent. As this period is only slightly greater than the proposed 14-day interval between applications of insecticide A, the

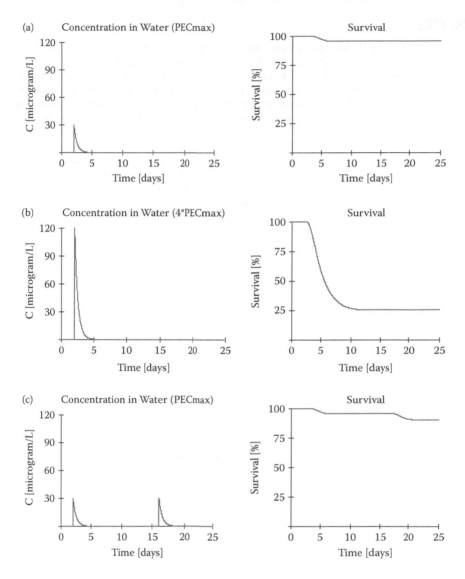

FIGURE CS7.1E Use of the TDM to simulate results of the mesocosm study for doses equivalent to the PEC$_{max}$ (a) and 4 times the PEC$_{max}$ (b). (c) TDM predictions for survival of *Gammarus* in response to 2 successive exposures at the PEC.

TK/TD modeling provides evidence that the combined effects on the most sensitive endpoint from repeat doses at the PEC$_{max}$ would be twice that of a single dose (i.e., there is no carryover toxicity from 1 expo-sure to the next). Model predictions for survival of *Gammarus pulex* in response to 2 doses of insecticide A at the PEC$_{max}$ are shown in Figure CS7.1-Ec. (Case study based on Ashauer et al. 2007c.)

NOTES

1. For example, ModelMaker (www.modelkinetix.com) or the free software OpenModel (http://www.nottingham.ac.uk/environmental-modelling/OpenModel.htm).
2. At present, it is not possible to make recommendations on what will be appropriate in specific cases; the key element is to evaluate how robust are the parameters that result from the fitting procedure; additional testing should be undertaken when parameter fits are not sufficiently robust (e.g., multiple-parameter combinations give equally good fits to the data).
3. Note 1 provides examples.
4. Work is ongoing to investigate the extent of variation in TK/TD parameters between arthropod species for a single pesticide (P.J. van den Brink, personal communication). In the absence of generic information, additional work would be required to demonstrate that parameters for *Gammarus* are representative if 2 or more species showed high sensitivity.

8 Model Ecosystem Approach in the Risk Assessment of Plant Protection Products[*]

CONTENTS

8.1 GENERAL INTRODUCTION

Freshwater microcosms and mesocosms are frequently used as research tools in aquatic ecotoxicology. They are bounded systems that are constructed artificially with samples from, or portions of, natural freshwater ecosystems or that consist of enclosed parts of natural freshwaters. Although these model ecosystems usually are characterized by a reduction in size and complexity when compared with natural ecosystems, they have to include an assemblage of organisms representing several trophic levels, and this assemblage should be in "dynamic equilibrium" with its ambient environment (see, e.g., Brock et al. 1995; Caquet et al. 2000). The terms "microcosm" and "mesocosm" are used more or less interchangeably when referring to model ecosystems. Following the definitions proposed by Crossland et al. (1993), microcosms are experimental systems containing less than 15 m^3 water volume or

---

[*] Box 5 and Box 6 in the decision scheme; see Figure 2.1 in Chapter 2 or Appendix 4.

experimental streams less than 15 m in length, while mesocosms are larger. However, in Europe, indoor experimental ecosystems are often referred to as microcosms and outdoor experimental ecosystems as mesocosms.

The diversity of types of aquatic model ecosystems is large. A major division is that between "generic" and "semirealistic" freshwater model ecosystems. The generic model ecosystems do not mimic any natural ecosystem in particular but rather exhibit some basic properties common to all ecosystems, such as species interaction, production, decomposition, and nutrient cycling. These systems are intended to contain only certain defined species and defined abiotic properties chosen by the experimenter, and they are relatively simple and readily standardized (Taub 1969; Metcalf et al. 1971; Kersting 1984). Most aquatic model ecosystems used in ecotoxicology and in the risk assessment of pesticides are of the semirealistic type in that they attempt to mimic real ecosystems. They can be classified according to the type of natural freshwater ecosystem that they represent and whether they are situated indoors or outdoors. In outdoor model ecosystems, a distinction can be made between constructed systems (e.g., concrete tanks in which sediment and water are introduced and that serve as experimental ponds) or enclosed parts of existing ecosystems (e.g., cylinders and artificially lined limnocorrals). The most frequently used freshwater model ecosystems in pesticide risk assessment are those that mimic shallow, static freshwater habitats (see Brock and Budde 1994), but ecotoxicological experiments with pesticides in artificial streams are also common (e.g., Schulz and Liess 2000; Hose et al. 2002; Schulz et al. 2002; Dabrowski, Bollen, Bennett, et al. 2005; Dabrowski, Bollen, and Schulz 2005; Heckmann and Friberg 2005; Beketov et al. 2008).

Differences in model ecosystem size and complexity are reported to have a profound effect on the enclosed community. Important aspects of ecosystem and community functions may be controlled by keystone organisms too large or mobile to be confined in experiments that are smaller than the ecosystem of concern, such as large predatory fish. In addition, problems in interpreting micro/mesocosm experiments may be caused by inadequate or erroneous scaling of sediment–water interactions and potential artifacts associated with containerization (wall effects and water renewal times). Furthermore, large freshwater ecosystems are usually characterized by diverse habitats differing in abiotic and biotic properties (e.g., the pelagic or the littoral zone of lakes), while most artificial aquatic ecosystems usually simulate 1 of these habitats only (Brock, Maltby, et al. 2008).

Belanger (1997) analyzed data from over 150 studies using model stream ecosystems ranging in size from 0.2 to 540 m long, 0.05 to 4.3 m wide, and 1.5 to 8×10^5 L volume. He concluded that, although larger systems could be sampled more intensively and were more likely to contain fish, there was no relationship between test system size and the species richness of invertebrate, algal, or protozoan assemblages. Few studies have compared assemblages in model streams and natural streams, but those that have indicated that assemblages in model streams can be representative of the natural streams from which they are derived (Belanger et al. 1995; Wong et al. 2004). Model ecosystems that simulate lentic aquatic ecosystems, however, usually contain species characteristic for deeper parts of freshwater ponds and often lack the species assemblages typical for littoral zones (Williams P et al. 2002). However, it appears from several model ecosystem experiments with insecticides that threshold levels for effects due to short-term exposure regimes may be very similar between lentic test systems that considerably differ in complexity, at least when they contain enough representatives of sensitive taxonomic groups, in this case arthropod populations (Van Wijngaarden, Brock, and Van den Brink 2005; Brock et al. 2006).

8.2   ROLE OF MICRO- AND MESOCOSMS IN PESTICIDE RISK ASSESSMENT

Registration schemes for pesticides require notifiers to assess the potential ecological risks of their products using a tiered approach. In the EU regulatory framework, a number of approaches may be used to address these concerns (Campbell et al. 1999; European Union 2002). They include freshwater micro- and mesocosm studies. Besides the aim of micro- and mesocosm studies to simulate natural conditions and exposing these systems to environmentally realistic pesticide exposure regimes, these studies normally follow experimental designs to demonstrate causality between treatment and effects and can identify concentration–effect relationships. Due to confounding factors, causality between pesticide exposure and effects is more difficult to demonstrate in-field monitoring studies. The advantage of micro- and mesocosm studies over the other types of experimental higher-tier studies (e.g., additional laboratory toxicity tests to construct species sensitivity distributions [SSDs]; laboratory population studies) is their ability to integrate more or less realistic exposure regimes with the assessment of endpoints at higher levels of biological integration and to study intra- and interspecies interactions and indirect effects. They also allow assessment of latency of effects and population and community recovery.

Micro- and mesocosm studies yield substantial amounts of data over many endpoints. As a consequence, this can make them difficult to interpret. In the case of the early model ecosystem studies conducted in the 1970s and 1980s for the registration of pesticides, an additional complication arose from the fact that many were not optimally designed for the purpose of risk evaluation. The systems generally were comprised of large ponds that contained fish that often led to very high variability between and within treatments.

Because of these difficulties, in 1992 the USEPA (US Environmental Protection Agency) dropped the requirement for mesocosm studies for pesticide registration. However, in Europe the knowledge and experience gained was used for further development and harmonization of micro- and mesocosm studies. In the early 1990s, this resulted in several guidance documents on the conduct of micro- and mesocosm experiments, including the workshop documents of SETAC Europe (Arnold et al. 1991) and EWOFFT (European Workshop on Fresh Water Field Tests; Crossland et al. 1993; Hill et al. 1993). Later workshop documents such as HARAP (Higher-Tier Aquatic Risk Assessment for Pesticides; Campbell et al. 1999) and CLASSIC (Community-Level Aquatic System Studies–Interpretation Criteria; Giddings et al. 2002) were more focused on the ecological interpretation of studies and on the implementation of the data into risk assessment. As more experience was gained through the conduct and design of micro- and mesocosm studies (see reviews of Brock, Lahr, et al. 2000; Brock, Van Wijngaarden, et al. 2000; and Van Wijngaarden, Brock, and Van den Brink 2005), their interpretation was aided by the development of software tools that facilitated multivariate statistical analysis of the data (Van den Brink and Ter Braak 1999).

In recent years, discussions shifted toward subjects like the need for the identification of protection goals and whether all types of water should receive the same level of protection (see, e.g., Crane and Giddings 2004; Brock et al. 2006; Van den Brink 2006). A key concern has been the awareness of inconsistencies in both the way the same mesocosm data are interpreted and the uncertainty (assessment) factors applied by regulatory experts in different EU member states. RIVM (Dutch National Institute for Public Health and the Environment) in the Netherlands has produced a guidance document on how mesocosm data should be presented and evaluated in a uniform and transparent manner (De Jong et al. 2008). In addition, Brock et al. (2006) and De Jong et al. (2008) proposed a refinement of the "Effect classes" used to categorize the results of micro/mesocosm experiments (see the following list), which are outlined in the *Guidance Document on Aquatic Ecotoxicology in the Context of Directive 91/414/EEC* (EC 2002).

Effect class 1 (no treatment-related effects demonstrated; NOEC$_{micro/mesocosm}$*)*
   No (statistically and ecologically significant) effects observed as a result of the treatment. Observed differences between treatment and controls show no clear causal relationship.
Effect class 2 (slight effects)
   Effects reported as "slight," "transient," or other similar descriptions. It concerns a short-term or quantitatively restricted response of 1 or a few sensitive endpoints, usually observed at individual samplings only.
Effect class 3A (pronounced short-term effects [<8 weeks], followed by recovery)
   Clear response of sensitive endpoints, but full recovery of affected endpoints within 8 weeks after the first application or, in case of delayed responses and repeated applications, the duration of the effect period is less than 8 weeks and followed by full recovery. Effects observed at some subsequent sampling instances.
Effect class 3B (pronounced effects and recovery within 8 weeks after the last application)
   Clear response of sensitive endpoints in micro/mesocosm experiment repeatedly treated with the test substance, but full recovery of affected endpoints within 8 weeks after the last application. The total effect period may be longer than 8 weeks because of possible responses in the treatment period.
Effect class 4 (pronounced effect in short-term study)
   Clear effects (e.g., large reductions in densities of sensitive species) observed, but the study is too short to demonstrate complete recovery within 8 weeks after the (last) application.

Effect class 5A (pronounced long-term effect followed by recovery)
    Clear response of sensitive endpoints, effect period longer than 8 weeks, and recovery did not yet occur within 8 weeks after the last application, but full recovery is demonstrated to occur in the year of application.
Effect class 5B (pronounced long-term effects without recovery)
    Clear response of sensitive endpoints (>8 weeks after last application), and full recovery cannot be demonstrated before termination of the experiment or before the start of the winter period.

8.3 SELECTING DOSING REGIMES IN MICRO/ MESOCOSM EXPERIMENTS

In the CLASSIC workshop (Giddings et al. 2002), exposure regime and dosing were already recognized as fundamental issues in the experimental design of micro- and mesocosm studies. The delegates recommended an exposure–response experimental design with preferably 5 or more concentrations and at least 2 replicates per concentration. Preferably, the lowest test concentration should not result in treatment-related responses, while the highest concentration tested should result in pronounced effects on several measurement endpoints. This allows the derivation of Effect class 1, 2, and 3 concentrations and puts in perspective the possibly more subtle treatment-related responses caused by intermediate concentration levels. This implies that the selected exposure concentrations should always be guided by lower-tier effect information (e.g., single-species toxicity tests) and the expected field exposure regime of the substance under evaluation (e.g., risks due to short- or long-term exposure).

Before designing a micro/mesocosm test for regulatory purposes, it is important to evaluate the possible exposure regimes in aquatic ecosystems that may result from normal agricultural use of the plant protection product of concern and to identify the relevant exposure regimes that should be addressed in the effect assessment (see Chapter 4). If the expected and relevant field exposure regime is characterized by a single high pulse (e.g., due to drift application) or by repeated pulses that are both toxicologically and ecologically independent, a single-application experimental design is an appropriate exposure regime to study in the micro/mesocosm experiment. The pulse duration, however, should either be equal to or larger in the micro/mesocosm experiment than that predicted for the field or it should be easy to extrapolate concentration–response relationships for shorter peaks to that of broader peaks. The latter may be the case if the time to onset of effect is very short for relevant organisms in single-species toxicity tests, when validated toxicokinetic and toxicodynamic (TK/TD) models can be used for the active substance and organisms of concern or when other empirical information is available on related substances with a similar toxic mode of action. For example, a positive correlation is reported for the toxicity of organophosphorous insecticides to arthropods and for the dissipation rates of these insecticides from water when temperature increases. In other words, at higher temperatures a shorter pulse of exposure may result in similar direct toxic effects when compared with a broader pulse of exposure under colder circumstances. This may explain the phenomenon that in single-application micro/mesocosms experiments performed with chlorpyrifos in different geographical regions or different ambient temperatures, similar threshold concentrations were observed when expressed in terms of its peak concentration (Van Wijngaarden, Brock, and Douglas 2005; Daam 2008; Lopéz-Mancisidor et al. 2008). Since changes in predicted exposure regimes due to changes in pesticide use patterns or changes in assumptions underlying the exposure scenarios cannot be excluded, several participants of the ELINK (EU Workshop on Linking Aquatic Exposure and Effects in the Registration Procedure of Plant Protection Products) workshop suggested always to adopt a repeated (e.g., 2 or 3 times) application exposure when designing a micro/mesocosm experiment for regulatory purposes.

If the expected exposure regime in the field triggers concerns due to long-term exposure, either a long-term exposure regime should be adopted in the micro/meso-cosm experiment or it should be easy to extrapolate population/community-level responses due to short-term exposure to that of a long-term exposure regime. In a micro/mesocosm experiment that aims to derive concentration–response relationships for long-term exposure, preferably a more or less constant pesticide concentration is maintained for at least the duration of the chronic toxicity test that triggered the micro/mesocosm test.

If the expected exposure regime in the field consists of several short pulses and the time frame between pulses is relatively short (e.g., <7 to 14 days) it may be an option to adopt a repeated exposure regime in the micro/mesocosm experiment. During the CLASSIC and ELINK workshops, however, there was some disagreement about whether a multiple-application experimental design should be avoided. The selection of a representative application scenario will not always be straightforward since use patterns of the pesticide and exposure events may vary considerably (e.g., due to variation in weather conditions and crop type). A pragmatic option would be the selection of a more or less regular multiple exposure regime (e.g., weekly application) on the basis of the predicted exposure concentrations for the most relevant exposure scenario. The number of applications has to be considered carefully in relation to the expected biological effects — but should be as low as possible — guided by the responses observed in the toxicity tests that triggered the micro/mesocosm study and by biological information of the species potentially at risk.

8.4 THRESHOLD CONCENTRATIONS FOR EFFECTS IN MICRO/ MESOCOSM EXPERIMENTS AND LINKING EXPOSURE AND EFFECTS IN THE RISK ASSESSMENT (BOX 5 IN FIGURE 2.1; SEE CHAPTER 2 OR APPENDIX 4)

Given the natural variability in the structure and function of freshwater communities, it is reasonable to question the spatiotemporal extrapolation of results of model ecosystem experiments with pesticides (see, e.g., Solomon et al. 2008). Since most model ecosystems enclose parts of — or have been seeded with components of — natural communities, the geographical location of micro- and mesocosms, and the time of the year in which these experiments are performed will determine their species composition and hence potentially their sensitivity. Within the context of the risk assessment for pesticides, a question at stake is how unique such test systems are with respect to their ecological threshold levels for toxic effects.

It appears from the model ecosystem experiments performed with the nonpersistent insecticides chlorpyrifos (single applications) and lambda-cyhalothrin (repeated applications) that concentrations specifically in the range of "no effect" to "slight and transient effects" are relatively consistent. For chlorpyrifos, the Effect class 1 to 2 concentrations based on peak exposure ranged from 0.1 to 0.3 µg/L ($n = 6$), while that was from 2.7 to 10 ng/L for lambda-cyhalothrin ($n = 6$) (Brock et al. 2006). Also Van Wijngaarden, Brock, and Van den Brink (2005) reported similar Effect class 1 to 2 concentrations for different model ecosystems treated once with azinphos-methyl or repeatedly with the fast-dissipating esfenvalerate. Furthermore, lake enclosure studies exploring effects of a single application of pentachlorophenol to plankton communities in spring, summer, autumn, and winter indicated that threshold levels for effects (Effect class 1 concentration based on peak exposure) varied little with season (24 to 54 µg/L) (Willis et al. 2004). Note that observed ranges in Effect class 1 to 2 concentrations are partly due to differences in experimental design between studies, particularly the treatment levels selected. In chlorpyrifos studies that selected the same treatment levels, the Effect class 1 concentrations were similar (= 0.1 µg/L). The same was observed for Effect class 2 concentrations in micro/mesocosm studies with lambda-cyhalothrin (Brock et al. 2006).

Whether the robustness in ecological threshold concentrations is also the case for more or less constant, chronic exposure regimes needs to be investigated. Data available for the herbicide atrazine suggest not only a larger variability in class 1 to 2 effect concentrations between experiments but also a larger number of studies is available ($n = 9$; range 2 to 25 µg/L) (Brock et al. 2006). There appears to be no information on other pesticide-treated model ecosystems comparing threshold concentrations for direct toxic effects as a result of chronic exposure. There is, however, evidence that threshold concentrations for chronic exposure to the surfactant C_{12}TMAC (dodecyl/trimethyl ammonium chloride) do not differ much between model stream experiments (range in $NOEC_{ecosystem}$ values 180 to 300 µg/L; $n = 5$) (Versteeg et al. 1999). Similarly, only a 3-fold difference in threshold concentrations was reported for long-term exposure to copper (adjusted to 50 mg/L $CaCO_3$ hardness) derived from 1 lentic mesocosm and 6 artificial stream studies conducted in the United States and Europe (Versteeg et al. 1999).

The short-term or long-term regulatory acceptable concentration (RAC) representative for the threshold level of effects in the field may be derived by applying an uncertainty factor (UF) (for spatiotemporal extrapolation) to the Effect class 1 to 2 concentration from the micro/mesocosm experiment. The height of this UF (which may be 1) should, among others, depend on the relevance of the tested assemblage for the species potentially at risk, the other higher-tier information available (e.g., toxicity data for additional test species and other micro/mesocosm experiments), and known variability in Effect class 1 to 2 concentrations for related compounds with a similar toxic mode of action. For a discussion on this topic, refer to the work of Brock et al. (Brock et al. 2006; Brock, Maltby, et al. 2008; Brock, Solomon, et al. 2008).

An important question at stake when evaluating concentration–response relationships in micro/mesocosm experiments is what constitutes the ecotoxicologically relevant concentration (ERC). According to our decision scheme, the peak concentration of the pesticide in the relevant matrix (water, sediment) usually is the appropriate predicted environmental concentration (PEC) to evaluate risks due to short-term exposure. Consequently, in the assessment of the threshold level of effects due to short-term exposure, the Effect class 1 to 2 concentrations should be expressed in terms of the nominal or measured or predicted peak concentration in the micro/ mesocosms of concern. In repeated-application studies, the peak concentration may occur immediately after the last application if the compound does not dissipate completely from the water column between applications. In that case, adopting the nominal treatment level to express the Effect class 1 to 2 concentration can be considered a conservative approach.

To evaluate risks due to long-term exposure, either the peak concentration or a time-weighted average (TWA) concentration of the pesticide in the relevant matrix (water, sediment) may be an appropriate PEC. As discussed, the selection of the length of the TWA time window is based on ecotoxicological considerations (e.g., acute–chronic [A/C] ratio, time-to-onset-of-effect information, length of the most sensitive life stage of the organisms at risk) and should be guided by the length of the relevant chronic toxicity tests that triggered the micro/mesocosm experiment. If the TWA approach is considered appropriate (for criteria, see Section 3.4 in Chapter 3), the majority of the ELINK workshop participants agreed to adopt a default time window of 7 days for the TWA estimate of the long-term PEC if no scientific arguments are provided to shorten or lengthen this default time window. Note that for a worst-case approach the time window for the TWA effect estimate in the micro/mesocosm study should not be smaller than the selected TWA time window for the PEC estimate in the field. In addition, the time window for the TWA effect estimate in the micro/mesocosm experiment should not be larger than the period in which the exposure remains more or less constant or, in case of a relatively fast-dissipating substance, the application period of the pesticide in the micro/mesocosm study. The application period is the period during which repeated-pulse applications occur. For example, when a 7-day time window is adopted for the PEC, the Effect class concentrations derived from a mesocosm experiment characterized by 3 weekly treatments can be expressed in terms of a TWA concentration that is 7 days or longer and 21 days or less if in the test systems the pesticide is not very persistent. Note that in repeated-application studies, the highest 7-day TWA concentration may be measured later in the application period if the active substance does not completely dissipate between applications.

In case the TWA approach is deemed not appropriate in the long-term risk assessment and consequently the PEC_{max} is used as the field exposure estimate, the Effect class concentrations derived from a mesocosm experiment simulating long-term exposure may be expressed in terms of the nominal, peak, or average concentration measured or calculated during the application period (or the period in which the exposure remains more or less constant in the micro/mesocosm test). Adopting the nominal or measured or calculated peak concentration may be realistic if it can be demonstrated that the dissipation from water in the mesocosm experiment overall is less fast than, or does not deviate much from, that in the relevant field scenarios. In that case, and if the concentration builds up due to repeated treatments, adopting the nominal concentration during the application period can be considered as a more conservative approach than adopting the measured or predicted peak concentration. Adopting the average concentration during the application period can be considered an even more conservative approach if the active substance does not build up due to repeated applications. For a more detailed discussion on this topic, see the work of Boesten et al. (2007).

Case Studies 8.1 and 8.2 outline refinement of the first-tier RAC.

Case Study 8.1   ELINKodinil (Fungicide): Refining the First-Tier RAC for Acute and Chronic Exposure by the Model Ecosystem Approach

This study concerns a relatively persistent fungicide, ELINKodinil, that is applied 3 to 5 times with a 7-day interval. In the relevant exposure scenario, both spray drift and drainage influence the exposure profile (causing a repeated-pulse exposure and a slight building up of the concentration between several pulses). The PEC_{max} for this fungicide was 5.0 µg as/L. The estimated 7-day TWA PEC was 3.4 µg as/L, and the 21-day TWA PEC was 2.1 µg as/L.

Of the basic set of standard test species, *Daphnia magna* (48-hour median effective concentration that affects 50% of the test organisms [EC50] = 428 µg as/L; 21-day no observed effect concentration [NOEC] = 20 µg as/L) and *Selenastrum capricornutum* (72-hour EC50 = 120 µg as/L) were relatively sensitive, while *Oncorhynchus mykiss* was less sensitive (96-hour median lethal concentration [LC50] = 957 µg as/L; 28-day NOEC = 102 µg as/L). In addition, a *Pimephales promelas* ELS (early life-stage toxicity) study resulted in a chronic NOEC of 86 µg as/L. The laboratory toxicity tests with *Daphnia* and fish concerned flow-through studies. Applying an UF (uncertainty factor) of 100 to the EC50 value of *Daphnia magna* resulted in a first-tier acute RAC of 4.28 µg as/L. Applying an AF of 10 to the NOEC value of *Daphnia magna* resulted in a first-tier chronic RAC of 2.0 µg as/L. These first-tier acute and chronic RACs were refined by the model ecosystem approach.

An outdoor microcosm experiment was performed in test systems that contained a diverse plankton and invertebrate community. Fish were not present in the test systems. The fungicide was applied 3 times at weekly intervals.

Based on treatment-related responses observed (Table CS8.1A), the overall $NOEC_{microcosm}$ was considered to be 6.4 µg as/L when expressed in terms of the peak concentration and 4.6 µg as/L and 2.8 µg as/L when expressed in terms, respectively, of the 7-day TWA and 21-day concentration during the application period. These values are higher than the calculated PEC_{max} (5.0 µg as/L), 7-day TWA PEC (3.4 µg as/L), and 21-day TWA PEC (2.1 µg as/L), indicating that both the acute and chronic risks are acceptable for invertebrates and primary producers, at least if it is assumed that the threshold level as observed in the microcosm experiment is representative for surface waters in the agricultural landscape and no additional UF is required. Since fish were not tested in the microcosm experiment, the $NOEC_{microcosm}$ should be compared with the laboratory toxicity data for fish. Based on the laboratory toxicity data provided, the acute first-tier RAC

TABLE CS8.1A

Overall Results of the Outdoor Microcosm Experiment (Expressed in Terms of Effect Classes)

	Treatment Level				
Nominal treatment level (µg as/L)	2.0	4.0	8.0	16	32
Measured peak concentration (µg as/L)	3.1	6.4	11.5	23.4	48.1
Highest 7-day TWA concentration (µg as/L)	2.4	4.6	9.3	19.8	40.8
Highest 21-day TWA concentration (µg as/L)	1.1	2.8	7.4	15.8	32.9
Effect Class Most Sensitive Endpoint[a]					
Phytoplankton community (PRC)	1	1	2	3A	3A
Phytoplankton populations	1	1	3A	3A	3A
Periphyton community (PRC)	1	1	3A	3B	5A
Periphyton populations	1	1	3A	3A	5A
Zooplankton community (PRC)	1	1	3A	3A	3B
Zooplankton populations	1	1 to 2	3A	3B	3B
Macroinvertebrates community (PRC)	1	1	2	3A	5B
Macroinvertebrates populations	1	1	2	3A	5B
Community metabolism	1	1	1	2	3A

[a] Principal responce curves.

for fish is $957/100 = 9.57$ µg as/L, while the first-tier chronic RAC for fish is $86/10 = 8.6$ µg as/L. These RACs for fish are higher than the NOEC$_{microcosm}$ and the calculated PECs, so the results of the NOEC$_{microcosm}$ value can be used to derive the overall RAC. Note, however, that the RACs for fish are lower than the peak and TWA exposure concentration of the microcosms that received a nominal concentration of 8.0 µg as/L and in which Effect class 3A responses were observed.

Case Study 8.2   ELINKalerate (Pyrethroid Insecticide): Refining the First-Tier RAC for Acute Exposure by the Model Ecosystem Approach and Additional Laboratory Tests

The most sensitive standard test species for the pyrethroid insecticide ELINKalerate was *Daphnia magna* (48-hour semistatic EC50 = 90 ng as/L), followed by the fish *Oncorhynchus mykiss* (96-hour flow-through LC50 = 970 ng as/L). The reported EC50 value for the green alga *Pseudokirchneriella subcapitata* was more than 10 000 ng as/L (above water solubility of as). Applying an AF of 100 to the toxicity value for *Daphnia* results in a first-tier RAC of 0.9 ng as/L. The calculated PEC$_{max}$ for the relevant ditch exposure scenario was 8 ng as/L, and the exposure profile consisted of 5 pulsed exposures, of which two had a more or less similar height (interval 10 days), while the height of the other pulses was considerably lower (<3 ng as/L). The first-tier RAC was refined by an experiment in enclosures placed in a macrophyte-dominated experimental ditch.

The insecticide was applied 3 times at weekly intervals at nominal concentrations of 0, 1, 3, 10, 30, and 100 ng as/L (3 replicate enclosures per treatment). Concentrations were confirmed analytically. The enclosures contained a diverse invertebrate community. At the community level, the NOEC of the enclosure study (based on multivariate analysis) was 10 ng as/L. At this treatment level, the only population-level response observed was a minor and transient (Effect class 2) effect on larvae of the insect *Chaoborus flavicans*. At the treatment level of 30 ng as/L, short-term but pronounced effects on *Chaoborus flavicans* (Effect class 3B) were observed, while the macrocrustacean *Gammarus pulex* showed long-term effects without recovery (Effect class 5B). In the 30-ng as/L enclosures, some other arthropods (including daphnids) showed short-term effects (Effect class 2). At the highest treatment level, pronounced treatment-related effects could be demonstrated for several insects, macrocrustaceans, and microcrustaceans. For most arthropods that showed a treatment-related response, the maximum effect already could be observed 4 days after the first treatment. During the first week after insecticide application, recorded water temperatures in the enclosures varied between 17 °C and 19 °C.

Although the overall $NOEC_{community}$ of the enclosure study was 10 ng as/L (based on the nominal treatment level), it was argued that this concentration could not be used as an RAC for short-term exposure because in the macrophyte-dominated enclosures the rate of dissipation of the insecticide was approximately twice as fast (time until 50% dissipation [DT50] = 24 hours) as that predicted for the relevant ditch exposure scenario (DT50 = 48 hours). For that reason, additional toxicity tests were performed with larvae of the phantom midge *Chaoborus flavicans* and the macrocrustacean *Gammarus pulex* in flow-through test systems that simulated pulsed exposure regimes characterized by different dissipation rates. The tests were performed at 18°C. Starting with peak concentration of 10 ng as/L, the following dissipation DT50s were simulated in the test vessels: 6, 12, 24, 48, 96, and 192 hours. In each test vessel, 10 individuals of either *Chaoborus* or *Gammarus* were incubated, and the test lasted 96 hour. Each exposure profile was tested in triplicate. The percentage affected individuals of the test organisms in the test vessels is given in Table CS8.2A.

The results of the refined exposure study with the 2 most sensitive species from the field enclosure experiment indicate that the effects observed were similar between test vessels treated with 10 ng as/L and a simulated dissipation DT50 of 24 and 48 hours, respectively. For that reason, it was decided that the overall NOEC of the field enclosure study of 10 ng as/L could be used to derive a higher-tier RAC for short-term exposure.

An important message of this case study is that results of micro/mesocosm experiments that did not simulate the appropriate exposure profile still may be useful for the risk assessment if results of additional laboratory experiments or

TABLE CS8.2B
Percentage Affected Individuals of *Chaoborus* and *Gammarus* in Laboratory Flow-Through Systems That Simulated Different Dissipation DT50s of the Pyrethroid Insecticide and That Started with 10 ng as/L

	Simulated Dissipation DT50 (%)					
	6 hours	12 hours	24 hours	48 hours	96 hours	192 hours
Chaoborus flavicans	10	10	20	20	20	30
Gammarus pulex	0	0	0	0	10	20

simulation studies (e.g., with TD/TK models) with the most sensitive species from these micro/mesocosm experiments allow reinterpretation of the exposure–response relationships observed. Such an approach certainly is more cost-effective than performing a new micro/mesocosm experiment.

8.5 FACTORS THAT AFFECT RECOVERY OF SENSITIVE POPULATIONS IN MICRO/MESOCOSM EXPERIMENTS (BOX 6 IN FIGURE 2.1; SEE CHAPTER 2 OR APPENDIX 4)

An advantage of microcosm and mesocosm studies is that information can be obtained about recovery of disturbed populations and ecosystem functions. The phenomenon of recovery is particularly important when exposure to a pesticide declines due to physical (e.g., hydrological), physicochemical (e.g., hydrolysis), or biological (e.g., bacterial breakdown) processes that result in the disappearance of the pesticide.

When defining recovery, a distinction between actual and potential recovery can be made. "Actual (or ecological) recovery" implies the return of the perturbed measurement endpoint (e.g., species composition, population density, dissolved oxygen concentration) to the window of natural variability in the ecosystem of concern or to the level that is not significantly different from that in control or reference systems of a microcosm or mesocosm study. This does not mean, however, that we should consider endpoints as recovered if the statistical difference primarily disappears due to an increase in variability in control test systems. "Potential (or ecotoxicological) recovery" is defined as the potential for recovery to occur following disappearance of the pesticide to a concentration at which it no longer has adverse toxic effects on the measurement endpoints of interest. However, if a substance shows fast dissipation from water due to partitioning to the sediment, certain sediment-dwelling organisms may still be exposed. When studying population responses to pesticide stress, it may be convenient to make a distinction between internal and external recovery. Internal recovery depends on surviving individuals in the stressed ecosystem or on a reservoir of resting propagules (e.g., seeds and ephippia) not affected by the pesticide. In contrast, external recovery depends on the immigration of individuals from neighboring ecosystems by active or passive dispersal (Caquet et al. 2007; Hanson et al. 2007; Brock, Solomon, et al. 2008).

When the pesticide degrades rapidly or its bioavailability decreases below a critical threshold concentration, the recovery rate of affected populations largely depends on the life-cycle characteristics of the affected species of concern and the presence of uncontaminated refuges in the surrounding landscape. Important life-cycle properties are the number of generations per year and related reproductive strategies, the presence of relatively insensitive (dormant) life stages, and the capacity of organisms to migrate actively from one site to another. Recovery of affected populations from pesticide stress may be rapid if

- The substance is not persistent, the exposure regime is short term or pulsed, and the time between pulses is long enough for recovery
- The physicochemical environment and ecologically important food web interactions are not altered by the pesticide or quickly restored
- The generation time of the populations affected is short
- There is a ready supply of propagules of eliminated populations through pesticide-resistant resting stages, by active immigration by mobile organisms, or through passive immigration by, for example, wind and water transport

When invertebrate populations are at risk and the acceptability of effects is based on the potential for recovery of affected populations, we should consider whether the populations in the test system (e.g., microcosm) are representative of the semivoltine, univoltine, or multivoltine populations occurring in the field. Semivoltine species need more than a year to complete their life cycle and univoltine species a year, while multivoltine species have more generations per year. For a discussion on this topic, see also the working group report on ecological scenarios in Chapter 15. The information on life-cycle characteristics may be used to better interpret the population responses observed in micro/mesocosm experiments, such as when extrapolating the experimental results by means of metapopulation models (see Chapter 9).

It follows from this discussion that the time of year during which the exposure to a pesticide takes place may determine the extent of biological recovery in a specific micro/mesocosm experiment. Pulsed exposures can have greater impacts on recovery patterns if they occur during critical life stages (particularly in the case of univoltine insects) or if they occur in autumn when lower dispersal activities and lack of winter reproduction may delay recovery until the following spring. In experimental ponds and enclosures, it appeared that application of chemicals at different times of year may induce different recovery patterns of the same invertebrate species (Hanazato and Yazuno 1990; Willis et al. 2004; Van Wijngaarden et al. 2006).

The number of generations per year of certain invertebrate species may vary with latitude and consequently with temperature and the length of the growing season (Niemi et al. 1990). For example, in colder regions, the same insect species may be univoltine, while in warmer regions this species may have more generations per year (multivoltine). Consequently, when recovery is taken into account in the assessment of acceptable concentrations, differences between latitudes may be of importance when extrapolating micro/mesocosm data from temperate to colder regions.

Estimates of recovery potential can be made either through focused micro- and mesocosm tests or modeling studies. It should be noted, however, that most micro/ mesocosm experiments on the impact of pesticide stress focus on the responses of dominant populations, which are often characterized by a relatively short life cycle (e.g., algae, invertebrates). Microcosm and mesocosm studies are generally less suitable to study the recovery of populations of larger organisms with a long life span (such as vertebrates). In addition, the duration of many published microcosm and mesocosm studies is too short to be able to derive the recovery period of sensitive semivolitine and univoltine invertebrate populations. Another point of attention in the interpretation and extrapolation of responses observed in micro/mesocosm studies is that most experiments utilize isolated test systems. This means that eliminated populations with a limited dispersal capacity cannot rapidly recolonize these test systems. For these organisms, observations on their recovery in isolated microcosms and mesocosms should be considered as a worst-case scenario.

8.6   CAN NOEAECS DERIVED FROM AQUATIC MICRO/ MESOCOSM EXPERIMENTS BE EXTRAPOLATED IN SPACE AND TIME? (BOX 6 IN FIGURE 2.1; SEE CHAPTER 2 OR APPENDIX 4)

The use of the no observed ecological adverse effect concentration (NOEAEC) implies that some treatment-related responses are acceptable, for example, if the sensitive endpoint shows a fast recovery. It should be noted that a NOEAEC value based on recovery will not be used to derive a long-term norm concentration (AA-EQS, annual average environmental quality standard) within the context of the Water Framework Directive (WFD; EC 2000), but that such an NOEAEC might be used to derive an RAC within the context of 91/414. If, for example, an Effect class 3A is considered the NOEAEC of a specific mesocosm experiment, the question at stake is whether this NOEAEC can be extrapolated to the variable field. When compared with the threshold level of effects (e.g., Effect classes 1 and 2), a larger variability in responses generally is observed for different types of pesticides studied at exposure concentrations well above their threshold concentration for direct toxic effects. For example, in indoor plankton-dominated microcosms simulating Mediterranean conditions and treated with chlorpyrifos, long-term algal blooms (indirect effect) were observed after treatment with 1 μg/L, while no algal blooms could be detected in similar microcosms simulating temperate conditions and treated with the same concentration (Van Wijngaarden, Brock, and Douglas 2005). Furthermore, in more complex outdoor experimental ditches after treatment with 0.9 μg/L chlorpyrifos, no indirect effects could be demonstrated (Van den Brink et al. 1996). Also, Roessink et al. (2005) concluded that, at higher concentrations of lambda-cyhalothrin (>10 ng/L), the magnitude and duration of effects differed between plankton-dominated enclosures (with a community characterized by short-lived organisms) and structurally more complex macrophyte-dominated enclosures. This indicates that once clear effects are caused by short-term exposure to pesticides, the variability in the rate of recovery may be relatively high between test systems. Consequently, if in the risk assessment an NOEAEC based on recovery is used, either an UF or a model approach for spatiotemporal extrapolation seems to be needed to derive an RAC based on an Effect class 3 NOEAEC. To scientifically underpin the size of this UF, information is needed on the variability in, for example, Effect class 3 responses of different micro/mesocosm studies treated with the same pesticide. Ideally, this type of information is needed for single- and repeated-pulse exposures as well as for pesticides that differ in toxic mode of action. Furthermore, when deriving an RAC on the basis of an Effect class 3 NOEAEC, for example, it should be carefully evaluated whether the populations that show recovery in the micro/mesocosm tests are representative for the populations potentially at risk in the field (see the ELINK WG report in Chapter 15).

Since microcosm/mesocosm tests are limited by the constraints of experimentation, in that usually only a limited number of recovery scenarios can be investigated, modeling approaches may provide an alternative tool for spatiotemporal extrapolation of recovery rates as observed in microcosm and mesocosm experiments. These modeling approaches should take into account the spatiotemporal dynamics in exposure concentrations. Life history and individual-based (meta)population models, which may be spatially explicit, provide the mathematical frameworks to do this. For examples of these models, see Chapter 9 and the ELINK WG report in Chapter 14 as well and the report of the LEMTOX workshop (Forbes et al. 2009).

8.7 LINKING EXPOSURE AND EFFECTS WHEN RECOVERY IS TAKEN INTO ACCOUNT (BOX 6 IN FIGURE 2.1; SEE CHAPTER 2 OR APPENDIX 4)

If the derived RAC value from a single- or multiple-application micro/mesocosm test is based on an Effect class 3 concentration (e.g., by application of an AF for spatiotemporal extrapolation), an appropriate risk assessment can only be performed by also plotting the threshold level for effects (e.g., based on lower-tier data or Effect class 1 to 2 concentrations) on the predicted field exposure profile. If in the appropriate field scenario the pulses are lower than the RAC value based on Effect class 3 concentrations but higher than the threshold level for direct toxic effects, the interval between successive peaks should be carefully considered. If the interval between peaks is smaller than the relevant recovery time of the sensitive populations of concern, these peaks should be considered as ecologically dependent. On the basis of this information, the total period of possible effects can be estimated for the exposure profile of the active ingredient in the field scenario of concern. Alternatively, population or community models may be the appropriate tools to evaluate whether the peaks are dependent or independent from an ecological point of view. These models may be based on empirical data for related compounds with a similar fate profile and toxic mode of action.

Case Study 8.3 refines the first-tier RAC.

Case Study 8.3   ELINKuron (Herbicide): Refining the First-Tier RAC for Acute Exposure by an NOEAEC Derived from a Mesocosm Experiment

This study concerns a herbicide, ELINKuron, for which in laboratory toxicity tests for algae are an order of magnitude more sensitive than *Lemna, Daphnia magna,* and rainbow trout. The first-tier risk assessment procedure identified possible risks for algal populations in particular. In a mesocosm experiment, the herbicide was applied once, and treatment-related effects on phytoplankton, periphyton, zooplankton, and community metabolism (DO [dissolved oxygen], pH, alkalinity) were investigated. The test systems contained a high enough diversity of different algal populations. The most sensitive endpoint concerned green algae in the phytoplankton and periphyton. Based on the treatment-related responses of the most sensitive measurement endpoint (densities of green algae), a threshold level for toxic effects (Effect class 1) and an NOEAEC (based on an Effect class 3A) could be identified. Full recovery of sensitive populations of green algae could be observed 5 weeks after herbicide application. As an RAC value representative for the threshold level ($RAC_{\text{Effect class 1}}$), the $NOEC_{\text{microcosm}}$ was used. The RAC value taking into account short-term effects ($RAC_{\text{Effect class 3A}}$) was derived by applying an AF of 3 to the NOEAEC. Both RAC values were

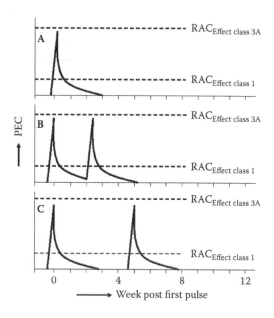

FIGURE CS8.3A Comparison of exposure profiles and regulatory acceptable concentrations that that do ($RAC_{\text{Effect class 3A}}$) and do not ($RAC_{\text{Effect class 1}}$) exhibit short-term effects followed by recovery.

plotted on the predicted exposure profile. As an example, 3 different exposure profiles (panels A, B, and C) are presented in Figure CS8.3A.

If, from a regulatory point of view, a short-term effect that does not last longer than 8 weeks is considered acceptable and the actual recovery of the most sensitive endpoint (abundance of green algae) was observed 5 weeks after herbicide application in the mesocosm study, then the exposure profile in panel A (single pulse) is acceptable. In this single-pulse exposure scenario, the peak concentration is slightly lower than the $RAC_{Effect\ class\ 3A}$, and the concentration drops below the $RAC_{Effect\ class\ 1}$ within a week. Consequently, the total period of effect will be approximately 5 weeks.

If the 2 pulses in the exposure profile of panel B are toxicologically independent, it seems logical to assume that recovery of sensitive green algae can be expected 5 weeks after the second pulse. The total period of effect, however, will be between 7 and 8 weeks since the 2 pulses are not independent from an ecological point of view. Since in panel B the total period of effect is expected to be less than 8 weeks, this pulsed exposure profile is considered acceptable.

The pulsed exposure profile given in panel C, however, is probably not acceptable from a regulatory point of view since the total period of effect is likely to be approximately 10 weeks (5 weeks after the first pulse and 5 weeks after the second pulse).

9 Ecological Models in the Risk Assessment of Plant Protection Products*

CONTENTS

9.1 INTRODUCTION

Ecological modeling may be used for the extrapolation of measured data from known scenarios to other (new) scenarios. This is usually done to reduce experimental effort, to predict outcomes of untested situations, and to check (by comparing model output with real data) to which extent ecological complexity has been understood. However, ecological models can also be valuable in ecotoxicology, for example, to optimize experimental design or to focus further research efforts on processes yet not sufficiently well understood. The use of population and ecosystem

---

* Boxes 4, 5, and 6 in the decision scheme; see Figure 2.1 in Chapter 2 or Appendix 4.

models in risk assessment has been explored in some recent publications, including that of Barnthouse (2008), and has been the focus of a workshop (Forbes et al. 2009). However, in the context of pesticide regulations, although modeling approaches are already used extensively in environmental exposure assessment (see Chapter 4), ecological modeling has hitherto not played a major role in the risk assessment of plant protection products.

Nonetheless, there are some developments in the use of models in ecotoxicological effects assessment, with the approach mainly focused on the extrapolation of results of laboratory toxicity tests (on survival, growth, reproduction, or behavior), to the population level. In the context of ELINK (EU Workshop on Linking Aquatic Exposure and Effects in the Registration Procedure of Plant Protection Products), the main need is to be able to use modeling to extrapolate effects within the same hierarchical level and, more specifically, to be able to use modeling for linking exposure and effects. Generally, measured toxicity data (mortality or sublethal effects at the individual level, populations and community effects or recovery) need to be applied to untested (and usually more complex) exposure patterns. There are various tools available, such as the ecotoxicologically relevant concentration (ERC) approach (Boesten et al. 2007); toxicokinetic and toxicodynamic models (see Chapter 7); population models; community, food web, or ecosystem models; and empirical models. An overview of ecological modeling approaches, including examples, is given in Section 9.2. Several of these models were considered in detail by ELINK Working Group (WG) 5, and the outputs from its work and discussions are provided as a WG report, "Extrapolation Methods in Aquatic Effect Assessment of Time-Variable Exposures to Pesticides," in Chapter 14. Only the most important points are covered here to avoid repetition. According to the decision scheme developed within the ELINK workshops, there are various steps (see Boxes 4 to 6 in Figure 2.1 in Chapter 2) at which ecological modeling may help to refine the acute or chronic effect thresholds or to address recovery in an appropriate manner. Further details are included in Section 9.3. To successfully design and apply modeling approaches for a refined risk assessment, it may well be required to conduct additional experimental work; further details are included in Section 9.4.

9.2   OVERVIEW OF MODELING APPROACHES

This section focuses on the following ecological modeling tools:

1) Population models
2) Community, food web, or ecosystem models
3) Empirical models

An overview of the different types of models within each of these three categories is provided, followed by examples for each type.

9.2.1 Population Models

Population models generally aim to describe the dynamics of one population. Depending on the properties of the population to be modeled, different types of population models may be distinguished and are listed in this section. The order used indicates an increasing degree of complexity based on the level of detail included in the approach. Unstructured models, demographic or matrix models, and individual-based models (IBMs), which are described first, all focus on temporal patterns.

Example 9.1: Mechanistic Algae Model (Weber et al. 2007)

The mechanistic algae model is a differential equation model with the state variables algae biomass and internal and external phosphate concentration. Temperature, light, and toxicant concentrations serve as input variables with an impact on the algal growth. Using this model, predictions are made of algal growth over time. Extensions of the model are planned for different algal species, incorporation of competition, and modeling of several peak exposures.

Example 9.2: Population Recovery Model (Barnthouse 2004)

The population recovery model approach uses the simple logistic growth model to estimate recovery times of different aquatic taxa depending on magnitude of effect (here reduction of abundance) and the intrinsic population growth rate. The model assumes a constant intrinsic growth rate of the population and constant carrying capacity of the environment.

9.2.1.1 Unstructured Models

Unstructured models describe the population with state variables like population abundance or density N. In choosing such variables, it is assumed that individuals can be treated as "nearly" identical. This means that when individuals differ little, these differences are lost by aggregation or averaging over those differences. Most of traditional theoretical population ecology consists of unstructured models.

9.2.1.2 Demographic or Matrix Models

Matrix population models are a specific type of population model that uses matrix algebra. These types of models are usually used to make predictions, including age or stage classes of the organisms under investigation.

Example 9.3: Age-Classified Matrix Model for Medaka (Lin et al. 2005)

The potential impacts of chemicals on medaka (*Oryzias latipes*) may be quantified in terms of reduction of population growth rate (lambda). An age-classified population matrix model (daily time step) was developed to combine life-cycle survivorship and fecundity data obtained from individual-level responses of medaka exposed to chemicals into population-level responses defined by the parameter lambda. Thereafter, from the resulting lambdas, different approaches for establishing population-level PNEC values were proposed and examined.

9.2.1.3 Individual-Based Models

Individual-based models are simulations based on the overall consequences of local interactions of members of a population; therefore, they include properties of individuals within populations. These models typically consist of an environment, or framework, in which the interactions occur and a number of individuals defined in terms of their behaviors (procedural rules) and characteristics. In an IBM, the characteristics of each individual are tracked through time. Some IBMs are also spatially explicit, meaning that the individuals are associated with a location in geometrical space. Some spatially explicit IBMs also exhibit mobility, by which the individuals can move around in their environment.

Example 9.4: *Daphnia* Model (Koh et al. 1997)

The perturbation in dynamics of a *Daphnia magna* population caused by the combined effects of two environmental stressors (temperature and dissolved oxygen) and a toxicant (a nonpolar organic lipophilic chemical) on individual physiology was investigated in a model setting, with the primary objective to determine indicators of stress at the population level. The methodology addressed stressor effects on physiological processes via investigations of individual-based population models.

In addition to the population models covered so far, metapopulation models and spatially explicit models both predict temporal and spatial patterns and dynamics.

9.2.1.4 Metapopulation Models

A metapopulation is a set of local populations connected by migrating individuals. Local populations usually inhabit isolated patches of resources, and the degree of isolation may vary depending on the distance among patches. Metapopulation models consider local populations as individuals. Dynamics of local populations are either not considered at all or are considered in a much abbreviated way. Most metapopulation models are based on colonization–extinction equilibrium. Metapopulation model approaches were introduced by Levins (1969), followed later, for example, by MacArthur and Wilson (2001). MacArthur and Wilson considered immigration of organisms (e.g., birds) from a continent to islands in the ocean. Angeler and Alvarez-Cobelas (2005) proposed an environmental risk assessment scheme for temporary wetlands based on island biogeography theory, life history strategies of temporary pond organisms, and landscape structure. This is an example of how the biological traits of aquatic organisms might be integrated in the risk assessment. These types of approaches have also been suggested in probabilistic risk assessments (Schulz et al. 2008).

Example 9.5: Metapopulation Model (Spromberg et al. 1998)

This example is of a toxicant-dosed metapopulation model to explore the range of possible dynamics of populations in contaminated field sites. This single-species metapopulation model is discrete and deterministic and incorporates dose–response curves and biotic growth rates to describe the effects of contamination on a metapopulation. The distribution of the chemical contamination is assumed limited to one of several patches. Both persistent and degradable toxicants can be modeled; however, results were not compared to real data.

9.2.1.5 Spatially Explicit Models

Spatially explicit models aim at providing predictions, which contain information on the spatial distribution and dynamics of populations dependent on impacting parameters such as toxicants. They may also be used in landscape management and to understand medium- and long-term species dynamics (e.g., through the introduction of a new species).

Example 9.6: CHARISMA (Van Nes et al. 2003)

CHARISMA is an individual-based spatially explicit model designed to simulate the growth of one or more competing species of submerged macrophytes. The model behavior of CHARISMA becomes rather complex and puzzling if more than one species is studied. This most likely reflects the complexity of nature but poses a challenge for calibrating and understanding the model. To control this complexity, a flexible design, facilitating the elimination of feedback loops and processes to trace the causes of the complex behavior of the model, has been used. The resulting "realistic, yet transparent" model appeared useful in many respects for bridging the existing gap between theory and reality.

Example 9.7: MASTEP (van den Brink, Baveco, et al. 2007)

MASTEP is an IBM model for assessing spatial and temporal effects of pesticides on metapopulations. It assumes a spatially explicit exposure (derived from FOCUS [Forum for the Co-ordination of Pesticide Fate Models and Their Use] scenarios). Dose–effect relationships were taken from mesocosm experiments, and individuals (in this case, *Asellus aquaticus*) were explicitly modeled, including their life cycle, in an individual-based approach. The focus of the model is on recolonization of disturbed populations based on movement patterns of individuals (active or passive by organismic drift) in a structured (1 × 1 m cells) environment.

9.2.2 COMMUNITY, FOOD WEB, OR ECOSYSTEM MODELS

Community, food web, or ecosystem models aim to describe the dynamics of communities, the fate and transfer of contaminants within food webs, or the interaction of communities with their abiotic environment. In most cases, they are combinations of unstructured models (differential equation system). However, it is also possible to use a different model complexity for different populations (e.g., MICMOD model; Swartzman & Rose 1984). Abiotic variables (e.g., nutrients, oxygen levels, detritus) may be included when community models are extended to the ecosystem level. Food web models emphasize the shifts in abundance that result from multiple trophic interactions and the transfer of energy and biomass through a food web. This framework includes top-down and bottom-up processes, trophic cascades, and more complex interactions across multiple trophic levels (Gotelli and Ellison 2006). Compounds with a high bioconcentration factor (BCF) are considered with respect to modeling the transfer of residues through the food web. Bartell et al. (1999) and Relyea and Hoverman (2006) proposed to use the framework of food web theory including density- and trait-mediated indirect effects to integrate ecology and ecotoxicology to allow a more general conceptualization of how pesticides affect aquatic communities.

Example 9.8: Food Web Model with Nutrients, Insecticides, and Recovery (Traas et al. 2004)

A microcosm experiment that addressed the interaction between eutrophication processes and contaminants was used to derive a food web model. Both direct and indirect effects of nutrient additions and a single insecticide application on biomass dynamics and recovery of functional groups were modeled. Direct toxicant effects on sensitive species could be predicted reasonably well using concentration–response relationships from the laboratory with representative species. Introducing recolonization scenarios in this specific model simulated dose-dependent recovery.

9.2.3   Empirical Models or "Data-Mining Models"

Empirical models aim to use existing data for extrapolation (e.g., by regression) although they are not based on a mechanistic understanding, for example, on the differential equation relations to be included. As almost all models use empirical data, it might be better to separate those explicitly built on large data sets as "data-mining models."

Example 9.9: PERPEST (Van den Brink, Roelsma, et al. 2002c)

A database of microcosm and mesocosm studies containing substance properties (mode of action, MoA), exposure pattern, and effect class per test concentration and group was set up for the PERPEST model. Using artificial intelligence (case-based reasoning), predictions are made regarding which effects will occur at the population or community level in a given ecological and exposure scenario.

In Figure 9.1, the extrapolation relationships are summarized, and the role of a number of the mentioned modeling approaches is included in this scheme. It becomes evident that while ecological relevance increases from top to bottom, the mechanistic understanding decreases.

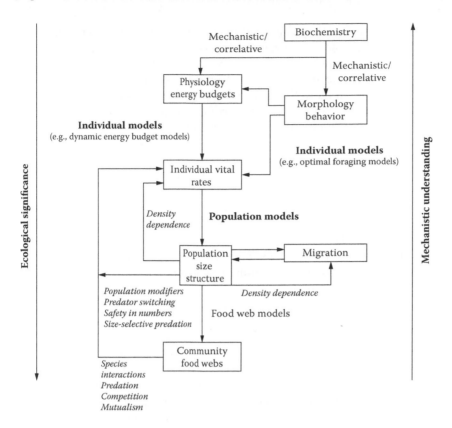

FIGURE 9.1 Extrapolation relationships, the role of ecological models, and the consequences for ecological significance and mechanistic understanding (Maltby et al. 2001). Empirical models are not included in this figure. Reprinted with permission from *Ecological Variability: Separating Natural from Anthropogenic Causes of Ecosystem Impairment.* Copyright 2001, Society of Environmental Toxicology and Chemistry (SETAC), Pensacola, FL, USA.

9.3 ECOTOXICOLOGICAL MODELING IN THE CONTEXT OF LINKING EXPOSURE AND EFFECTS

9.3.1 REFINING ACUTE OR CHRONIC EFFECT THRESHOLDS (BOX 5 DECISION SCHEME IN FIGURE 2.1; SEE CHAPTER 2 OR APPENDIX 4)

Modeling approaches may be used to refine acute or chronic effect thresholds. In acute risk assessment, this may be particularly helpful when the duration of a pulsed exposure (or combined pulses) is longer than the duration of the standard acute aquatic toxicity test. In this case, modeling may be used to predict the effects of longer-exposure pulsed durations (or multiple pulses) on the population dynamics. In the context of chronic risk assessment, modeling may be used in a similar way when predicted or measured (time-variable) exposure durations in the field exceed the duration of the appropriate chronic effects study. The main goal here is to predict and better characterize the potential effects of exposure regimes beyond the timelines of those regularly employed in chronic effect tests. Given that sufficient knowledge on the dose–effect relationship is present, models can help to reduce considerably the experimental effort yet provide substantial information for variable exposure situations (see also Chapter 7). Depending on the physicochemical properties of the pesticide, food web models may be appropriate to use as well.

In both acute and chronic risk assessment procedures, modeling is undertaken potentially in combination with micro/mesocosm experiments. In principal, any of the modeling approaches described may be helpful depending on the specific case; however, unstructured models, demographic or matrix models, and IBMs with a focus on temporal patterns are most relevant. Usually, the question to be answered in this context refers to reactions of populations of selected species to untested exposure situations. As the selected species are usually those with ecological characteristics (i.e., dose–effect relationships) that are relatively well understood (e.g., *Daphnia*, algae, amphipods), these models may be developed further in due course into tools ready to use in pesticide risk assessment. However, it is important to ensure that a sufficient level of complexity in exposure profile is considered (see Chapter 4).

9.3.2 ADDRESSING RECOVERY (BOX 6 DECISION SCHEME IN FIGURE 2.1; SEE CHAPTER 2 OR APPENDIX 4)

In addition to issues discussed, recovery is considered to be a key element in effects assessments using micro/mesocosm studies, from which no observed ecologically adverse effect concentrations (NOEAECs) may be derived. Recolonization potential from external sources may also be relevant for some groups or ecological traits. Modeling approaches could therefore be used in combination with data from micro/mesocosm studies to answer a number of questions, including the following:

- Do the species or ecological traits showing recovery and determining an NOEAEC in the micro/mesocosm experiment conclusively represent those typical in the field with regard to both similar taxonomic groups and differing taxonomic groups or ecological traits?
- Is the recovery potential obtained in the micro/mesocosm experiment as a result of the system setup (orientation of control and exposed units) representative of field conditions?
- How do we estimate the true potential for external recovery (recolonization) of many of the species affected in micro/mesocosm studies (e.g., species with relatively long life cycles and without a terrestrial life stage such as amphipods and isopods, for which recovery is prohibited by the short duration of the experiment or the exposure scenario)?

Any of the models described in Section 9.2 may be used here; however, metapopulation models and spatially explicit models do have the advantage of being able to cover both temporal and spatial aspects. Spatial aspects are of particular importance when recovery or recolonization processes in dynamic environments are to be considered. However, it is worth emphasizing that the models to be applied in these circumstances are mostly designed on a case-by-case basis, taking into account factors such as the spatiotemporal population dynamics of the ecological system and considering other additional field-based experimental or monitoring data (see Section 9.4, Example 9.2). It is likely that further experimentation and field-monitoring data are required before such models become more effective tools for use in pesticide risk assessment. This caveat, however, should not exclude the use of these types of models in risk assessment, provided the assumptions on which they are based are sufficiently well justified by data. To do so, additional experiments may be required to produce data needed for the setup or validation of modeling approaches. Examples are described in Section 9.4.

9.4   RELATED EXPERIMENTS

9.4.1   REFINING ACUTE OR CHRONIC EFFECT THRESHOLDS (BOX 5 DECISION SCHEME IN FIGURE 2.1; SEE CHAPTER 2 OR APPENDIX 4)

The additional experimental work required to support the setup of modeling approaches for refining acute or chronic effect thresholds should mainly focus on one question: What are the potential additional or differing effects of exposure scenarios not covered by the exposure regimes in regular acute or chronic aquatic toxicity tests? Thus, studies providing further insight into the importance of extended or more frequent (pulsed) exposure are needed. These types of studies are described in the work of Ashauer et al. (2007a, 2007c; also see Chapter 7). There may also be a need for micro/mesocosm studies to address spatially heterogeneous exposure scenarios for relevant species (Dabrowski, Bollen, Schulz 2005).

> **Example 9.10: LIMPACT (Neumann, Baumeister, et al. 2002; Neumann, Liess, et al. 2002)**
>
> Measured pesticide exposure data and macroinvertebrate community composition as well as temporal dynamic data from field surveys are used as input variables for the LIMPACT approach. A knowledge-based expert system was then set up to estimate the potential pesticide exposure in the field based on macroinvertebrate data. A heuristic knowledge base was developed that contains 921 diagnostic rules with scores to either accept or reject a diagnosis.

Example 9.11 outlines a stream microcosm study evaluating the individual and combined effects of elevated flow, suspended particle, and pesticide concentration on mayfly nymphs. The presence of suspended particulates, not surprisingly, reduces the toxicity of cypermethrin. This can be predicted by models if basic toxicity and sorption data are available. Interestingly, a significant antagonistic interaction was also present in the cypermethrin and elevated flow rate setup. In contrast to the "classical" assumption of increasing effects of chemicals at increasing flow rates in flow-through test systems (due to higher exposure), the opposite appears to be the case for mayflies. Mayfly nymphs react to high flow rate conditions by reducing their activity and organism drift and by positioning themselves underneath rocks. These behavioral responses lead to reduced exposure to the chemical, which results in a reduced toxicological effect. These types of reactions are only predictable in models if the models are based on data from sets of experiments using combinations of relevant biological response variables.

Example 9.11: Targeted Micro/Mesocosm Study (Dabrowski, Bollen, Schulz, 2005.) Reprinted with permission from *Combined effects of discharge, turbidity and pesticides on mayfly behaviour: experimental evaluation of spray-drift and runoff scenarios. Environ Toxicol Chem* 24:1395–1402. Copyright 2005, Society of Environmental Toxicology and Chemistry (SETAC), Pensacola, FL, USA

The targeted micro/mesocosm study example comes from a study in which the effects of the pyrethroid insecticide, cypermethrin (CYP), increased flow speed (Flow), and increased suspended particles (Part) on drift behavior and activity of mayfly nymphs (*Baetis harrisoni*) were investigated individually and in combination in replicated (*n* = 5) laboratory stream microcosm experiments. The case study showed that abiotic interactions among chemicals, particles, and flow rates may result in complex behavioral reactions in aquatic organisms. It indicated that mayflies reacted actively in response to flow conditions as the reduced drift was associated with reduced activity and passivity in response to pesticide exposure as increased drift was observed, although the activity was reduced. These responses in turn can act to protect the organisms and greatly affect the extent of toxic effects of pesticides such as the

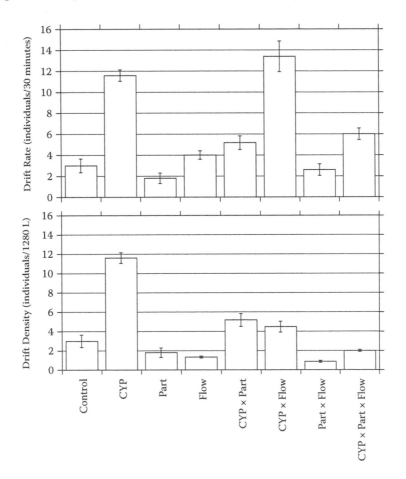

tested pyrethroid cypermethrin. It may be concluded therefore that testing under field-relevant conditions yields valuable information for a realistic description of potential adverse effects and is important to gain relevant input data for appropriate approaches to model the ecological responses to pesticide stress. Mean (± standard error, n = 5) of drifting mayflies (Baetis harrisoni) expressed as drift rate (top) and as drift density (bottom) in 30-min treatments either with the pyrethroid cypermethrin (CYP), suspended particles (Part), or flow increase (Flow) and their combinations.

9.4.2 ADDRESSING RECOVERY (BOX 6 DECISION SCHEME IN FIGURE 2.1; SEE CHAPTER 2 OR APPENDIX 4)

A wide range of additional studies may be needed to support modeling approaches to better address recovery in risk assessment. Targeted experiments on the recovery potential of selected species are needed to address recovery in a temporal context. Micro/mesocosm experiments may also be used to derive relevant input parameters for models such as population growth rates (Barnthouse 2004). This is the case for species with a rather short life cycle, such as algae or zooplankton species. For many of the aquatic invertebrates and most vertebrates, ecological information needs to be gathered and classified to be able to define ecological traits. To do so, species-specific information such as migration behavior, reproductive potential, life cycle strategy, generation time, and so on needs to be collected (Furse et al. 2006). When this information is not available, appropriate laboratory or field studies may need to be conducted. Generally, this type of information will be useful for considering recovery in both temporal and spatial contexts. This also means that further classification and understanding of landscape features, such as refuges, spatial distribution of contaminated and uncontaminated patches, and so on is needed (Brown et al. 2006), which may be derived using landscape images and geographical information systems (GISs).

Micro/mesocosm experiments may tend to overestimate the toxicity for certain species that only have aquatic life stages and a relatively long generation cycle (e.g., amphipods and isopods). This is largely because exposure regimes used in these experiments comprise homogeneously dosed systems that are atypical of natural water bodies in the field. The lack of refuges and the relatively short duration of these experiments often lead to strong effects on the mentioned invertebrate groups, with no recovery occurring within the experimental phase. To tackle this problem, targeted field-monitoring studies (Schulz 2005; Schulz and Liess 1999a, 1999b) may be a solution. Example 9.12 shows that even the results of well-established in situ bioassays using freshwater amphipods may consistently overestimate the toxicity for these organisms as far as transient nonpoint source pollution is concerned. In any spatially explicit modeling approach, the type of reactions described in Example 9.12 may be considered. In addition, the quantification of refugia in the field (e.g., by means of GIS and aerial landscape photographs) may be used in this context. The following conclusions may be drawn with regard to the use of ecological modeling in the assessment of pesticides:

- Ecological modeling is considered to be a helpful tool in terms of contextualizing risk.
- Many of the existing examples are still snapshots.

- Guidance regarding when and how to use ecological modeling is still needed, specifically concerning the validity of input data and model parameterization; see the LEMTOX workshop results for details (Forbes et al. 2009).
- Quality control of the modeling results (e.g., through appropriate model validation using "new" independent data sets) is needed to allow a more thorough assessment of the overall value of modeling.

Example 9.12: Targeted Field-Monitoring Study (Schulz and Liess 1999b)

A validation study to test for the toxicological and ecological relevance of an in situ bioassay with *Gammarus pulex* L. (Amphipoda) and *Limnephilus lunatus* Curtis (Trichoptera) was conducted in an agricultural stream system. Both tested species showed sensitive reactions to transient insecticide input, but the effects on their population dynamics were different: While larval densities of *L. lunatus* were decreased due to elevated mortality, an active drift and avoidance behavior enabled *G. pulex* to leave contaminated sites temporarily and thus to occur at high densities in contaminated streams. During runoff events, *G. pulex* migrated from the potentially contaminated headstream section (site 4) into the uncontaminated tributary (site C), which can be regarded as a refuge and source for recolonization. Hence, although the bioassay is valuable for identifying insecticide input events, supplementary field studies are recommended for a correct ecological interpretation of the results and further understanding of the complexity of reactions under field conditions. If the reaction of amphipods to transient chemical stress under field conditions should be implemented in a population model, these types of avoidance reactions need to be considered. Mean (±standard error, *n* = 5) of drifting mayflies (*Baetis harrisoni*) expressed as drift rate (top) and as drift density (bottom)

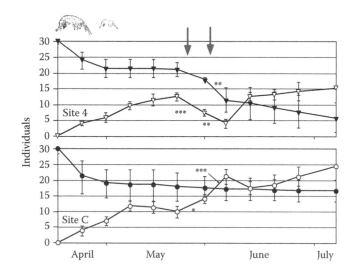

in 30-min treatments either with the pyrethroid cypermethrin (CYP), suspended particles (Part), or flow increase (Flow) and their combinations. Mean survival rate ($n = 4$) of adult *Gammarus pulex* (solid circles) and number of immigrant juvenile *G. pulex* (open circles) in the in situ bioassay at the potentially contaminated stream site 4, 30 m upstream of the confluence with the uncontaminated control tributary (site C). Arrows indicate times of runoff-related insecticide inputs. Asterisks indicate significant (analysis of variance [ANOVA], Fisher's PLSD; $*p < 0.05$; $**p < 0.01$; $***p < 0.001$) changes in the number of organisms between sampling dates.

Mean ($\pm$ standard error, $n = 5$) of drifting mayflies (Baetis harrisoni) expressed as drift rate (top) and as drift density (bottom) in 30-min treatments either with the pyrethroid cypermethrin (CYP), suspended particles (Part), or flow increase (Flow) and their combinations. Mean survival rate ($n = 4$) of adult Gammarus pulex (solid circles) and number of immigrant juvenile G. pulex (open circles) in the in situ bioassay at the potentially contaminated stream site 4, 30 m upstream of the confluence with the uncontaminated control tributary (site C). Arrows indicate times of runoff related insecticide inputs. Asterisks indicate significant (analysis of variance [ANOVA], Fisher's PLSD; $*p < 0.05$; $**p < 0.01$; $***p < 0.001$) changes in the number of organisms between sampling dates.

10 Ecological Field Data in the Risk Assessment of Plant Protection Products*

CONTENTS

10.1 INTRODUCTION

Ecological field data are defined in the following as data, generated from real ecosystem sampling, that allow the description of the physical, chemical, and biological characteristics of the aquatic ecosystems of concern. In the following, these data are referred to as "ecological data." Ecological data aim at providing a picture of the properties of aquatic ecosystems or water bodies that we intend to protect by the risk assessment performed. As an example, knowledge of typical assemblages of benthic macroinvertebrates in small streams resembling the streams to be encountered close to treated fields can make an evaluation of higher-tier tests far easier to the risk assessor. The reasoning is also valuable for fish and plants.

---

* Box 5 and Box 6 in decision scheme; Figure 2.1 in Chapter 2 or Appendix 4.

By their nature, ecological data sets may be particularly useful in higher-tier risk assessments by means of model ecosystems and ecosystem models (Boxes 5 and 6 in Figure 2.1). At these steps of the risk assessment, the role of ecological data is to reduce uncertainty in the risk assessment by incorporating ecological knowledge of the aquatic ecosystems we aim to protect. But, ecological data may also be useful at earlier steps of the risk assessment (i.e., before additional laboratory studies are generated) since field-monitoring data may provide useful information about the relevant species to be tested (e.g., to construct species sensitivity distributions [SSDs]) or about ecologically relevant experimental conditions. It has to be noted, however, that in most cases all the uncertainties that find a solution with the ecological data set are supposed to be covered in the first-tier risk assessment by the "realistic worst-case" work hypothesis for both fate and effects assessment and by the assessment factors that are used.

Reality cannot be fully deduced from model ecosystems (i.e., microcosm and mesocosm studies) or ecosystem models. Indeed, since model ecosystems usually are built from samples taken from real freshwater ecosystems, they represent a limited-scale water body. These smaller test systems do not take into account all the possible internal feedback mechanisms (e.g., effects of large predators) or interactions with the surrounding environment (e.g., water, nutrient and organic matter fluxes) that play a role in larger-scale ecosystems. On the other hand, model ecosystems are tools that can provide experimental proof of exposure–response relationships, while this proof is more difficult to obtain by means of observations in real freshwater ecosystems because of possible confounding factors and lack of replication. For this reason, the information provided by model ecosystem studies does not have the same status as information provided by field monitoring (see also Liess et al. 2005 for further considerations on this topic).

The following text proposes some guidance to generate and use real ecological data in the aquatic risk assessment. Examples of real ecological data sets that were gathered by one of the ELINK (EU Workshop on Linking Aquatic Exposure and Effects in the Registration Procedure of Plant Protection Products) working groups (WGs) are presented in the report, "Ecological Characterization of Water Bodies," in Chapter 15 for Dutch ditches, German and Spanish streams, and English ponds.

10.2   GENERATING ECOLOGICAL DATA

10.2.1   Link with FOCUS Scenarios

To facilitate the risk assessment, ecological data should ideally provide a picture of the typical physical, chemical, and biological properties of water bodies in agricultural landscapes (e.g., pond, ditch, and stream). This information can be used not only to improve the pond, ditch, and stream scenarios that currently are used for the exposure assessment according to FOCUS (Forum for the Co-ordination of Pesticide Fate Models and Their Use; 2001) but also to adapt these scenarios for the effect assessment to better link the exposure and effect assessment. Consequently, the selected water bodies for field monitoring should ideally be comparable, in type and size, to the water bodies in which the exposure estimates are calculated (e.g., a

TABLE 10.1

Overview of the 10 Scenarios Defined[a]

Name	Mean Annual Temperature (°C)	Annual Rainfall (mm)	Topsoil	Organic Matter (%)	Slope (%)	Water Bodies	Weather Station
D1	6.1	556	Silty clay	2.0	0 to 0.5	Ditch, stream	Lanna
D2	9.7	642	Clay	3.3	0.5 to 2	Ditch, stream	Brimstone
D3	9.9	747	Sand	2.3	0 to 0.5	Ditch	Vreedepeel
D4	8.2	659	Loam	1.4	0.5 to 2	Pond, stream	Skousbo
D5	11.8	651	Loam	2.1	2 to 4	Pond, stream	La Jailliere
D6	16.7	683	Clay loam	1.2	0 to 0.5	Ditch	Thiva
R1	10.0	744	Silt loam	1.2	3	Pond, stream	Weiherbach
R2	14.8	1402	Sandy loam	4	20[b]	Stream	Porto
R3	13.6	682	Clay loam	1	10[b]	Stream	Bologna
R4	14.0	756	Sandy clay loam	0.6	5	Stream	Roujan

[a] From FOCUS 2001.
[b] Terraced to 5%.

ditch, pond, or stream in which predicted environmental concentrations in surface water [PECsw] are estimated with FOCUS tools). The selected water bodies should also preferably be typical of an agricultural landscape rather than aquatic ecosystems of pristine, or even protected, areas to ascertain that the data provide a realistic picture of the biodiversity in agricultural areas. However, particular care should be taken to avoid the use of monitoring data from very stressed and overpolluted water bodies when constructing the scenarios (e.g., water bodies subject to bad agricultural practices or other sources of pollution) as they would not adequately represent the level of environmental protection expected from a risk assessment. Similarly, in the case of products assessed for specific use (e.g., in forests), the corresponding field database should represent forest water bodies as much as possible.

The main climate and soil or landscape characteristics relevant for the water bodies in the current FOCUS scenarios are listed in Table 10.1. It is acknowledged that these scenarios are based on a set of assumptions that do not necessarily correspond to real situations. In addition, the set of assumptions is not exhaustive, so influencing factors may be missed. As a consequence, it is important to note that in selecting water bodies to generate real ecological data, it is not necessary to fulfill all the specific scenario characteristics mentioned in Table 10.1. In other words, the aim is not to find streams in the area of Porto with a mean annual temperature of 14.8 °C, an annual rainfall of 1402 mm, and so on. It is more important to address the key characteristics that probably exert an important influence on the behavior of active substances or their degradation products (e.g., pH, depth of the water column, water flow, biomass of plants, organic matter content of the sediment), on the freshwater communities present (e.g., pH, nutrient status, water flow, organic matter content of the sediment, biomass macrophytes), and thus also on the potential ecological risk of pesticide residues. However, any deviation from the values reported in the existing FOCUS scenarios that could be observed in the selected water bodies should be interpreted with regard to their consequences into the risk assessment. As an example, if the candidate streams of a region in the south of Europe would display a mean annual temperature of 19.0 °C, thus being above the temperature for the FOCUS scenarios, the possible consequences on both the fate of the active substance and on the composition and sensitivity of the assemblages should be discussed.

10.2.2   DESCRIPTION OF THE WATER BODIES

To select representative water bodies that can be linked to the FOCUS scenarios, it is necessary to collect the available information on the geographical, geological, physicochemical, and biological characteristics of water bodies in European agricultural landscapes and preferably on the land use of the surroundings, particularly with regard to cropped area. The representativeness of the selected water bodies for a specific type of freshwater ecosystem in the country or landscape of concern should also be described. Such information is, for example, available for Dutch ditches, German streams, and Spanish streams in the WG report in Chapter 15. This information helps to address the uncertainty related to the extrapolation of higher-tier risk assessments to the actual freshwater ecosystems of concern. Whether the water bodies are natural or made by humans also should be reported, together with the age of the system in the latter case. Furthermore, the description should include information on important management activities, such as weed cutting, dredging, flow regulation, and fishing.

10.2.3   DESCRIPTION OF ABIOTIC FACTORS

Several abiotic factors influence the species composition of aquatic ecosystems, the exposure, and the susceptibility of the populations to all kind of stressors:

- water body depth
- nature of the sediment substrate and its thickness
- water flow (linked to the slope)
- pH and variation
- temperature and variation
- organic matter content
- light intensity

Information on the typical abiotic factors associated with benthic invertebrate assemblages have, for example, been published in Tachet et al. (2000). Abiotic factors also strongly correlate with certain species traits of the organisms present, such as voltinism, which is directly linked to temperature. These abiotic factors may also affect bioavailability and exposure to the chemical substance of concern. Water column and sediment depth condition the dilution of the active substance in the water column and its partitioning between the water column and the sediment. The quality and quantity of suspended or sediment-associated organic debris matter in particular may affect the bioavailability of hydrophobous substances. Also, substances subject to photolysis or hydrolysis may behave in different ways, depending on the prevailing light and pH conditions in the water body of concern.

Inclusion of all these parameters as inputs for surface water exposure modeling would probably increase the realism of output data. However, it would also move the focus from worst-case exposure estimates (relevant for generic risk assessments) to exposure estimates more relevant for local situations. The latter imply a higher workload, depending on the number of local situations that need to be assessed. Furthermore, the development of improved or new FOCUS scenarios that integrate more input parameters also increases the workload.

Table 10.2 attempts to describe the potential influence of abiotic factors on FOCUS output and the possible consequences for the risk assessment. All the parameters listed in Table 10.2, except conductivity, may be used either qualitatively or quantitatively in a refined risk assessment. Even for qualitative use, it is recommended to get consolidated data. The data should be reported together with the description of the data collection period and the spatial scale of the area where sampling was done. Examples may be found in the WG report in Chapter 15.

10.2.4 DESCRIPTION OF BIOTIC FACTORS

Biotic factors correspond to the faunal and floral assemblages encountered in the described water bodies, that is, to species composition and related species traits. Fauna and flora inventories can inform us about the general "ecosystem health" of the system, its current developmental stage, and the presence of relevant taxa (trophic and functional groups) from a risk assessment point of view. Ideally, fauna and flora inventories should be made at the most detailed level. However, it may not be possible to describe the whole assemblage at the species level. The workload for such a level of detail may be discussed in light of its significance for the risk assessment. For example, Maltby et al. (2005) showed that species sensitivities toward insecticides are more similar within than between phylogenetic taxa. Particularly, arthropods were demonstrated to be more sensitive to insecticides than other taxonomic groups of invertebrates. As a consequence, for the risk assessment of insecticides it may be wise to collect data in greater taxonomical detail for arthropods, while for other taxa a rather high level of identification (e.g., family or order) may suffice. Similarly, for herbicides it may be wise to put more energy in the identification of algae and macrophytes than in invertebrates.

Defining the species or taxa of interest not only depends on abundance criteria, but also species traits and developmental stages of the species in the assemblage may influence the overall sensitivity and vulnerability to pesticide stress (see FOCUS 2007b for a review). As a consequence, the ecological traits of the potentially sensitive taxa (e.g., voltinism, dispersal) should be described when possible to facilitate spatiotemporal extrapolation.

Generic data on voltinism, dispersal, and other ecological traits are available in several databases (see, e.g., Tachet et al. 2000), but real information from the aquatic systems sampled should always be considered as more valuable since voltinism of a given species may be different (e.g., in a German stream near Hamburg and in

TABLE 10.2

Possible Influence of Water Body Abiotic Factors on Exposure Estimates and Proposals to Address This in the Risk Assessment

Parameter	Influence on Exposure Estimates	Use in Risk Assessment
Dimensions of the water bodies	Influences partitioning process and dynamics of concentrations in water column and sediment	In principle, a quantitative parameter but to be used in a qualitative manner to avoid underestimation of local effects. Use the information to inform the risk assessment in terms of realism or representativity of a worst case
Water flow	Influences dynamics of concentrations in water column due to dilution and discharge	See above
Conductivity	Influence probably not large	Less useful in exposure assessment. Useful water quality parameter to check the condition of the test system
pH	Affects degradation rate (e.g., hydrolysis); substance dependent. May also affect adsorption in case of weak acid and thereafter partition to sediments and dissipation rate in the total system	Quantitative parameter. Can be easily taken into account through laboratory studies determining DT50 or DT50 dissipation as a function of pH
Organic matter	Influences bioavailability; substance dependent	Quantitative parameter. Can be taken into account manually into TOXSWA[a]
Temperature	Influences degradation rate of substances and sensitivity and growth rate of species	Quantitative parameter. Included in TOXSWA
Suspended sediment	Influences adsorption; might be important but difficult to take quantitatively into account	Can be taken into account manually into TOXSWA
Organic matter content upper sediment layer	Influences sorption and desorption; organic matter content of upper sediment is season dependent	Can be taken into account manually into TOXSWA

[a] A toxic substance in surface water model.

French streams as reported in the Tachet database). The analysis of data from the abundance and ecological aspects may help in proposing (focal) species to be considered as leading the risk assessment for a specific purpose (scenario, area, toxic mode of action).

Ideally, information on different functional groups (also including nonsensitive taxa) should be gathered to allow identification of indirect effects and shifts in food web interactions due to pesticide stress (Relyea and Hoverman 2006). In doing so, the ecological data collected can be used to provide "target images" of aquatic systems that may be encountered in agricultural landscapes and thus may be used as generic ecological data independently of the nature of the risks that are assessed.

10.2.5 REAL ECOLOGICAL DATA SETS PRESENTED IN WG REPORT "ECOLOGICAL CHARCTERIZATION OF WATER BODIES" AND GENERAL RECOMMENDATION FOR FUTURE DATA SET

Examples of ecological data collection and reporting are provided in Chapter 15. An attempt was made to describe all 3 water body types in which PECsw are calculated in FOCUSsw (forum for the coordination of pestcide fast models and their use in surface waters). For this purpose, data were collected for Dutch ditches, German and Spanish streams, and English ponds. Data were also collected to describe Mediterranean wetlands. As can be seen in Chapter 15, the data that could be collected within the time frame of the ELINK working group differed in detail and presentation and reflect that they were collected for different purposes. All these ecological data sets have advantages and limitations that are discussed in the corresponding sections of Chapter 15, but all brought information that could be used to refine the risk assessment.

There is still a need for additional ecological data sets. Additional data for each type of water body can be used for comparative purposes to study uncertainties for spatiotemporal extrapolation. Furthermore, ecological data sets for ditches, streams, and ponds should be collected for different climatic zones in Europe to be used in a generic way as a handbook of "ecological scenarios" to which risk assessors could refer in evaluating refined risk assessments. It is proposed that the generation of these real ecological data sets should be based on a FOCUS-based distinction:

- One per water body type (streams, ditches, and ponds)
- One per European region to distinguish north, south, and central Europe

Once the ecological data sets become available, their geographical representativeness could be tested by collecting information of surrounding locations to know whether they sufficiently represent the aquatic systems relevant for the risk assessment (e.g., as described in FOCUS scenarios) or if there is a need for additional data sets (e.g., Do Hungarian streams resemble German streams from a risk assessment point of view?). Work sharing is needed to ensure the harmonization of data collection and of their use. The integration of the collected real ecological data from surface waters in agricultural landscapes into a European database is recommended.

10.3 USE OF REAL ECOLOGICAL DATA
TO REFINE THE RISK ASSESSMENT

In the following, examples are given on some key issues for effect and fate assessments, which are based on the data provided in real ecological data sets. Other examples could be found from the data sets presented in the WG report presented in Chapter 15 and from future data sets. The aim is to illustrate how such data could be used quantitatively or qualitatively in the refined risk assessment. In some cases, the data set itself does not provide ready-to-use information to refine the risk assessment but rather provides the basis for additional and simple experiments that may provide the required information. Real ecological data sets may also bring a sound basis for ecological modeling.

10.3.1 USE TO REFINE EFFECTS ASSESSMENT

Ecological data may be used to evaluate whether the ecotoxicological data set contains taxa or species that were identified as representative or focal species in the water bodies of concern. Thus, the availability of a real ecological data set may address the uncertainty with regard to the level of protection that is achieved by the risk assessment.

Example 10.1

An ecotoxicological data set may contain multiple laboratory tests, the results of which are further analyzed into an SSD. It may also contain a micro- or meso-cosm study. In the ELINKmethrin case study (Appendix 3), a refinement of the effect assessment used an SSD, and a mesocosm study focused on invertebrates. The species used in the SSD could be compared with the assemblage data from real ecological data. ELINKmethrin is an insecticide, and elements for comparison could be found in the real data set for the Dutch ditch (see Chapter 15). Tables 15.12 and 15.13 of Section 15.3 in Chapter 15 provide an inventory of arthropods of the macroinvertebrate community in macrophyte-dominated ditches. These tables could be used to check whether the most frequent arthropod taxonomic groups in Dutch ditches were covered by the species used to construct the SSD for ELINKmethrin. It appeared from this comparison that representatives of several key taxa were represented in the SSD. This means that the results of the SSD could be considered ecologically relevant for Dutch ditches.

With respect to Example 10.1, it has to be noted that the output of an SSD is influenced by species selection, and criteria of use may be different at the EU and national levels. Real ecological data may then also help in refining risk assessment via eliminating some groups of the SSD as not relevant for the risk assessment.

This kind of assemblage data could also be used to evaluate the representativeness of the species present in mesocosms. Again, in the case of ELINKmethrin, there was a good deal of correspondence between the taxa tested in the mesocosm compared to those expected in the Dutch ditch assemblage. This analysis thus reinforced the confidence in the ecological relevance of the ecotoxicological data set in the case of Dutch ditches.

10.3.2   USE TO REFINE EXPOSURE ASSESSMENT

Ideally, the aquatic system selected for ecological data set collection should be as similar as possible to the water bodies for which exposure calculations are performed.

Example 10.2

In estimating PECsw for ELINKmethrin (Appendix 3), the rules for selecting input parameters in FOCUSsw simulations meant the selection of values that do not perfectly reflect the fate of the substance in the system. A DT50 (period required for 50% disappearance) of 1000 days is used (default value) due to strong partition properties. The substance quickly partitions to the sediment, and it is hydrolytically labile with a DT50 of 0.5 day at pH 9.0. In the German streams studies (Chapter 15), pH ranged from 7.0 to 8.5, the median pH derived from the review of Dutch ditch data was 7.9, and in Spanish streams the pH ranged from 7.5 to 8.5, meaning that hydrolysis of ELINKmethrin could be a more or less important factor in substance dissipation. The DT50 default value of 1000 days to simulate the fate of ELINKmethrin in water probably overestimates exposure in the water, at least for water body types with low water flow. Exposure estimates could in such cases be revisited, looking at both hydrolysis and partition to the sediment. Simple experiments performed in controlled systems with and without sediment could be undertaken under varying pH conditions to derive DT50 in the water column and in the total systems. The pH values selected could be in the range of pH values recorded under real aquatic systems to generate more realistic input data for FOCUS modeling. The consequence of this may be a significantly lower DT50 value, leading to faster degradation, and probably lower than expected amounts of chemical being adsorbed to the sediment.

Refining PEC calculations through, for example, the use of tailored DT50, pH, and rate of water flow may lead to less-conservative exposure assessments.

In Example 10.2, the rate of decline of the pesticide is influenced by pH. It is therefore reasonable that account should be taken of the pH of the water bodies predominating in real-world ecological landscapes when refining potential exposure concentrations derived from modeling. However, note should be taken of the fact that fluctuations in abiotic (physicochemical) and biotic (microbiological) characteristics of the water body do occur in space and time, which may influence the fate and behavior of active substances (particularly in lentic systems) with a consequent effect on exposure concentrations.

Example 10.3

Another illustration is given based on high sorption properties. The substance ELINKenzuron is quite lipophilic and has a Koc of 25 000. This suggests that sorption to macrophytes may play a role in the dissipation of ELINKenzuron. Information on the biomass of macrophytes present in the water body could be used to refine the modeling via adsorption to macrophytes. Information on the biomass corresponding to macrophytes was available from the ecological data set for Dutch ditches (Chapter 15). The proportion of substance being adsorbed as a function of plant biomass would not be a difficult information to generate. Used as an input into

FOCUS modeling, it is anticipated that this adsorption would reduce the height of the exposure peaks, specifically to the water bodies to protect. Although macrophytes may influence exposure estimates in aquatic systems, basic data on sorption and desorption of the active substance to macrophytes usually is not available in the dossier, so it is difficult to take this into account. Degradation studies with and without macrophytes and including sorption or desorption measurements may allow quantification of the phenomenon when relevant.

Care should therefore be taken to assess the appropriateness of the refinement in terms of its perceived contribution to a reduction in exposure in water bodies associated with agricultural landscapes. In Example 10.3, the fraction of a substance adsorbed to macrophytes reduces the exposure concentration in the water column. However, account should be taken of the potential release of residues into the water body when plants decay or those that might be bioavailable to phytophagous species or decomposers.

10.3.3   Use to Discuss Seasonal Influence on Exposure in the Field

Example 10.4

A potential example of where seasonality of exposure and its likelihood to co-occur with presence of the relevant taxa was provided by the ELINKsulfuron case study. The risk assessment for ELINKsulfuron (Appendix 3) highlighted a potential concern for aquatic plants. Plant growth may be highly seasonal, and abundance and growth follow the vegetation period. Review of the exposure profile for D4 (pond) indicated that the maximum peak of exposure occurs on the first of January, in the middle of the winter for the climate at D4. In the case of effects directed specifically on the growth of plants, it may be useful to discuss the time of occurrence of exposure peaks in line with the main periods of plant growth. The consequences of the expected exposure peaks on plant growth over the next spring season may, for example, be estimated through modeling.

10.3.4   Use as Input for Ecological Modeling

Ecological modeling may be very useful as additional information for a refined risk assessment. In particular, ecological modeling may allow a reduction in the uncertainties remaining in extrapolating a risk assessment to the wide range of situations to be covered (see Chapter 9 and the WG report in Chapter 14 for further details). Modeling may bring answers on untested situations or untested species. This is the case when questions remain on the species of concern within time (i.e., What happens after the study has finished? What if the exposure pattern is different?) or for species that should be accounted for in the risk assessment but were not present in the mesocosm studies.

An ecological data set may be useful at different levels:

- As a reference situation with regard to expected population levels reached as an output of models
- As a data resource for model parameterization for tested species but with ecological traits being specific to a region

- As a data resource for model parameterization for untested species
- As a data resource for model parameterization to simulate effects in specific water body types

Example 10.5

Effects of a herbicide to algae under conditions specific for Spain could, for example, be simulated based on the mechanistic algae model of Weber et al. (2007) (see Chapter 9 for a description). In this model, conditions in Spain would be simulated through inputs for temperature and light from data for the streams in Spain. Such information would then be used to discuss the expected level of risk under these particular cases. As another example, recovery potential after an exposure could be simulated for a specific case study based, for example, on a population recovery model (Barnthouse 2004), provided that the organisms present in the ecological data set appear in the model developed. Furthermore, spatially explicit models including a spatiotemporal component (e.g., based on a metapopulation model like MASTEP [metapopulation model for assessing spatial and temporal effects of pesticides]; Van den Brink, Baveco, et al. 2007), which are adapted to the life-cycle traits of the species as described in the ecological data set, may be used.

In all examples mentioned in the previous discussion, some further data than those included regularly in dossiers would be needed. The amount of data required depends obviously on the model used and more specifically on the complexity of the model and the respective model input parameters (Example 10.5). For the algae model (Weber et al. 2007), information on algal growth rates for different temperature, nutrient, and light conditions are required. For the recovery model (Barnthouse 2004), population growth rates of the species to be considered are needed. This may be derived from mesocosm experiments. However, other data sources are needed in case species not tested in the mesocosm need to be addressed. For the metapopulation model (Van den Brink, Baveco, et al. 2007), quite extensive information of the migration patterns (organismic drift, upstream migration) of the target species (in this case, *Asellus aquaticus*) are needed. These data can only be extracted from literature sources, if available, or from additional laboratory or field experimentation.

In principle, it is possible to use these different types of models in the refined risk assessment; however, their use will often be restricted to species on which a comparably high amount of additional ecological data are available. The data demand increases with the complexity of models: Models like the dynamic algae model are the easiest to use with few data other than those that are available in dossiers, more information is needed for recovery-type models, and even more (also spatial) information may be needed for spatially explicit metapopulation types of model. In this context, ecological data may be a source of information to be used in ecological models for each of the parameters that the models address.

11 References for Guidance Chapters (Part I)

Aldenberg T, Jaworska JS. 2000. Uncertainty of hazardous concentrations and fraction affected for normal species sensitivity distributions. Ecotoxicol Environ Saf 46:1–18.

Aldenberg T, Jaworska JS, Traas TP. 2002. Normal species sensitivity distributions in probabilistic ecological risk assessment. In: Posthuma L, Traas TP, Suter GW, editors, Species sensitivity distributions in risk assessment. Boca Raton (FL): CRC Press. p 49–102.

Angeler DG, Alvarez-Cobelas M. 2005. Island biogeography and landscape structure: integrating ecological concepts in a landscape perspective of anthropogenic impacts in temporary wetlands. Environ Pollution 138:420–424.

Arnold D, Hill IR, Matthiessen P, Stephenson R, editors. 1991. Guidance document on testing procedures for pesticides in freshwater mesocosms. Brussels: SETAC-Europe.

Ashauer R, Boxall ABA, Brown CD. 2006a. Predicting effects on aquatic organisms from fluctuating or pulsed exposure to pesticides. Environ Toxicol Chem 25(7):1899–1912.

Ashauer R, Boxall ABA, Brown CD. 2006b. Uptake and elimination of chlorpyrifos and pentachlorophenol into the freshwater amphipod *Gammarus pulex*. Arch Environ Contam Toxicol 51:542–548.

Ashauer R, Boxall ABA, Brown CD. 2007a. Modelling combined effects of pulsed exposure to carbaryl and chlorpyrifos on *Gammarus pulex*. Environ Sci Technol 41:5535–5541.

Ashauer R, Boxall ABA, Brown CD. 2007b. New ecotoxicological model to simulate survival of aquatic invertebrates after exposure to fluctuating and sequential pulses of pesticides. Environ Sci Technol 41(4):1480–1486.

Ashauer R, Boxall ABA, Brown CD. 2007c. Simulating toxicity of carbaryl to *Gammarus pulex* after sequential pulsed exposure. Environ Sci Technol 41(15):5528–5534.

Ashauer R, Brown CD. 2007. Comparison between FOCUS outputs for pesticide concentrations over time and field observations [Internet]. University of York report for Defra project PS2231, 48 p. Available from: <http://randd.defra.gov.uk/>.

Barber MC. 2003. A review and comparison of models for predicting dynamic chemical bioconcentration in fish. Environ Toxicol Chem 22:1963–1992.

Barnthouse LW. 2004. Quantifying population recovery rates for ecological risk assessment. Environ Toxicol Chem 23(2):500–508.

Barnthouse LW. 2008. Population-level ecological risk assessment. Boca Raton (FL): Taylor & Francis Group.

Bartell SM, Lefebvre G, Kaminski G, Carreau M, Campbell KR. 1999. An ecosystem model for assessing ecological risks in Québec rivers, lakes, and reservoirs. Ecol Modell 124:43–67.

Beketov MA, Schäfer RB, Marwitz A, Paschke A, Liess M. 2008. Long-term stream invertebrate community alterations induced by the insecticide thiacloprid: effects concentrations and recovery dynamics. Sci Total Environ 405:96–108.

Belanger SE. 1997. Literature review and analysis of biological complexity in model stream ecosystems: influence of size and experimental design. Ecotoxicol Environ Saf 52:150–171.

Belanger SE, Meiers EM, Bausch RG. 1995. Direct and indirect ecotoxicological effects of alkyl sulfate and alkyl ethoxysulfate on macroinvertebrates in stream mesocosms. Aquat Toxicol 33:65–87.

Boesten JJTI, Köpp H, Adriaanse PI, Brock TCM, Forbes VE. 2007. Conceptual model for improving the link between exposure and effects in the aquatic risk assessment of pesticides. Ecotoxicol Environ Saf 66:291–308.

Boxall ABA, Brown CD, Barrett KL. 2002. Higher-tier laboratory methods for assessing the aquatic toxicity of pesticides. Pest Manage Sci 58:637–648.

Brock TCM, Arts GHP, Maltby L, Van den Brink PJ. 2006. Aquatic risks of pesticides, ecological protection goals and common aims in EU legislation. Integr Environ Assess Manag 2:e20–e46.

Brock TCM, Bos AR, Crum SJH, Gylstra R. 1995. The model ecosystem approach in ecotoxicology as illustrated with a study on the fate and effects of an insecticide in stagnant freshwater microcosms. In: Koch B, Niessner R, editors, Immunochemical detection of pesticides and their metabolites in the water cycle. Weinheim (DE): Deutsches Forschungsgemeinschaft VCH. p 167–185.

Brock TCM, Budde BJ. 1994. On the choice of structural parameters to indicate responses of freshwater ecosystems to pesticide stress. In: Hill IA, Heimbach F, Leeuwangh P, Matthiesen P, editors, Freshwater field tests for hazard assessment of chemicals. Boca Raton (FL): Lewis Publishers. p 19–56.

Brock TCM, Lahr J, Van den Brink PJ. 2000. Ecological risks of pesticides in freshwater ecosystems. Part 1: Herbicides. Wageningen (NL): Alterra. Report no. 088.

Brock TCM, Maltby L, Hickey CW, Chapman J, Solomon K. 2008. Spatial extrapolation in ecological effect assessment of chemicals. In: Solomon KR, Brock TCM, De Zwart D, Dyer SD, Posthuma L, Richards SM, Sanderson H, Sibley PK, Van den Brink PJ, editors. Extrapolation practice for ecotoxicological effect characterization of chemicals. Boca Raton (FL): SETAC Press and CRC Press, p 223–256.

Brock TCM, Solomon K, Van Wijngaarden R, Maltby L. 2008. Temporal extrapolation in ecological effect assessment of chemicals. In: Solomon KR, Brock TCM, De Zwart D, Dyer SD, Posthuma L, Richards SM, Sanderson H, Sibley PK, Van den Brink PJ, editors, Extrapolation practice for ecotoxicological effect characterization of chemicals. Boca Raton (FL): SETAC Press and CRC Press, p 187–221.

Brock TCM, Van Wijngaarden RPA, Van Geest GJ. 2000. Ecological risks of pesticides in freshwater ecosystems. Part 2: Insecticides. Wageningen (NL): Alterra. Report no. 089.

Brown CD, Dubus IG, Fogg P, Spirlet M, Reding M-A, Gustin C. 2004. Exposure to sulfosulfuron in agricultural drainage ditches. Pest Manage Sci 60:765–776.

Brown CD, Turner N, Hollis J, Bellamy P, Biggs J, Williams P, Arnold D, Pepper T, Maund S. 2006. Morphological and physico-chemical properties of British aquatic habitats potentially exposed to pesticides. Agric Ecosys Environ 113:307–319.

Campbell PJ, Arnold DJS, Brock TCM, Grandy NJ, Heger W, Heimbach F, Maund SJ, Streloke M. 1999. Guidance document on higher-tier aquatic risk assessment for pesticides (HARAP). Brussels: SETAC-Europe.

Caquet T, Hanson ML, Roucaute M, Graham DW, Lagadic L. 2007. Influence of isolation on the recovery of pond mesocosms from the application of an insecticide. II. Benthic macroinvertebrate responses. Environ Toxicol Chem 26:1280–1290.

Caquet T, Lagadic L, Sheffield SR. 2000. Mesocosms in ecotoxicology. 1. Outdoor aquatic systems. Rev Environ Contam Toxicol 165:1–38.

Crane M, Giddings JM. 2004. Risk around the world: "Ecological acceptable concentration" when assessing the environmental risks of pesticides under European Directive 91/414/EEC. Human Ecol Risk Assess 10:733–747.

Crawford CG. 2004. Sampling strategies for estimating acute and chronic exposures of pesticides in streams. J Am Water Res Assoc 40:485–502.

Crossland NO, Heimbach F, Hill IR, Boudou A, Leeuwangh P, Matthiessen P, Persoone G. 1993. European Workshop on Freshwater Field Tests (EWOFFT), summary and recommendations. Workshop held in Potsdam, Germany, June 25 to 26, 1992.

Daam M. 2008. Influence of climatic factors and microcosm complexity on the fate and effects of pesticides [dissertation]. Aveiro (PT): Universidade de Aveiro.

Dabrowski JM, Bollen A, Bennett ER, Schulz R. 2005. Pesticide interception by emergent aquatic macrophytes: potential to mitigate spray-drift input in agricultural streams. Agric Ecosyst Environ 111:340–348.

Dabrowski JM, Bollen A, Schulz R. 2005. Combined effects of discharge, turbidity and pesticides on mayfly behaviour: experimental evaluation of spray-drift and runoff scenarios. Environ Toxicol Chem 24:1395–1402.

De Jong FMW, Brock TCM, Foekema EM, Leeuwangh P. 2008. Guidance for summarizing and evaluating aquatic micro- and mesocosm studies. Bilthoven (NL): RIVM. RIVM Report 601506009.

[EC] European Commission. 1997. Commission proposal for a council objective establishing annex VI to directive 91/414/EEC concerning the placing of plant protection products on the market. Off J Eur Comm C 240:1–23.

[EC] European Commission. 2000. Directive 2000/60/EC of the European Parliament and of council of 23 October 2000 establishing a framework for community action in the field of water policy. Off Eur Comm L 327/1.

[EC] European Commission. 2002. Guidance document on aquatic ecotoxicology in the context of the Directive 91/414/EEC. Brussels: European Commission. Health and Consumer Protection Directorate-General. SANCO/3268/2001 rev, 4 (final).

[EC] European Commission. 2003. Technical guidance document on risk assessment in support of Commission Directive 93/67/EEC on risk assessment to new notified substances and Commission Regulation (EC) No. 1488/94 on risk assessment for existing substances and Directive 98/8/EC of the European Parliament and the Council concerning the placing of biocidal products on the market. Luxembourg (Luxembourg): Office for Official Publications of the European Communities.

[EFSA] European Food Safety Authority. 2005. Opinion of the Scientific Panel on Plant Health, Plant Protection Products and Their Residues on a request from the EFSA related to the evaluation of dimoxystrobin. EFSA J 178:1–45.

[EFSA] European Food Safety Authority. 2005. Opinion of the Scientific Panel on Plant Health, Plant Protection Products and Their Residues on a request from the EFSA related to the assessment of the acute and chronic risk to aquatic organisms with regard to the possibility of lowering the assessment factor if additional species were tested. EFSA J 301:1–45.

[EFSA] European Food Safety Authority. 2006. Opinion of the Scientific Panel on Plant Health, Plant Protection Products and Their Residues on a request from the EFSA related to the aquatic risk assessment for cyprodinil and the use of a mesocosm study in particular. EFSA J 329:1–77.

[EFSA] European Food Safety Authority. 2006. Opinion of the Scientific Panel on Plant Protection Products and Their Residues on a request from EFSA on the Final Report of the FOCUS Working Group on Landscape and Mitigation Factors in Ecological Risk Assessment. EFSA J 437:1–30.

Forbes VA, Calow P. 2002. Species sensitivity distributions revisited: a critical appraisal. Human Ecol Risk Assess 8:473–492.

Forbes VE, Hommen U, Grimm V, Thorbek P, Heimbach F, Van den Brink P, Wogram J, Thulke H. 2009. Ecological models in support of regulatory risk assessment of pesticides: developing a strategy for the future. Integr Environ Assess Manag 5:167–172.

Forbes VE, Sibley RM, Calow P. 2001. Determining toxicant impacts on density-limited populations: a critical review of theory, practice and results. Ecol Appl 11:1249–1257.

[FOCUS] Forum for the Co-ordination of Pesticide Fate Models and Their Use. 2001. FOCUS surface water scenarios in the EU evaluation process under 91/414/EEC. Report of the FOCUS Working Group on Surface Water Scenarios. EC Document Reference SANCO/4802/2001-rev2.

[FOCUS] Forum for the Co-ordination of Pesticide Fate Models and Their Use. 2006. Guidance document on estimating persistence and degradation kinetics from environmental fate studies on pesticides in EU registration. Report of the FOCUS Working Group on Degradation Kinetics. EC Document Reference SANCO/10058/2005 v2.0, May 2005.

[FOCUS] Forum for the Co-ordination of Pesticide Fate Models and Their Use. 2007a. Landscape and mitigation factors in aquatic risk assessment. Volume 1. Extended summary and recommendations. Report of the FOCUS Working Group on Landscape and Mitigation Factors in Ecological Risk Assessment. EC Document Reference SANCO/10422/2005 v2.0, September 2007.

[FOCUS] Forum for the Co-ordination of Pesticide Fate Models and Their Use. 2007b. Landscape and mitigation factors in aquatic risk assesment. Volume 2. Detailed technical reviews. Report of the FOCUS Working Group on Landscape and Mitigation Factors in Ecological Risk Assessment. EC Document Reference SANCO/10422/2005 v2.0, September 2007.

Furse M, Hering D, Moog O, Verdonschot P, Johnson RK, Brabec K, Gritzalis K, Buffagni A, Pinto P, Friberg N, Murray-Bligh J, Kokes J, Alber R, Usseglio-Polatera P, Haase P, Sweeting R, Bis B, Szoszkiewic, K, Soszka H, Springe G, Sporka F, Krno I. 2006. The STAR project: context, objectives and approaches. Hydrobiologia 566:3–29.

Giddings JM, Biever RC, Helm RL, Howick GL, deNoyelles FJ. 1994. The fate and effects of Guthion (azinphos methyl) in mesocosms. In: Graney RL, Kennedy JH, Rogers JH, editors, Aquatic mesocosm studies in ecological risk assessment. Boca Raton (FL): Lewis Publishers. p 469–495.

Giddings JM, Brock TCM, Heger W, Heimbach F, Maund SJ, Norman S, Ratte H-T, Schäfers C, Streloke M, editors. 2002. Community-level aquatic system studies-interpretation criteria (CLASSIC). Pensacola (FL): SETAC. 44 p. ISBN 1-880611-49-X.

Giesy JP, Graney RL. 1989. Recent developments in and intercomparisons of acute and chronic bioassays and bioindicators. Hydrobiologia 188/189:21–60.

Gotelli NJ, Ellison A. 2006. Food-web models predict species abundances in response to habitat change. PloS Biol 4(10): 1869–1873.

Hanazato T, Yazuno M. 1990. Influence of time of application of an insecticide on recovery patterns of a zooplankton community in experimental ponds. Arch Environ Contam Toxicol 19:1856–1866.

Hanson ML, Graham DW, Babin E, Azam D, Coutellec M, Knapp CW, Lagadic L, Caquet T. 2007. Influence of isolation on the recovery of pond mesocosms from the application of an insecticide. I. Study design and planktonic community responses. Environ Toxicol Chem 26:1265–1279.

Heckman LH, Friberg N. 2005. Macroinvertebrate community response to pulse exposure with the insecticide lambda-cyhalothrin using in-streaam mesocosms. Environ Toxicol Chem 24:582–590.

Hendley P, Giddings J. 1999. Draft report of the Aquatic Workgroups of ECOFRAM (Ecological Committee on FIFRA Risk Assessment) — Aqex-ecofram-Peer01-may499.doc, p 450 [Internet]. Available from: <http://www.epa.gov/oppefed1/ecorisk/aquareport.pdf>.

Hill IR, Heimbach F, Leeuwangh P, Matthiessen P, editors. 1993. Freshwater field tests for hazard assessment of chemicals. Boca Raton (FL): Lewis Publishers.

Holland PT. 1996. Glossary of terms relating to pesticides (IUPAC Recommendations 1996). Pure Appl Chem 68:1167–1193.

Holvoet K, Seuntjens P, Mannaerts R, De Schepper V, Vanrolleghem PA. 2007. The dynamic water-sediment system: results from an intensive pesticide monitoring campaign. Water Sci Technol 55:177–182.

Hose GC, Lim RP, Hyne RV, Pablo F. 2002. A pulse of endosulfan-contaminated sediment affects macroinvertebrates in artificial streams. Ecotoxicol Environ Saf 51:44–52.

Hose GC, Van den Brink PJ. 2004. Confirming the species sensitivity distribution concept for endosulfan using laboratory, mesocosm and field data. Arch Environ Contam Toxicol 47:511–520.

Jager T, Kooijman SALM. 2005. Modelling receptor kinetics in the analysis of survival data for organophosphorus pesticides. Environ Sci Technol 39:8307–8314.

Jagers op Akkerhuis GAJM, Kjær C, Damgaard C, Elmegaard N. 1999. Temperature-dependent, time-dose-effect model for pesticide effects on growing, herbivorous arthropods: bioassays with dimethoathe and cypermethrin. Environ Toxicol Chem 18:2370–2378.

Jagers op Akkerhuis GAJM, Seidelin N, Kjaer C. 1999. Are we analysing knockdown in the right way? How independence of the knockdown-recovery process from mortality may affect measures for behavioural effects in pesticide bioassays. Pestic Sci 55:62–68.

Kallander DB, Fisher SW, Lydy MJ. 1997. Recovery following pulsed exposure to organophosphorus and carbamate insecticides in the midge, *Chironomus riparius*. Arch Environ Contam Toxicol 33:29–33.

Kersting K. 1984. Development and use of an aquatic micro-ecosystem as a test system for toxic substances. Properties of an aquatic microecosystem IV. Int Rev Hydrobiol 69:567–607.

Kooijman SALM, Bedaux JJM. 1996. The analysis of aquatic toxicity data. Amsterdam: VU University Press.

Lee JH, Landrum PF, Koh CH. 2002. Prediction of time-dependent PAH toxicity in *Hyalella azteca* using a damage assessment model. Environ Sci Technol 36:3131–3138.

Lepper P. 2002. Towards the derivation of quality standards for priority substances in the context of the Water Framework Directive. Identification of quality standards for priority substances in the field of water policy. Final Report. Schmitten (DE): Fraunhofer Institute. Report No. B4-3040/2000/30637/MAR/E1.

Levins R. 1969. Some demographic and genetic consequences of environmental heterogeneity for biological control. Bull Entomol Soc Am 15:237–240.

Liess M, Brown C, Dohmen P, Duquesne S, Hart A, Heimbach F, Krueger J, Lagadic L, Maund S, Reinert W, Streloke M, Tarazona JV, editors. 2005. Effects of pesticides in the field (EPIF). Brussels (Belgium): SETAC Press.

Lin BL, Tokai A, Nakanishi J. 2005. Approaches for establishing predicted-no-effect concentrations for population-level ecological risk assessment in the context of chemical substances management. Environ Sci Technol 39(13):4833–4840.

López-Mancisidor P, Fernández C, Marina A, Carbonell G, Tarazona JV. 2008. Zooplankton community responses to chlorpyrifos in mesocosms under Mediterranean conditions. Ecotoxicol Environ Saf 71:16–25.

Lye Koh H, Hallam T., Ling Lee H. 1997. Combined effects of environmental and chemical stressors on a model *Daphnia* population. Ecol Modell 103(1):19–32.

MacArthur RH, Wilson EO. 2001. The theory of island biogeography. Princeton (NJ): Princeton University Press.

Maltby L, Blake N, Brock TCM, Van den Brink PJ. 2002. Addressing interspecific variation in sensitivity and the potential to reduce this source of uncertainty in ecotoxicological assessments. London: UK Department for Environment, Food and Rural Affairs. London, DEFRA project code PN0932.

Maltby L, Blake N, Brock TCM, Van den Brink PJ. 2005. Insecticide species sensitivity distributions: the importance of test species selection and relevance to aquatic ecosystems. Environ Toxicol Chem 24:379–388.

Maltby L, Brock TCM, Van den Brink PJ. Accepted. Fungicide risk assessment for aquatic ecosystems: interspecies variation in sensitivity and its relation with toxic mode of action. Environ. Sci. Technol.

Maltby L, Kedwards T, Forbes VE, Grasman K, Kammenga JE, Munns WR Jr, Ringwood AH, Weis JS, Wood SN. 2001. Linking individual-level responses and population-level consequences. In: Baird DJ, Burton GA Jr, editors. Ecological variability: separating natural from anthropogenic causes of ecosystem impairment. Pensacola (FL): SETAC. p 27–82.

Metcalf RL, Sanga GK, Kapoor IP. 1971. Model ecosystems for the evaluation of pesticide biodegradability and ecological magnification. Environ Sci Technol 5:709–713.

Muir DCG, Hobden BR, Servos MR. 1994. Bioconcentration of pyrethroid insecticides and DDT by rainbow trout: uptake, depuration, and effect of dissolved organic carbon. Aquat Toxicol 29:223–240.

Naddy RB, Johnson KA, Klaine SJ. 2000. Response of *Daphnia magna* to pulsed exposures of chlorpyrifos. Environ Toxicol Chem 19:423–431.

Neumann M, Baumeister J, Liess M, Schulz R. 2002. An expert system to estimate the pesticide contamination of small streams using benthic macroinvertebrate as bioindicators, part 2: the knowledge base of LIMPACT. Ecol Indicators 2:391–401.

Neuman M, Liess M, Schulz R. 2002. An expert system to estimate the pesticide contamination of small streams using benthic macroinvertebrate as bioindicators, part 1: the database of LIMPACT. Ecol Indicators 2:379–389.

Nichols JW, Fitzsimmons PN, Whiteman FW. 2004. A physiologically based toxicokinetic model for dietary uptake of hydrophobic organic compounds by fish — II. Simulation of chronic exposure scenarios. Toxicol Sci 77:219–229.

Niemi GJ, DeVore P, Detenbeck N, Taylor D, Lima A, Pastor J, Yount JD, Naiman RJ. 1990. Overview of case studies on recovery of aquatic systems from disturbance. Environ Manag 14:571–587.

Nuutinen S, Landrum PF, Schuler LJ, Kukkonen JVK, Lydy MJ. 2003. Toxicokinetics of organic contaminants in *Hyalella azteca*. Arch Environ Contam Toxicol 44(4):467–475.

Parsons JT, Surgeoner GA. 1991. Acute toxicities of permethrin, fenitrothion, carbaryl and carbofuran to mosquito larvae during single- or multiple-pulse exposures. Environ Toxicol Chem 10:1229–1233.

Péry ARR, Bedaux JJM, Zonneveld C, Kooijmann SALM. 2001. Analysis of bioassays with time varying concentrations. Water Res 35(16):3825–3832.

Pieters BJ, Jager T, Kraak MHS, Admiraal W. 2006. Modelling response of *Daphnia magna* to pesticide pulse exposure under varying food conditions: intrinsic versus apparent sensitivity. Ecotoxicology 15:601–608.

Posthuma L, Traas TP, Suter GW, editors. 2002. Species sensitivity distributions in risk assessment. Boca Raton (FL): CRC Press.

Reinert KH, Giddings JM, Judd L. 2002. Effects analysis of timevarying or repeated exposures in aquatic ecological risk assessment of agrochemicals. Environ Toxicol Chem 21:1977–1992.

Relyea R, Hoverman P. 2006. Assessing the ecology in ecotoxicology: a review and synthesis in freshwater systems. Ecology Lett 9:1157–1171.

Roessink I, Arts GHP, Belgers JDM, Bransen F, Maund SJ, Brock TCM. 2005. Effects of lambda-cyhalothrin in 2 ditch microcosm systems of different trophic status. Environ Toxicol Chem 24:1684–1690.

Roessink I, Belgers JDM, Crum SJH, Van den Brink PJ, Brock TCM. 2006. Impact of triphenyltin acetate in microcosms simulating floodplain lakes. II. Comparison of species sensitivity distributions between laboratory and semi-field. Ecotoxicology 15:411–424.

Rozman KK, Doull J. 2000. Dose and time as variables of toxicity. Toxicology 144:169–178.

Schroer AFW, Belgers JDM, Brock TCM, Matser AM, Maund SJ, Van den Brink PJ. 2004. Comparison of laboratory single species and field population-level effects of the pyrethroid insecticide λ-cyhalothrin on freshwater invertebrates. Arch Environ Contam Toxicol 46:324–335.

Schulz R. 2005. Aquatic in situ bioassays to detect agricultural non-point source pesticide pollution: a link between laboratory and field. In: Ostrander GK, editor. Techniques in aquatic toxicology. Boca Raton (FL): CRC Press. p 427–448.

Schulz R, Liess M. 1999a. A field study of the effects of agriculturally derived insecticide input on stream macroinvertebrate dynamics. Aquat Toxicol 46:155–176.

Schulz R, Liess M. 1999b. Validity and ecological relevance of an active in situ bioassay using *Gammarus pulex* and *Limnephilus lunatus*. Environ Toxicol Chem 18:2243–2250.

Schulz R, Liess M. 2000. Toxicity of fenvalerate to caddisfly larvae: chronic effects of 1- versus 10-h pulse-exposure with constant doses. Chemosphere 41:1511–1517.

Schulz R, Stehle S, Elsaesser D, Matezki S, Müller A, Neumann M, Ohliger R, Wogram J, Zenker K. 2008. Geodata-based probabilistic risk assessment and management of pesticides in Germany, a conceptual framework. Integr Environ Assess Manag 5:69–79.

Schulz R, Thiere G, Dabrowski JM. 2002. A combined microcosm and field approach to evaluate the aquatic toxicity of azinphosmethyl to stream invertebrates. Environ Toxicol Chem 21:2172–2178.

Solomon KR, Brock TCM, De Zwart D, Dyer SD, Posthuma L, Richards SM, Sanderson H, Sibley PK, Van den Brink PJ, editors. 2008. Extrapolation practice for ecotoxicological effect characterization of chemicals. Boca Raton (FL): SETAC Press and CRC Press. p 105–134.

Spromberg JA, John BM, Landis WG. 1998. Metapopulation dynamics: indirect effects and multiple distinct outcomes in ecological risk assessment. Environ Toxicol Chem 17(8):1640–1649.

Suter GW, Traas TP, Posthuma L. 2002. Issues and practises in the derivation and use of species sensitivity distributions. In: Posthuma L, Traas TP, Suter GW, editors, Species sensitivity distributions in risk assessment. Boca Raton (FL): CRC Press, p 437–474.

Swartzman GL, Rose K. 1984. Simulating the biological effects of toxicants in aquatic microcosm systems. Ecol Modell 22:123–134.

Tachet H, Richoux P, Bournaud M, Usseglio-Polatera P. 2000. Invertebrates d'eau douce; systematique, biologie, ecologie. Paris: CNRS.

Taub FB. 1969. A biological model of a freshwater community: a gnotobiotic ecosystem. Limnol Oceanogr 14:136–142.

Traas TP, Janse JH, Van den Brink PJ, Brock TCM, Aldenberg T. 2004. A freshwater food web model for the combined effects of nutrients and insecticide stress and subsequent recovery. Environ Toxicol Chem 23(2):521–529.

Van den Brink PJ. 2006. Response to recent criticism on aquatic (semi-)field studies experiments: opportunities for new developments in ecological risk assessment of pesticides [Letter to the editor]. Integr Environ Assess Manag 2:202–203.

Van den Brink PJ, Baveco JM, Verboom J, Heimbach F. 2007. An individual-based approach to model spatial population dynamics of invertebrates in aquatic ecosystems after pesticide contamination. Environ Toxicol Chem 26(10):2226–2236.

Van den Brink PJ, Blake N, Brock TCM, Maltby L. 2006. Predictive value of species sensitivity distributions for effects of herbicides in freshwater ecosystems. Human Ecol Risk Assess 12:645–674.

Van den Brink PJ, Brock TCM, Posthuma L. 2002. The value of the species sensitivity distribution concept for predicting field effects: (non-)confirmation of the concept using semi-field experiments. In: Posthuma L, Traas TP, Suter GW, editors. Species sensitivity distributions in risk assessment. Boca Raton (FL): CRC Press. p 155–198.

Van den Brink PJ, Maltby L, Wendt-Rasch L, Heimbach F, Peeters F, editors. 2007. New improvements in the aquatic ecological risk assessment of fungicidal pesticides and biocides. Boca Raton (FL): SETAC, and Brussels.

Van den Brink PJ, Roelsma J, Van Nes EH, Scheffer M, Brock TCM. 2002. Perpest model, a case-based reasoning approach to predict ecological risks of pesticides. Envir Toxicol Chem 21(11):2500–2506. Example 9.9, p 124.

Van den Brink PJ, Sibley PK, Ratte HT, Baird DJ, Nabholz JV, Sanderson H. 2008. Extrapolation of effect measures across levels of biological organization in ecological risk assessment. In: Solomon KR, Brock TCM, De Zwart D, Dyer SD, Posthuma L, Richards SM, Sanderson H, Sibley PK, Van den Brink PJ, editors. Extrapolation practice for ecotoxicological effect characterization of chemicals. Boca Raton (FL): SETAC Press and CRC Press. p 105–133.

Van den Brink PJ, Ter Braak CJF. 1999. Principal response curves: analysis of time-dependent multivariate responses of biological communities to stress. Environ Toxicol Chem 18:138–148.

Van den Brink PJ, Van Wijngaarden RPA, Lucassen WGH, Brock TCM, Leeuwangh P. 1996. Effects of the insecticide Dursban 4 E (a.i. chlorpyrifos) in outdoor experimental ditches: II. Community responses and recovery. Environ Toxicol Chem 15:1143–1153.

Van Wijngaarden RPA, Brock TCM, Douglas MT. 2005. Effects of chlorpyrifos in freshwater model ecosystems: do experimental conditions change ecotoxicological threshold levels? Pest Manag Sci 61:923–935.

Van Wijngaarden RPA, Brock TCM, Van den Brink PJ. 2005. Threshold levels of insecticides in freshwater ecosystems; a review. Ecotoxicology 14:353–378.

Van Wijngaarden RPA, Brock TCM, Van den Brink PJ, Gylastra R, Maund SJ. 2006. Ecological effects of spring and late summer applications of lambda-cyhalothrin in freshwater microcosms. Arch Environ Contam Toxicol 50:220–239.

Versteeg DJ, Belanger SE, Carr GJ. 1999. Understanding single-species and model ecosystem sensitivity: data-based comparison. Environ Toxicol Chem 18:1329–1346.

Weber D, Preuß TG, Ratte HT, Bruns E, Görlitz G, Schäfer D, Reinken G, Ottermanns R, Claßen S, Agatz A. 2007. Modellierung der Effekte von Pflanzenschutzmitteln auf ausgewählte Phytoplankter — Modellentwicklung. Vortrag SETAC-Europe GLB annual meeting, Leipzig (DE).

Widianarko B, Van Straalen N. 1996. Toxicokinetics-based survival analysis in bioassays using non-persistent chemicals. Environ Toxicol Chem 15:402–406.

Williams P, Whitfield M, Biggs J, Fox G, Nicolet P, Shillabeer N, Sheratt T, Heneghan P, Jepson P, Maund S. 2002. How realistic are outdoor microcosms? A comparison of the biota of microcosms and natural ponds. Environ Toxicol Chem 21:143–150.

Williams WM, Singh P, Reinert K, Giddings J. 1999. RADAR risk assessment tool for aquatic exposure characterisation. Poster session. 20th annual meeting of the Society of Environmental Toxicology and Chemistry. Philadelphia.

Willis KJ, Van den Brink PJ, Green JG. 2004. Seasonal variation in plankton community responses of mesocosms dosed with pentachlorophenol. Ecotoxicology 13:707–720.

Wong DCL, Maltby L, Whittle D, Warren P, Dorn PB. 2004. Spatial and temporal variability in the structure of invertebrate assemblages in control stream mesocosms. Water Res 38:128–138.

Part II

Reports from ELINK Work Groups

Information presented in Part II of this book has been produced by a number of specific ELINK (EU Workshop on Linking Aquatic Exposure and Effects in the Registration Procedure of Plant Protection Products) work groups to aid the understanding of issues raised within the ELINK workshops. Views expressed are those of the authors, who are mentioned at the start of each section for which they were responsible. The authors of the different chapters or sections, however, invited the other work group members to comment on earlier versions of their report. Consequently, the work group reports can be considered as a consensus view of the corresponding work group members but not of all participants of the ELINK workshops.

12 A Novice's Guide to FOCUS Surface Water

CONTENTS

Warning! Information in this work group report is a simplified but very useful guide to exposure assessment using FOCUS (Forum for the Co-ordination of Pesticide Fate Models and Their Use) Surface Water modeling approaches.

12.1   PROBLEM FORMULATION

As outlined in Part I of this book, since the introduction of FOCUS Step 3 models and scenarios, exposure assessments at the EU level have become considerably more sophisticated. Furthermore, the implementation of risk mitigation measures will lead to FOCUS Step 4 calculations and the likelihood of more intensive interactions between exposure and effect assessment and risk management. The current framework builds on the guidance established in workshops such as HARAP (Higher-Tier Aquatic Risk Assessment for Pesticides; Campbell et al. 1999) and CLASSIC (Community-Level Aquatic System Studies–Interpretation Criteria; Giddings et al. 2002) that preceded the release of the FOCUS surface water framework, guidance, and modeling tools (FOCUS 2001, 2006).

Until recently, most higher-tier ecotoxicological experiments with pesticides only considered the impact of spray drift. However, since the FOCUS surface water framework addresses exposure not only from drift but also from drainage and runoff, a much wider range of exposure patterns may now need to be considered in aquatic risk assessments. As a consequence of the more complex exposure profiles generated by this framework, it is sometimes not clear whether the exposure regimes considered in current higher-tier ecotoxicological studies (e.g., complex mesocosm experiments) represent a "realistic worst case." Further, the current FOCUS scenarios may not take sufficient account of ecological properties of the aquatic systems that they intend to simulate.

Since it is impractical (for logistical as well as financial reasons) to study the full range of likely exposure regimes within laboratory or field-based effect assessments, it is important to consider how exposure regimes are established within ecotoxicological assessments and the most effective and meaningful basis for establishing exposure endpoints for use in risk assessments, particularly when considering risks from long-term exposure. These uncertainties in the current risk assessment may hamper the development and implementation of practical and appropriate risk mitigation procedures. To date, the interaction between the assessment of exposure and of ecotoxicological effects in the risk assessment procedure typically remains at a relatively low level of sophistication compared with the complexity of the individual exposure or effects assessments. To improve the potential to more effectively and meaningfully link exposure and effect assessments, the ELINK workshop was established with the objective of reviewing or providing guidance on a number of topics, namely:

• Reality check of ecosystem characteristics simulated by FOCUS surface water scenarios

- Use of higher-tier fate data (e.g., of semifield experiments) in the exposure assessment
- Clustering of exposure patterns emerging from FOCUS scenarios
- Reality check of exposure regimes studied in ecotoxicological experiments
- Proper use of peak and time-weighted average (TWA) concentrations in the different tiers of the effects assessment
- Extrapolation of effects of time-variable exposures

The focus of the workshop was not to improve existing test methods and exposure and effect models but to consider how the output of these methods and models can best be used in risk assessment and methods or tools that may need to be developed to enhance risk assessment. It should be kept in mind that, ideally, the new methods, tools, or frameworks for evaluation elaborated within this workshop should

- Be in accordance with the principles of the tiered approach
- Be usable for Annex I inclusion under Directive 91/414/EEC and national registration procedures
- Be easy to use in routine risk assessment and risk mitigation procedures
- Not result in substantial increases of staff and resources of regulatory authorities and companies

To establish a firm foundation on which subsequent discussions in this report will build, a "primer" has been prepared with the following overviews of the current aquatic risk assessment framework:

- Summary of conceptual framework and simple mechanistic description of model roles and processes
- What FOCUS scenarios represent and their context for regulatory decision making
- Capabilities and limitations of FOCUS models
- Summary of regulatory experiences with modeling assessments
- Summary of regulatory experiences with ecotoxicological assessments
- Reality check, regulatory context, and confidence overview

12.2   FOCUS SURFACE WATER SCENARIOS FOR NONMODELERS

Helmut Schaefer

While there is detailed knowledge of the setup, use, and interpretation of FOCUS surface water tools and scenarios in the e-fate modeling community, the objective of this document is to provide that information in a brief and comprehensive way for those who have to deal with the predicted environmental concentration in surface water (PECsw values) in aquatic risk assessments but have no special expertise in e-fate modeling.

12.2.1 ENTRY ROUTES

Following the data requirements laid down in Directive 91/414/EEC, aquatic exposure assessment should cover drift, runoff, drainage, and atmospheric deposition as potential entry routes of pesticides into surface water bodies. As atmospheric deposition is usually of minor relevance, the FOCUS surface water scenarios group did not consider this route. It has, however, subsequently been discussed in the FOCUS air group. Spray drift is based on the Ganzelmeier/Rautmann tables (Rautmann et al. 2001) and is always taken into account unless the formulated pesticide is not a foliar spray.

FOCUS has a number of steps with increasing levels of refinement. At lower tiers (Steps 1 and 2), runoff and drainage are combined and characterized with worst-case percentage loadings to water based on expert judgment. At higher tiers (Steps 3 and beyond), runoff and drainage are considered separately and never assumed to occur in parallel at the same scenario location (see Figure 12.1). Runoff is calculated using the PRZM (pesticide root zone model) simulation model, which was developed by the US Environmental Protection Agency (USEPA) and is used in the US authorization process for runoff estimations. Drainage is estimated with the MACRO model, developed in Sweden by Nick Jarvis (University of Uppsala). The actual fate of an active substance in the surface water body is calculated with the TOXSWA model, developed in the Netherlands by Paulien Adriaanse (Alterra).

12.2.2 WATER BODIES

Different water bodies (lakes, ponds, rivers, canals, streams, irrigation or drainage canals), either "static or slow moving," are required to be taken into account under

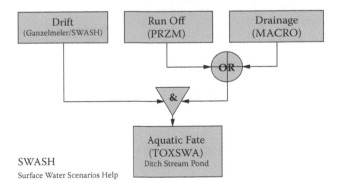

FIGURE 12.1 FOCUS SW Step 3 approach.

TABLE 12.1

Water Body Geometry and Residence Time

Type of Water Body	Width (m)	Total Length (m)	Average Water Depth (m)	Target Average Residence Time (days)
Ditch	1	100	0.3	5
Pond	30	30	1.0	50
Stream	1	100	0.3 to 0.5	0.1

91/414/EEC, but in the interests of standardization FOCUS defined 3 "representative" water bodies: ditch, pond, and stream, with standardized characteristics shown in Tables 12.1 to 12.3.

12.2.3 TIERED APPROACH

As indicated, a tiered scheme (see Figure 12.2) was developed, starting with simple extreme worst-case approaches and progressing to more detailed and less worst-case procedures. This was done with the intention to save time and effort for compounds that easily pass the aquatic risk assessment even under worst-case assumptions.

Step 1: Is extremely worst case and considers only the ditch scenario.
Step 2: Works on a "spreadsheet" type of basis, characterizing the input from drift, runoff, and drainage with percentages of applied amount entering the water body. As in Step 1, only a ditch is used as a standard water body.
Step 3: Drainage and runoff are calculated with process-driven simulation models, and 3 different water bodies (ditch, pond, and stream) are implemented.

TABLE 12.2

Characteristics of Water and Sediment

Characteristic	Value
Concentration of suspended solids in water column ($mg \cdot L^{-1}$)	15
Sediment layer depth (cm)	5
Organic carbon content (%)	5
Dry bulk density ($kg \cdot m^{-3}$)	800
Porosity (%)	60

TABLE 12.3
"Water Body–Entry Route" Combination

Type of Water Body	Drift	Runoff	Drainage
Ditch	X		X
Pond	X	X	X
Stream	X	X	X

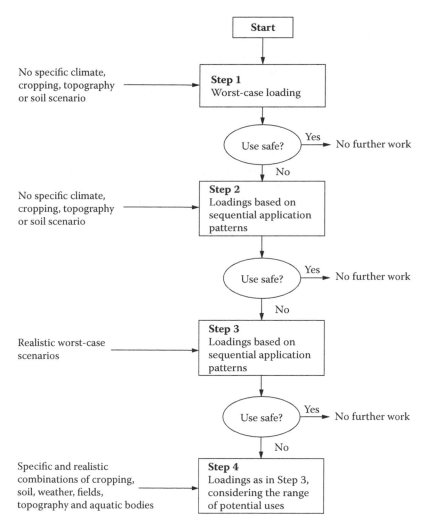

FIGURE 12.2   Tiered approach in FOCUS surface water exposure assessment.

Six drainage scenarios and 4 runoff scenarios covering different soil–climate conditions throughout Europe were defined.

Step 4: Is not part of the FOCUS surface water scenarios report, and no standards are defined for doing Step 4 calculations. However, the FOCUS landscape modeling and mitigation report gives detailed advice on which refinements can in principle be considered at Step 4 (FOCUS 2007a, 2007b).

12.2.3.1 Some Details of Step 2 Calculations

Step 2 works on a spreadsheet basis to characterize drift, runoff, and drainage in terms of defined percentages of the applied chemical entering the water body. Loadings are defined as 1 application or a series of individual applications (depending on the intended use), resulting in drift to the water body, followed by a runoff/drainage event occurring 4 days after the last application. Between applications and the runoff or drainage event, degradation in soil is considered. Drift entries are based on the Ganzelmeier/Rautmann tables (Rautmann et al. 2001). Minimum distances assumed to surface water (not as a mitigation measure) are 1 m for low crops (cereals, potatoes, etc.) and 3 m for high crops (orchards, grapes, etc.). Combined runoff or drainage entries, as a percentage of residues on soil, are based on region and season (see Table 12.4).

The receiving water body is a 30 cm deep ditch 1 m wide underlain by a 5 cm deep sediment with 5% organic carbon content. Partitioning between water and sediment is based on adsorption coefficients derived in a standard batch equilibrium adsorption experiment with agricultural soils. Following a drift event, equilibrium between water and sediment is established within 24 hours. The "runoff/drainage" entry is distributed between water and sediment at the time of loading according to the adsorption coefficient of the compound. Degradation in water and sediment is based on the DT50 (period required for 50% disappearance) derived in water–sediment studies.

TABLE 12.4
Input into Surface Water via Runoff or Drainage at Step 2

Region, Season	Soil Residue (%)
North Europe, October to February	5
North Europe, March to May	2
North Europe, June to September	2
South Europe, October to February	4
South Europe, March to May	4
South Europe, June to September	3
No runoff	0

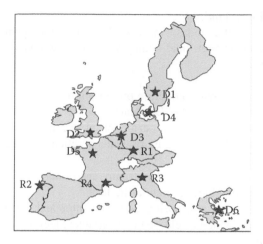

FIGURE 12.3   Location of the weather stations used for FOCUS surface water scenarios.

12.2.3.2   Some Details of Step 3 Calculations

To cover different soil–climate (pedoclimatic) conditions throughout Europe, 6 drainage scenarios (D1 to D6) and 4 runoff scenarios (R1 to R4) were defined (Figure 12.3).

The scenarios were characterized as follows:

D1

   Climate: Cool with moderate precipitation.

   Representative site: Lanna, Sweden.

   Soil type: Slowly permeable clay with field drains. Seasonally waterlogged by groundwater.

   Surface water bodies: Field ditches and first-order streams.

   Landscape: Gently sloping to level land.

   Crops: Grass, winter and spring cereals, and spring oilseed rape.

D2

   Climate: Temperate with moderate precipitation.

   Representative site: Brimstone, United Kingdom.

   Soil type: Impermeable clay with field drains. Seasonally waterlogged by water perched over impermeable massive clay substrate.

   Surface water bodies: Field ditches and first-order streams.

   Landscape: Gently sloping to level land.

   Crops: Grass, winter cereals, winter oilseed rape, field beans.

D3

   Climate: Temperate with moderate precipitation.

   Representative site: Vredepeel, Netherlands.

   Soil type: Sands with small organic carbon content and field drains. Subsoil waterlogged by groundwater.

   Surface water bodies: Field ditches.

Landscape: Level land.

Crops: Grass, winter and spring cereals, winter and spring oilseed rape, potatoes, sugar beet, field beans, vegetables, legumes, maize, pome or stone fruit.

D4

Climate: Temperate with moderate precipitation.

Representative site: Skousbo, Denmark.

Soil type: Light loam, slowly permeable at depth and with field drains. Slight seasonal waterlogging by water perched over the slowly permeable substrate.

Surface water bodies: First-order streams and ponds.

Landscape: Gently sloping, undulating land.

Crops: Grass, winter and spring cereals, winter and spring oilseed rape, potatoes, sugar beet, field beans, vegetables, legumes, maize, pome or stone fruit.

D5

Climate: Warm temperate with moderate precipitation.

Representative site: La Jaillière, France.

Soil type: Medium loam with field drains. Hard, impermeable rock at depth. Seasonally waterlogged by water perched over the impermeable substrate.

Surface water bodies: First-order streams and ponds.

Landscape: Gently to moderately sloping, undulating land.

Crops: Grass, winter and spring cereals, winter and spring oilseed rape, legumes, maize, pome or stone fruit, sunflowers.

D6

Climate: Warm Mediterranean with moderate precipitation.

Representative site: Thiva, Greece.

Soil type: Heavy loam over clay with field drains. Seasonally waterlogged by groundwater.

Surface water bodies: Field ditches.

Landscape: Level land.

Crops: Winter cereals, potatoes, field beans, vegetables, legumes, maize, vines, citrus, olives, cotton.

R1

Climate: Temperate with moderate precipitation.

Representative site: Weiherbach, Germany.

Soil type: Free-draining light silt with low organic matter content.

Surface water bodies: First-order streams and ponds.

Landscape: Gently to moderately sloping, undulating land.

Crops: Winter cereals, winter and spring oilseed rape, sugar beet, potatoes, field beans, vegetables, legumes, maize, vines, pome or stone fruit, sunflowers, hops.

R2

Climate: Warm temperate with very high precipitation.

Representative site: Porto, Portugal.

Soil type: Free-draining light loam with relatively high organic matter content.

Surface water bodies: First-order streams.

Landscape: Steeply sloping, terraced hills.

Crops: Grass, potatoes, field beans, vegetables, legumes, maize, vines, pome or stone fruit.

R3

Climate: Warm temperate with high precipitation.

Representative site: Bologna, Italy.

Soil type: Free-draining calcareous heavy loam.

Surface water bodies: First-order streams.

Landscape: Moderately sloping hills with some terraces.

Crops: Grass, winter cereals, winter oilseed rape, sugar beet, potatoes, field beans, vegetables, legumes, maize, vines, pome or stone fruit, sunflower, soybean, tobacco.

R4

Climate: Warm Mediterranean with moderate precipitation.

Representative site: Roujan, France.

Soil type: Free-draining calcareous medium loam over loose calcareous sandy substrate.

Surface water bodies: First-order streams.

Landscape: Moderately sloping hills with some terraces.

Crops: Winter and spring cereals, field beans, vegetables, legumes, maize, vines, pome or stone fruit, sunflower, soybean, citrus, olives.

Due to the weather-related drainage and runoff water inputs at step 3, the actual water depth and residence time vary with time and the weather file that is selected. Some examples are given in Tables 12.5 to 12.8. These tables show that significant deviations from the desired residence times as presented in Table 12.1 do occur. Also noteworthy are the variations of water depths in the stream scenarios (Table 12.6). Minimum water levels are kept in the simulation model TOXSWA by artificially introducing weirs. Drift as a function of distance is described with regression equations, which are based on the Ganzelmeier/Rautmann tables. Minimum distances

TABLE 12.5
Drainage/Ditch Scenarios

Scenario	Min/Max Discharge (L/s)	Min/Max Water Depth (m)	Min/Max Monthly Residence Time (Days)
D1	0.008 to 3.88	0.30 to 0.32	0.4 to 45.5
D2	0.001 to 11.5	0.30 to 0.35	0.65 to 250
D3	0.08 to 0.71	0.30 to 0.31	0.70 to 4.4
D6	0.04 to 12.8	0.30 to 0.36	0.24 to 8.1

TABLE 12.6
Drainage/Stream Scenarios

Scenario	Min/Max Discharge (L/s)	Min/Max Water Depth (m)	Min/Max Monthly Residence Time (Days)
D1	0.38 to 131	0.31 to 0.82	0.017 to 0.93
D2	0.007 to 388	0.30 to 1.40	0.022 to 50.2
D4	1.23 to 85.2	0.31 to 0.68	0.017 to 0.29
D5	0.86 to 218	0.29 to 0.92	0.012 to 0.39

TABLE 12.7
Drainage/Pond Scenarios

Scenario	Min/Max Discharge (L/s)	Min/Max Water Depth (m)	Min/Max Monthly Residence Time (Days)
D4	0.025 to 0.40	1.00 to 1.01	87.7 to 283
D5	0.026 to 0.90	1.00 to 1.01	46.8 to 405

TABLE 12.8
Runoff/Pond Scenario

Scenario	Application Season	Min/Max Discharge (L/s)	Min/Max Water Depth (m)	Min/Max Monthly Residence Time (Days)
	Spring	0.1 to 1.4	1.00 to 1.01	108 to 157
R1	Summer	0.1 to 1.6	1.00 to 1.01	85 to 157
	Autumn	0.1 to 1.6	1.00 to 1.01	85 to 157

to water bodies (not as a mitigation measure) were defined and depend on crop and water body type. Minimum distances can range from 1 m for low crops near a ditch to 6 m for high crops near a pond. The drift percentage is averaged over the width of the water body, starting with the minimum distance mentioned.

The fate in soil needs to be considered in the context of runoff and drainage as entry routes. The degradation in the soil is quantified with rates from either laboratory or field degradation experiments. Sorption to soil is characterized with adsorption coefficients derived in standard batch equilibrium adsorption experiments with agricultural soils. Runoff from the soil surface is calculated with the PRZM simulation model. Within this model, the estimation of runoff water is based on the so-called curve number approach, which considers the soil type, soil moisture, crop type, soil management, and soil permeability. Furthermore, soil erosion is estimated with the modified universal soil loss equation (MUSLE), which considers the runoff volume and peak flow as well as slope, slope length, soil erodibility, soil cover, and soil management. Pesticides enter surface water bodies via runoff water and eroded soil (both processes are often summarized as "pesticide runoff"). The amount of pesticide entering the water body is calculated from the concentration in soil and soil pore water at the day of the runoff event. PRZM calculations are run over several years; however, only a 12-month period is used for the exposure assessment. As some worst-case assumptions were already included, the weather conditions for this period were intended to represent a median case concerning the predicted daily and annual runoff water volume. As hydrological flows vary with season, different weather files were used for different application seasons (see Table 12.9).

For the drainage calculations, the MACRO model is used. Drainage is assumed to be associated with preferential flow, that is, bypass flow-through macropores (e.g., cracks or wormholes) in the soils. Macropore flow is calculated assuming gravity flow and a power law function for conductivity. Drainage water flow occurs under saturated conditions in the soil layers containing the drainage pipes. The amount of pesticide drained depends on the concentration in this soil layer. MACRO calculations are run over several years; however, only a 16-month period is used for the exposure assessment. Again, the weather conditions for this period were intended to represent a median case concerning the amount of drainage water (see Table 12.10).

In all cases, instantaneous mixing in the water body is assumed; that is, there are no vertical or horizontal concentration gradients in the water body. Fate in the surface water body is characterized by partitioning between water and sediment as well as degradation in the water and sediment phase. Partitioning is quantified with adsorption coefficients derived in a standard batch equilibrium adsorption experiment with agricultural soils. It is implemented as a diffusion process; thus, it is kinetically limited (not instantaneous). Parameters characterizing the degradation in water and sediment are normally taken from water–sediment studies. These studies are conducted without light exposure. For compounds that photolytically degrade, this leads to an overestimation of the persistence in water. Photolysis is not explicitly

---

TABLE 12.9
Selected Weather Years for Runoff

	Selected Year for Each Application Season		
Scenario	Spring (March to May Application)	Summer (June to September Application)	Autumn (October to February Application)
R1	1984	1978	1978
R2	1977	1989	1977
R3	1980	1975	1980
R4	1984	1985	1979

TABLE 12.10

Selected Weather Years for Drainage (Values in Parentheses Refer to Median Value)

Scenario	Selected Weather Year	Precipitation (mm)	Drainage (mm)
D1	1982	538 (556)	136 (130)
D2	1986	623 (642)	212 (230)
D3	1992	693 (747)	264 (274)
D4	1985	692 (659)	115 (98)
D5	1978	627 (651)	182 (177)
D6	1986	733 (683)	259 (263)

implemented in the TOXSWA model, which calculates the fate in surface water bodies at Step 3. Therefore, specific steps need to be undertaken if photolysis needs to be included. These could consist of calculating a lumped degradation covering both microbial and photolytic degradation while considering the light conditions of the European region of interest and the timing of application.

Except for strongly sorbing or rapidly degrading compounds, fate in the surface water typically plays only a minor role in the exposure assessment.

Runoff or drainage entries into surface water are driven by storm events. To ensure the realistic worst-case character of the calculation, the so-called "pesticide application timer" (PAT) was implemented. The objective of the tool is to find an application date in a given time window that fulfills the following 2 criteria:

1) At least 10 mm of rainfall in the 10 days following application
2) Less than 2 mm of rain each day in a 5-day period, starting 2 days before application, extending to 2 days following the day of application

If no appropriate date can be found, first criteria 1) and then criteria 2) are relaxed. In case of multiple applications, this procedure is done for each individual application.

12.2.4   RESULTS OF FOCUS SW CALCULATIONS

The final result of the FOCUS sw calculations are edge-of-field concentrations in water and sediment. As mentioned, instantaneous mixing in the water body (no horizontal or vertical concentration gradients) is assumed. Following the requirements laid down in 91/414/EEC, the maximum as well as TWA concentrations for periods of 1, 2, 4, 7, 14, 21, 28, and 42 days after last application are reported. For Step 2 calculations, the time-weighted concentration is directly derived for the periods after the last application. For Step 3, averages are calculated using the moving time window method over the evaluation period of 12 (runoff) or 16 (drainage) months. These should build the basis for the acute and chronic aquatic risk assessment.

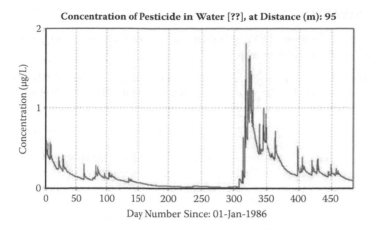

FIGURE 12.4 Concentration–time course calculated with TOXSWA.

12.2.5 INTERPRETATION OF RESULTS

At Step 3, very detailed concentration–time courses are reported over the evaluation period of 12 or 16 months (see Figure 12.4). For several reasons, these should only be seen as illustrative and not used "as is" in the aquatic risk assessment. As described, runoff or drainage entries are event driven. Around the day of application, the PAT ensures that a realistic worst-case situation is addressed. However, for later events this is not checked in detail. Representative weather files were identified based on test runs for maize (runoff) or winter cereals (drainage). No such checks are done for other crops. For runoff, selected weather files are said to be representative for certain applications seasons (Table 12.9). Runoff events for a spring-applied compound might still occur in summer; whether the weather file that was representative for a spring application will still be representative for a runoff event in summer is unknown. Thus, the concentration–time courses should be treated with care. For higher-tier aquatic risk assessments, it is recommended to use this information only in an aggregated way (e.g., typical return interval period, typical duration of runoff or drainage event).

In the foreword to the FOCUS surface water report, it is stated that "it can be concluded that passing 1 (one) of the proposed surface water scenarios would be sufficient to achieve Annex I listing within the framework of 91/414/EEC." However, consider that passing just a single scenario might be seen as insufficient for Annex I inclusion by some member states and will definitely lead to a specific provision like "in this overall assessment Member States should pay particular attention to the protection of aquatic ecosystems" in the publication of Annex I listing in the *Official Journal*.

12.3 CAPABILITIES AND LIMITATIONS OF MODELS

Neil Mackay

Each of the models used at Step 3 is summarized here, building on the overview of the operational framework provided in the preceding section.

12.3.1 SWASH

The following definition of SWASH (Surface Water Scenarios Help software) is provided from the FOCUS home page:

> SWASH is an overall user-friendly shell, encompassing a number of individual tools and models, involved in Step 3 calculations for the FOCUS Surface Water Scenarios. Its main functions are:
>
> - Maintenance of a central pesticides properties database for use in MACRO, PRZM and TOXSWA
> - Preparation of other input for MACRO, PRZM and TOXSWA, notably application patterns, application methods and dosages
> - Creation of projects, containing all Step 3 FOCUS runs required for use of a pesticide on a specified crop
> - Calculation of spray drift deposition onto ditch, stream and pond like waterbodies
> - Provision of an overview of crop and waterbody combinations in each scenario, of the extent of each scenario and of the installed versions of each model, including its shell and database.

Capabilities and Associated Constraints Indicate the Scope and Flexibility of the SWASH Software

Operational Capability	Constraint
25 of the most common cropping situations are represented	Less frequently encountered crops may be approximated by finding a similar cropping regime. Further refinements may become necessary at Step 4 to better reflect crop transpiration demands, cropping cycles, foliar interception, irrigation needs, and so on.
Diverse association of crops and scenarios is represented	Some crops are associated with almost all scenarios (e.g., winter cereals), but there are a number associated with just 1 or 2 scenarios (e.g., citrus, cotton, hops, soybeans, olives, tobacco).
Representation of application interval	Intervals that result in applications that span years are not immediately accepted. For example, there are usage regimes in which an autumn application to an autumn crop is followed by a second application in spring to the same crop. To overcome this requires definition of an application window with • the spring application date acting as the initiation of the application window • the modified interval defined as the number of Julian days between the spring application data and the autumn application date • the final date in the application window based on the Julian day for initiation of application window plus the modified application interval plus 30 days (or 30 days times the number of applications minus 1) More complex application patterns in which intervals vary between more frequent applications cannot be accurately represented at Step 3.
Representation of application techniques	Up to 4 application regimes may be represented, depending on crop type (aerial application, granular application, soil incorporation and air blast or ground spray). It is necessary to choose the closest equivalent application technique

and refine at Step 4. In some cases, there is no sensible approximation at Step 3, and it then becomes necessary to amend assumptions to generate meaningful results (e.g., ground spray applications beneath the canopy or between rows of pome fruit or vines cannot be represented in these cropping scenarios). It becomes necessary to amend drift assumptions in TOXSWA input files to overcome this limitation. It may also be necessary to replace foliar interception assumptions in some cases.

Representation of drift	SWASH generates Step 3 assumptions and directly generates underlying definitions to enable generation of MACRO, PRZM, and TOXSWA input files. Drift representation at Step 3 is based on a set of crop- and water body-specific worst-case assumptions regarding proximity of crop to water and deposition interface area. SWASH will not directly generate Step 4 representations of mitigation options such as no-spray buffers or vegetated filter strips. The FOCUS drift calculator embedded within SWASH will facilitate definition of appropriate drift rates to take forward into Step 4, but input files must be adapted independently.

12.3.2   MACRO

The following definition of MACRO is provided from the FOCUS home page:

The model calculates coupled unsaturated–saturated water flow in cropped soil, including the location and extent of perched water tables, and can also deal with saturated flow to field drainage systems. The model accounts for macropore flow, with the soil porosity divided into two flow systems or domains (macropores and micropores) each characterized by a flow rate and solute concentration. Richards' equation and the convection–dispersion equation are used to model soil water flow and solute transport in the soil micropores, while a simplified capacitance type-approach is used to calculate fluxes in the macropores. Exchange between the flow domains is calculated using approximate, physically-based, expressions based on an effective aggregate half-width. Additional model assumptions include first-order kinetics for degradation in each of four "pools" of pesticide in the soil (micro- and macropores, solid/liquid phases), together with an instantaneous sorption equilibrium and a Freundlich sorption isotherm in each pore domain.

Capabilities and Associated Constraints Indicate the Scope and Flexibility of the MACRO Software

Operational Capability	Constraint
Representation of application techniques	As discussed when considering SWASH, a number of application techniques may be represented as standard within FOCUS. The interception potential at the timing of application is not reported and can only be determined by creating batch files for running and checking PAR files for ZFINT parameter definition prior to executing the batch. Representation of incorporation?

Pesticide application timer (PAT)	In both MACRO and PRZM, algorithms have been embedded that use a set of rules to define application dates. A description of these rules and potential associated limitations is provided in a separate entry relating to the PAT tool.
Representation of drainage networks	Drainage networks have been defined on the basis of observation and expert judgment. The most robust representations of these networks are mainly related to arable field study sites in northern Europe. Representation of drainage networks for perennial crops is based on highly pragmatic adoption of arable networks at the same scenario locations.
Representation of kinetics	In the currently available version of FOCUS MACRO degradation kinetics is limited to SFO (simple first-order) kinetic formats.
Representation of sorption	In the currently available version of FOCUS MACRO sorption is simulated as simple equilibrium sorption. A non-FOCUS version of MACRO that represents nonequilibrium sorption is available and may be introduced as a revision to FOCUS MACRO in the future.
Representation of degradation pathways	MACRO has a limitation in only being capable of simulating 1 parent → metabolite linkage at a time with a maximum of 2 primary metabolites. For more complex degradation pathways involving longer metabolite chains or degradation of parent to multiple metabolites, it becomes necessary to simulate this through a series of simulations, generally with pragmatic representation of pathways. Accurate representation of pathways requires a highly complex and time-consuming workaround strategy.
Timescales for simulation	PECsw profiles generated by TOXSWA are based on 12 months (runoff scenarios) or 16 months (drainage scenarios). The simulation framework in MACRO was therefore based on this constraint (and in recognition of the relatively long simulation run times) and involved 5 years of hydrological and chemical "warm-up" simulation followed by the simulation of the reported period for preparation of MACRO → TOXSWA m2t files. It is not possible at Step 3 to undertake longer time series runs. Meteorology files provided as standard with FOCUS drainage scenarios also do not allow for long-term simulations at Step 4.

12.3.3 PRZM

The following definition of PRZM is provided from the FOCUS home page:

The first official release of PRZM was published in 1984 by the US EPA although beta versions were available beginning in 1982. The German PELMO model was developed based on this PRZM1. An upgraded version PRZM2 was issued as part of the RUSTIC package and later released as a stand-alone model by the Center for Exposure Assessment Modeling (CEAM) of the US EPA (October 1994). In the mid-1990s the runoff routines were upgraded as part of the work of the FIFRA Exposure Modeling Work Group and the FIFRA Environmental Model Validation Task Force (FEMVTF) to produce version 3.12. This version also included more flexibility with application techniques, the ability to make degradation a function of soil temperature, and output which is more user friendly. Version 3.12 is also the version that has been used by the FIFRA Environmental Model Validation Task Force in its program to compare model predictions with actual data from runoff and leaching field studies. The actual PRZM version 3.12 beta was released officially in March 1998 by CEAM. For use in the FOCUS ground water and surface water scenarios an enhanced PRZM3 (version 3.20, FOCUS release) was developed. In addition to the capabilities of version 3.12 it has the option of using the normalized Freundlich isotherm, the ability to make the degradation rate a function of soil moisture, and the capability to consider increasing sorption with time. The new PRZM 3.20 executable is a truly Windows based, 32bit PRZM3 code. The program coding was conducted by Waterborne Environmental.

Capabilities and Associated Constraints Indicate the Scope and Flexibility of the PRZM Software

Operational Capability	Constraint
Pesticide application timer (PAT)	In both MACRO and PRZM, algorithms have been embedded that use a set of rules to define application dates. A description of these rules and potential associated limitations is provided in a separate entry relating to the PAT tool.
Representation of runoff	In the absence of hourly weather data, it is necessary within PRZM to represent storm duration on translation of standard daily PRZM output into an hourly format utilized by TOXSWA. The following description is provided in the FOCUS surface water report: daily runoff and erosion time series output files (*.ZTS) are automatically post-processed into a series of hourly runoff and erosion values by assuming a peak runoff rate of 2 mm/hr in output files designated as *.P2T (for PRZM to TOXSWA). Thus, an 18 mm daily precipitation event is entered into TOXSWA as a nine hour runoff loading of 2 mm/hr. This pragmatic translation may not represent peak shape and duration accurately.
Representation of kinetics	In the currently available version of FOCUS PRZM degradation kinetics is generally limited to SFO kinetic formats. However, in PRZM it is possible to represent biphasic degradation (e.g., hockey stick kinetic formats). However, this is a necessarily simplistic representation of this scheme and should only be carried out with care as it may not accurately represent degradation — particularly where multiple applications are carried out.
Representation of sorption	In the currently available version of FOCUS PRZM sorption is simulated as simple equilibrium sorption.
Representation of degradation pathways	PRZM is more flexible than MACRO and has the capability of simulating the following options: Parent → up to 2 primary metabolites Parent → primary metabolite → secondary metabolite For more complex degradation pathways involving longer metabolite chains or degradation of parent to multiple metabolites, it becomes necessary to simulate this through a series of simulations, generally with pragmatic representation of pathways.

12.3.4 Pesticide Application Timer

The following definition of the PAT is provided from the FOCUS surface water report:

Pesticide losses in both surface runoff and subsurface drainage flow are "event-driven" and therefore very strongly dependent on the weather conditions immediately following application, in particular the rainfall pattern. It was therefore considered necessary to develop a procedure which would help to minimize the influence of the user choice of application date on the results of FOCUS surface water scenario calculations, at the same time as retaining some degree of flexibility in simulated application timings to allow realistic use patterns for widely different compounds. A Pesticide Application Timing calculator (PAT) was developed to achieve this dual purpose. PAT is incorporated in the shell programs for both MACRO and PRZM, and is also available as a stand-alone program.

Initially, the pre-set criteria state that there should be at least 10 mm of rainfall in the ten days following application and at the same time, there should be less than 2 mm of rain each day in a five day period, starting two days before application, extending to two days following the day of application. PAT then steps through the "application window" to find the first day which satisfy these requirements. For multiple applications, the procedure is carried out for each application, respecting the minimum interval specified between applications.

Depending on the rainfall pattern in the application window defined by the user, it is quite possible that no application day exists which satisfies the two basic criteria defined above. In this case, the criteria are relaxed and the procedure repeated until a solution is found, as follows:

- The 5-day period around the day of application is reduced first to a 3-day period (one day either side of the application day), and then if there is still no solution, to just the day of application. Relaxing these criteria makes the resulting leaching estimates potentially more conservative.
- If PAT still fails to find a solution, then the second criteria is relaxed, such that 10 mm of rain is required to fall in a 15-day period following application, rather than 10 days. Relaxing these criteria makes the leaching estimates less conservative.
- If a solution is still not forthcoming (for example, for dry periods, such that the total rainfall during the entire application window is less than 10 mm), then the minimum rainfall requirement is reduced 1 mm at a time, to zero.
- If PAT still fails to find a solution (this will be the case if the application window is very wet, with more than 2 mm of rain every day), then the amount of rain allowed on the day of application is increased 1 mm at a time, until a solution is found.
- Note: If multiple applications occur within the application window, it is important to make the window as large as possible (but still in agreement with the GAP) in order to prevent PAT from unnecessarily relaxing the precipitation rules.

As noted, the PAT scheme will relax the criteria used to identify an application date to find a solution. Under some circumstances, it will identify extreme rainfall events that will result in a potentially unrealistically challenging basis for risk assessment. The challenge context of events that drive the selection of application dates by the PAT can only be determined by subsequently interpreting weather data files. For example, it may be appropriate to better understand the temporal context of the rainfall event. Is the chosen event, for example, the worst-case seasonal, annual, or even long-term event?

The FOCUS framework selects meteorologically and hydrologically "representative" years for the basis of the reported 12-month (runoff) or 16-month (drainage) simulation carried out by TOXSWA. This approach is a relatively simple and crude strategy to make the simulation framework more easily managed by SWASH, MACRO, and PRZM. A more robust approach would require higher temporal resolution analysis of the relationship between application timing windows and return intervals between individual storm events over longer time periods. This can only be carried out at present through independent analysis of weather data and Step 4 simulations.

12.3.5   TOXSWA

The following definition of TOXSWA is provided from the FOCUS home page:

> TOXSWA describes the behavior of pesticides in a water body at the edge-of-field scale, i.e. a ditch, pond or stream adjacent to a single field. It calculates pesticide concentrations in the water layer in horizontal direction only and in the sediment layer in both horizontal and vertical directions. TOXSWA considers four processes: 1) transport, 2) transformation, 3) sorption, and 4) volatilization. In the water layer, pesticides are transported by advection and dispersion, while in the sediment, diffusion is included as well. The transformation rate covers the combined effects of hydrolysis, photolysis and biodegradation and it is a function of temperature. It does not simulate formation of metabolites. Sorption to suspended solids and to sediment is described by the Freundlich equation. Sorption to macrophytes is described by a linear sorption isotherm but this feature is not used for the FOCUS scenarios. Pesticides are transported across the water–sediment interface by diffusion and by advection (upward or downward [see FOCUS scenarios]).

Capabilities and Associated Constraints Indicate the Scope and Flexibility of the TOXSWA Software

Operational Capability	Constraint
Representation of application techniques	As discussed, the definition of drift percentages is defined directly by SWASH and may not accurately represent the actual usage situation. Overriding these definitions can be achieved through careful manipulation of files.
Representation of upstream catchment	The presence and implications of usage in an upstream catchment for stream scenarios is very simplistically represented through the assumption of a further 20% loading (e.g., 20% of upstream catchment is treated).
Representation of degradation processes	Chemicals may undergo degradation via a wide range of processes in water–sediment systems. In general, these can also be represented in water–sediment systems that are the basis for definition of bulk water and sediment-phase degradation rates. One important exception is the inability to represent photolysis as such studies are conducted in the dark. Unlike other models such as EXAMS, there is no ability within TOXSWA to simulate the implications of photolysis. For some compounds, this is the primary route of degradation in the water phase, and as a consequence, simulations in TOXSWA may be unrealistically conservative.
Representation of kinetics	In the currently available version of FOCUS TOXSWA degradation kinetics is limited to SFO kinetic formats.

Representation of sorption	In the currently available version of FOCUS TOXSWA sorption is simulated as simple equilibrium sorption.
Representation of degradation pathways	At present, FOCUS TOXSWA does not have the capability of simulating formation of metabolites in water bodies. Instead, simulations of parent and any metabolites are carried out completely independently. This sometimes requires workarounds for metabolites that result solely from degradation of parent in water–sediment systems.
Timescales for simulation	For practical reasons, at the time of the development of the FOCUS framework a constraint was imposed that TOXSWA would carry out simulations for a 12-month (runoff) or 16-month (drainage) period. Longer-term simulations can only be conducted at Step 4 and are then limited by 1) the temporal reporting window of the MACRO → TOXSWA m2t and PRZM → TOXSWA p2t files and 2) the availability of long-term weather data to support MACRO simulations.

12.4 WHAT FOCUS SCENARIOS REPRESENT AND THEIR CONTEXT FOR REGULATORY DECISION MAKING

Neil Mackay

12.4.1 INTRODUCTION

The FOCUS Surface Water Working Group was given the following objective:

Develop scenarios that can be used as a reliable input for modeling in the EU registration process as proposed by the FOCUS Surface Water Working Group in the step by step approach proposed in their report.

It was specifically pointed out in the background to this objective that

The existence of standard scenarios will make a uniform procedure for assessing the PECsw of plant protection products in surface water possible.

In developing the 6 drainage and 4 runoff scenarios, the FOCUS Surface Water Working Group was aware that it would be necessary to carefully outline what the scenarios were intended to represent and, perhaps more important, what they did not. The following text is abstracted from the introduction to the FOCUS surface water report (FOCUS 2001):

The contamination of surface waters resulting from the use of an active substance is represented by ten realistic worst-case scenarios, which were selected on the basis of expert judgment. Collectively, these scenarios represent agriculture across Europe, for the purposes of Steps 1 to 3 assessments at the EU level. However, being designed as "realistic worst case" scenarios, these scenarios do not mimic specific fields, and nor are they necessarily representative of the agriculture at the location or the Member State after which they are named. Also they do not represent national scenarios for the registration of plant protection products in the Member States. It may be possible for a Member State to use some of the scenarios defined also as a representative scenario to be used in national authorizations but the scenarios were not intended for that purpose and specific parameters, crops or situations have been adjusted with the intention of making the scenario more appropriate to represent a realistic worst case for a wider area.

The purpose of the standard scenarios is to assist in establishing relevant Predicted Environmental Concentrations (PECs) in surface water bodies which — in combination with the appropriate end points from ecotoxicology testing — can be used to assess whether there are safe uses for a given substance. The concept of the tiered approach to surface water exposure assessment is one of increasing realism with Step 1 scenarios representing a very simple but unrealistic worst case calculation and Step 3 scenarios presenting a set of realistic worst cases representative of a range of European agricultural environments and crops.

12.4.2   AGRICULTURAL RELEVANCE OF SCENARIOS

The FOCUS Surface Water Working Group undertook an assessment of the relevance of individual scenarios to soil and climate conditions associated with agricultural land in Europe (a "pedoclimatic relevance" assessment). It was established that scenarios were considered representative of the proportions of agricultural land in the Europe Union (as constituted in 2001) (Table 12.11).

This summary demonstrates that the 10 surface water scenarios are considered representative of 32.9% of all the agricultural land in the European Union. However, in the regulatory context it is important to know how representative the scenarios are in terms of a worst case for pesticide movement to surface waters.

TABLE 12.11

Extent of the 10 Surface Water Scenarios within the European Union

(Table 3.5-1, FOCUS 2001)

Scenario	Area (km²)	Total Agricultural Land (%)
D1	15703	1.5
D2	8459	0.8
D3	8855	0.9
D4	44204	4.2
D5	15999	1.5
D6	36531	3.5
R1	75631	7.7
R2	6779	0.7
R3	22912	2.3
R4	95716	9.7

The FOCUS Surface Water Working Group acknowledged that because of the lack of comprehensive databases that characterize the environmental characteristics across the European Union, it is not possible to undertake a worst-case assessment in a rigorous, statistically based manner. Instead, available data sources were used to examine the extent of land with characteristics that are more challenging than those of the identified scenarios from the perspective of pesticide movement to surface water.

The FOCUS Surface Water Working Group recognized that it is not possible to scale the factors of soil, slope, rainfall or recharge, and temperature in terms of their relative contribution to an overall worst-case environmental combination. Any of the factors may be the most important depending on how each set of pesticide-specific application and physicochemical characteristics interacts with the rainfall patterns and volumes, soil, and slope characteristics of each scenario. Therefore, to simplify the worst-case assessments agricultural land was divided into temperature zones, and the proportions of land vulnerable to drainage and runoff were identified. These areas were then further investigated to quantitatively establish the areas of agricultural land with more challenging characteristics as follows:

Drainage
  • Texture
  • Average spring and autumn temperature
  • Annual recharge estimates
Runoff
  • Soil hydrological class (PRZM)
  • Average spring and autumn temperature
  • Annual precipitation

On this basis, the following conclusions were drawn:

Drainage: Scenario D2 combines an extreme worst-case soil with a worst-case recharge and represents a 98.8 percentile worst case for drainage within the worst-case temperature range. The extreme worst-case temperature range contains no extreme worst-case soils or any agricultural land with significantly larger recharge values than D2. The only drained land "worse than" D2 is thus the 1.2% of areas within the worst-case temperature range that have significantly larger recharge (see FOCUS 2001 report Table 3.5-4). These areas represent 0.7% of all drained land (1.2% of worst-case temperature drained agricultural land, which is 59% of all drained land). D2 thus represents a 99.3 percentile worst case for all drained agricultural land.

Run-off: Scenario R2 combines an extreme worst-case slope with an extreme worst-case rainfall, and it represents a 98.1 percentile worst case for runoff within the intermediate-case temperature range. There are no worse slopes under agriculture within all the runoff agricultural land in Europe. The only significantly worse areas of rainfall within all the agricultural runoff land occur in the intermediate temperature range, where they represent 0.9% of the agricultural runoff land. Worse-case runoff soils (hydrologic classes

C and D) occur within the worst- and extreme worst-case temperature land, but areas with more than 1402 mm of rainfall occupy only 1.3% of the total agricultural runoff land. The only agricultural runoff land worse than R2 is thus this 1.3% of agricultural runoff land and the 0.9% of areas within the intermediate-case temperature range that have significantly larger rainfall plus the 1% of areas within the intermediate-case temperature range with class C or D soils (see FOCUS 2001 report Table 3.5-5). These areas represent 2.0% of all runoff land (1.3% plus 1.9% of intermediate-case temperature agricultural runoff land, which is 34.5% of all agricultural runoff land). R2 thus represents a 98 percentile worst case for all agricultural runoff land.

A generalized overview of the areas considered represented or that have conditions more or less challenging than the scenarios developed are summarized in Figure 12.5. It is important to emphasize that these assessments apply only to the combination of general environmental characteristics that were used to identify the 10 surface water scenarios. More detailed statistics are available for each scenario as shown in tables extracted from the FOCUS Surface Water report and summarized

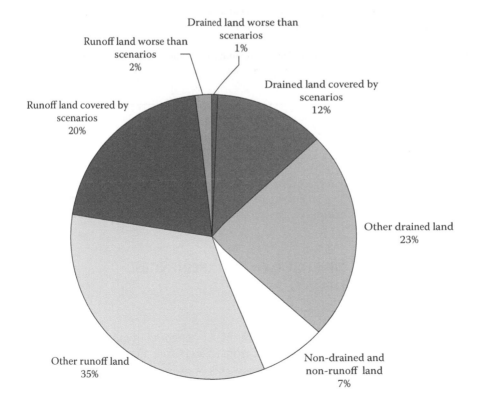

FIGURE 12.5 Overall assessment of the relevance of the 10 surface water scenarios to European Union agriculture (see Figure 3.5-4 in FOCUS 2001).

TABLE 12.12

Assessment of the Amount of European Agricultural Land Protected by Each Drainage Scenario (see Table 3.6-1 in FOCUS 2001)

Drainage Scenario	D1	D2	D3	D4	D5	D6
Area of drained land with a worse temperature expressed as a percentage of all drained and runoff land	1.2	6.6	6.6	6.6	29.6	34.7
Area of drained land with a worse soil expressed as a percentage of all drained and runoff land	0.9	0	1.3	17.9	0.4	0.4
Area of drained land with a worse recharge expressed as a percentage of all drained and runoff land	8.3	1.4	1.2	9.3	1.8	0.5
Total area of drained land with worse characteristics expressed as a percentage of all drained and runoff land	10.4	8.0	9.1	33.8	31.8	35.6
Total runoff land expressed as a percentage of all drained and runoff land	59	59	59	59	59	59
Total area of land unprotected expressed as a percentage of all runoff and drained land	69.4	67	68.1	92.8	90.8	94.6
Total area of land protected expressed as a percentage of all runoff and drained land	**30.6**	**33**	**31.9**	**7.2**	**9.2**	**5.4**
Total area of land protected expressed as a percentage of all runoff and drained land in northern European agriculture	**55.0**	**61.0**	**59.1**	**13.3**	**16.9**	**n.a.**
Total area of land protected expressed as a percentage of all runoff and drained land in southern European agriculture	**n.a.**	**n.a.**	**n.a.**	**n.a.**	**n.a.**	**11.7**

in Tables 12.12 and 12.13. On this basis, the FOCUS Surface Water Working Group concluded the following:

> Based on these results it is estimated that a favorable risk assessment for any single drainage scenario or any single runoff scenario should protect a significant area (at least >5%) of relevant European agriculture and thus should be adequate for achieving Annex 1 listing.

12.5 SUMMARY OF REGULATORY EXPERIENCES: FOCUS MODELING

James Hingston

This section provides a brief account of the ongoing regulatory experiences with the FOCUS surface water tools from an exposure modeling point of view. The approach of the FOCUS surface waters scenarios to the aquatic exposure assessment integrates the estimation of pesticide entry to surface water by spray drift, subsurface drainage, or surface runoff. It was confirmed by the European Commission during the working group (legislation) meeting of 3 to 4 July 2003 that the FOCUS surface water approach to the estimation of aquatic PECs would

TABLE 12.13

Assessment of the Amount of European Agricultural Land Protected by Each Runoff Scenario (see Table 3.6-2 in FOCUS 2001)

Runoff Scenario	R1	R2	R3	R4
Area of runoff land with a worse temperature expressed as a percentage of all drained and runoff land	13.1	24.7	24.7	45.8
Area of runoff land with a worse soil expressed as a percentage of all drained and runoff land	3.0	14.5	3.0	0.7
Area of runoff land with a worse rainfall expressed as a percentage of all drained and runoff land	6.6	0.4	4.1	2.8
Total area of runoff land with worse characteristics expressed as a percentage of all drained and runoff land	22.2	39.6	32.4	50.9
Total drained land expressed as a percentage of all drained and runoff land	39	39	39	39
Total area of land unprotected expressed as a percentage of all runoff and drained land	61.2	78.6	71.4	89.9
Total area of land protected expressed as a percentage of all runoff and drained land	**38.8**	**21.4**	**28.6**	**10.1**
Total area of land protected expressed as a percentage of all runoff and drained land in northern European agriculture	**71.9**	**n.a.**	**n.a.**	**n.a.**
Total area of land protected expressed as a percentage of all runoff and drained land in southern European agriculture	**n.a.**	**46.6**	**62.1**	**21.9**

need to be used by industry in their submissions for existing substances in List 3 of the review program designated under Council Directive 91/414/EEC. In addition, it was confirmed that this approach would be required for submissions for Annex I consideration of all new active substances from November 2003. The FOCUS models have therefore been used by industry and member states for several years, and a considerable amount of experience in their use in the regulatory context has been gained during this time.

It should be noted that the FOCUS approach was primarily developed as a tool for evaluating active substances for inclusion in Annex I of Directive 91/414/EEC, and although some EU member states will apply the existing FOCUS surface water scenarios at Annex III, there is to date no agreed use of these models at member-state level across the European Union. Therefore, this section deals with experiences with the FOCUS models during the EU Annex I listing peer review process only and should not be taken to represent the views of any individual member state.

Prior to FOCUS, surface water exposure assessments were typically based on spray drift only. This was because spray drift was the only route of entry to surface water that had an agreed exposure assessment scheme at the EU level. Such assessments only required knowledge of the application rate, the appropriate spray drift percentile (taken from either the Ganzelmeier or Rautmann data sets), and an estimate of the water phase dissipation rate. The use of the FOCUS models has resulted in a significant increase in the complexity of the input parameters required. For example, in place of a simple water-phase dissipation rate, FOCUS Step 3 ideally requires true degradation rates for both the water and sediment phases. The derivation of these individual values is complex and is dealt with in detail in Chapter 10 of the FOCUS degradation kinetics report (FOCUS 2006). Early regulatory experiences often involved detailed discussion of the derivation of these input parameters, particularly where water-phase dissipation rates were erroneously used instead of true degradation rates at Step 3. Use of a dissipation rate (which may include loss from the water phase via both degradation and partitioning to sediment) could result in an underestimate of persistence of the substance in the water phase since the FOCUS Step 3 models simulate sediment partitioning using separate expressions based partly on the substance's sorption properties. The use of a simple dissipation rate may therefore result in a double accounting of the partitioning process in the FOCUS simulations. As experience with the use of the models increased, it became apparent that for many substances the water or sediment phase DT50 values were not particularly sensitive input parameters, particularly for the ditch and stream water bodies, which typically give the highest PEC values and were of most concern for the purposes of the subsequent aquatic risk assessment. This is because the hydraulic retention times in these water bodies are often relatively short compared with the corresponding substance degradation rates; therefore, the principal mechanism for loss from such systems tends to be via outflow from the system rather than either degradation or partitioning to sediment.

Given the complexities of deriving acceptable input parameters for water- and sediment-phase degradation, applicants have chosen, with increasing regularity, to use simple worst-case default parameters. One common approach has been to use the whole system DT50 from the standard laboratory water–sediment study to represent degradation in the compartment where most degradation appears to occur, and a simple 1000-day default value for the nondegrading compartment. Such approaches are in line with the recommendations of the FOCUS degradation kinetics work group and are increasingly being accepted at the EU level for purposes of Annex I inclusion.

Given the significant increase in the complexity of the exposure assessment generated using the FOCUS surface water approach, the need for clear reporting of all input parameters and the assumptions made has become more apparent as ever more substances have progressed through the EU peer review process. One example of where clear reporting of the starting assumptions is necessary is with regard to the application windows selected for use with the PAT. Given the "event-driven" nature of many drain flow and runoff events simulated by FOCUS, there is still some uncertainty regarding the sensitivity of the models to the actual application dates selected by the models. Clear reporting of the assumptions made in setting application windows has proven important to increase confidence that the models have been parameterized appropriately.

With regard to the presentation of results, approaches between different applicants and member states has varied considerably from simple tables of maximum PEC values only for each scenario–water body combination up to full presentation of all actual and TWA concentrations for all scenarios at all steps. Given the huge increase in the number of exposure values generated by the FOCUS approach, it is not always clear that the exposure and effects assessments have used the appropriate endpoints in all assessments. Generally, a tiered approach has proven to be the most successful way of presenting the results in both the exposure and effects assessments. When the risk assessment for the most sensitive aquatic species results in acceptable toxicity exposure ratio (TER) values using the peak PEC values, it has been possible to simplify the presentation of results to just those values used in the risk assessment. Only when certain TER triggers are breached has there been a need to consider detailed results from all scenarios to identify those scenarios that result in safe uses.

A number of the issues related to handling the increased complexity of the FOCUS surface water outputs were first highlighted in the European Food Safety Authority (EFSA) PPR Panel opinion on dimoxystrobin (EFSA 2005; Question No. EFSA-Q-2004-81). This opinion highlighted the need for consistent approaches to characterizing both effects and exposure concentrations and included some of the first examples of attempts to match exposure and effects profiles for higher-tier effects studies (e.g., effects studies performed in the presence of sediment or the exposure profile from a fish early life stage study [ELS], for example). The opinion stressed the need for in-depth analysis of the full range of fate and behavior data when using tailored exposure regimes in effects studies. Unfortunately, this opinion was published after the submission date for the List 3a and 3b substances; therefore, many of the regulatory submissions supporting the substances that are only now being considered in detail by the EU peer review process do not necessarily have a full consideration of these issues. If such issues are only raised during the EU peer review process and require more detailed consideration of the FOCUS exposure outputs, this invariably delays the progress of substances through the Annex I inclusion process. It is hoped that as experience using these tools increases such issues will be identified at a much earlier stage to ensure the EU peer review process progresses in an efficient manner.

In some cases, the presentation of simple tabulated results of actual and TWA concentrations has highlighted the fact that such presentations are not necessarily able to reflect the greater complexity of the FOCUS surface water outputs. For example, tabulated results for actual PEC values from some runoff scenarios based on the standard FOCUS output time points (e.g., 0, 1, 2, 4, 7, 14, 21, 28, 42, 50, and 100 days) may miss significant peak, short-term exposure events due to runoff events occurring on days other than those selected in the standard FOCUS output. For example, an event occurring 30 days after the peak exposure would not be represented in the standard table of actual PEC outputs; however, these events may be indirectly represented in the variation of the corresponding TWA PEC value. As a result, the significance of such exposure events may not be fully considered during the aquatic risk assessment; however, the need to consider these issues in detail is largely dependent on the ecotoxicological context of the accompanying risk assessment and is not necessarily a universal concern in every case. Where the exposure assessment needs to be compared to higher-tier effects data such that the more detailed patterns of exposure have to be considered, additional presentation of the full graphical outputs has been provided in a small number of regulatory assessments. However, the detailed matching of exposure and effects patterns has received relatively little attention during the EU peer review process at this time, and it is hoped that work in other areas of this report will further aid this in future regulatory submissions.

One area with continuing uncertainty in linking the exposure and effects assessment is the risk assessment for sediment-dwelling organisms. Issues have arisen when the sediment dweller effects study is based on a water spiked study design (giving an effect concentration in milligrams per liter). In such circumstances, it is uncertain which is the most appropriate exposure endpoint to use in the subsequent risk assessment. Some applicants have converted the sediment PEC (based on mass of substance per dry weight of sediment) to an effective water-phase concentration for direct comparison with the effects endpoint. This has been achieved by simply assuming the mass of substance in the sediment phase instantaneously repartitioned back into the overlying water phase to generate a PECsw for use against the effects endpoint. In other cases, the effect endpoint has been compared directly with the peak PECsw from the FOCUS scenarios. However, where multiple exposure events occur (e.g., due to spray drift or runoff), it is clear that such an approach (based on surface water PEC alone) is not always acceptable. There has been some confusion over this point, and further guidance in this area is probably required.

In some cases, regulatory exposure assessments have deviated from the standard FOCUS approach on a case-by-case basis. One example of this is the calculation of potential accumulation of residues in sediment for very persistent substances. In some cases, simplistic approaches to calculating PECsed (predicted environmental concentration in sediment) using the sediment DT50 and an assumption of repeated annual applications has been performed outside the FOCUS framework (e.g., using simple MS Excel spreadsheets). This is comparable to the first-tier approaches that would have been used pre-FOCUS surface water and has sometimes been used to address possible weaknesses in the current FOCUS tools that may not be able to provide a sufficiently precautionary estimate of PECsed for very persistent substances. The principal weakness in the FOCUS models in this area appears to be the fact that the initial PECsed for the simulation period is always set to zero, even when the models indicate that significant residues could remain in sediment after 1 year and be subject to carryover to the following simulation period. The FOCUS models may not therefore be capable of assessing potential sediment accumulation satisfactorily using the standard assumptions at Step 3.

A further example of where modeling has deviated from the FOCUS approach would be in the derivation of appropriate exposure concentrations for use against effects endpoints derived for the formulated product. When the formulation has been shown to be significantly more toxic than the active substance alone, a separate exposure assessment has sometimes been performed based on spray drift only using the drift calculator in SWASH. This is because it is assumed that in the environment the individual components of a formulation will disperse separately, and it is unlikely that an intact formulation would be subject to processes such as drain flow or runoff. Therefore, the standard FOCUS surface water outputs that may be based on such exposure routes may not be appropriate for use against formulation toxicity data.

With regard to the implementation of risk mitigation measures to reduce exposure levels during FOCUS Step 4 assessments, at present the reduction of spray drift exposures using increased no-spray buffer zones is the only mitigation method that has received widespread agreement at the EU level. Other methods to reduce exposure, such as the incorporation of vegetated filter strips to mitigate runoff exposures, have not been formally agreed during the EU peer review process. The acceptability of runoff mitigation measures has been further complicated by the publication of the PPR Panel opinion on the FOCUS Landscape and Mitigation Factors work group report (Question No. EFSA-Q-2006-063; EFSA 2006), which questioned the validity of the reduction factors proposed by the FOCUS report as originally drafted. In general, at the time of the workshops the EU peer review process left issues such as the usefulness of runoff mitigation to be considered by individual member states on a case-by-case basis.

In general, despite the issues raised in this section, the overall regulatory experience with the FOCUS models to date has been very positive. The models have proven to be a useful and reliable tool for integrating multiple exposure routes into the aquatic risk assessment process. It is clear that additional guidance is needed to improve the link between the exposure and effects assessments, and many of these issues are addressed in this workshop report.

12.6   SUMMARY OF REGULATORY EXPERIENCES: USE OF FOCUSsw MODELING IN ECOTOX ASSESSMENT

Gary Mitchell and Steve Norman

The FOCUS surface water scheme incorporates 3 potential routes of entry to surface water (spray drift, runoff, and drainage) as a basis for exposure assessment at the EU level. In aquatic exposure assessment using FOCUSsw modeling (and therefore risk assessment), active substances applied as sprays can generally be differentiated between those where spray drift is the dominant potential route of input to surface water and those where runoff or drain flow is the major potential input route. In terms of environmental fate, the distinction between active substances (spray drift vs. runoff or drain flow dominated) is driven by rate of degradation in soil, solubility in water, and potential for adsorption to the soil particles.

Some experiences where spray drift or runoff or drain flow were the principal input routes are described next.

12.6.1   RISK ASSESSMENTS WHERE SPRAY DRIFT DOMINATES

For active substances that are applied as a spray and have a high potential for adsorption to soil particles (high Koc), the spray drift route of input usually dominates. A major group of active substances that fits this profile is the synthetic pyrethroid (SP) insecticides. From experience, profiles from Step 3 FOCUSsw (see Section 12.2) for adsorptive compounds such as SP insecticides give a distinct pulse of exposure for each spray drift event. The duration of this pulse is shortest for streams (dissipation driven primarily by advection) and longest for ponds (dissipation driven primarily by degradation and partitioning to sediment). Generally, because of the relative depth of the systems, the PECsw values for streams (and ditches) are much higher than for ponds; thus, stream and ditch PEC values tend to determine the outcome of the assessment. Microcosm and mesocosm experiments are typically conducted to determine the effects of SPs (for example) under more realistic conditions in terms of exposure and ecology. These studies are usually conducted in static test systems without flow. Hence, the use of the FOCUSsw PEC values for streams (and ditches) in comparison with micro/mesocosm-derived endpoints can be regarded as conservative; from experience, this approach has been accepted by regulators.

12.6.2 RISK ASSESSMENTS WHERE RUNOFF OR DRAIN FLOW DOMINATE

A combination of 1 or more properties of a compound (e.g., high solubility in water, low Koc, relatively long DT50 in soil) — or the method of application — may lead to runoff or drain flow being the major potential route of input to surface water. Runoff and drain flow are "event driven," the "event" being significant rainfall.

In the first tier of risk assessment, maximum PEC values are compared with endpoints from standard laboratory toxicity studies, which generally include constant exposure for a predetermined period. Where these maximum PEC values for FOCUS Steps 1, 2, or 3 have resulted in a TER greater than the relevant Annex VI 91/414 EEC trigger, a low risk is concluded. Only if a Step 3 TER occurs below the relevant trigger is it necessary to consider the Step 3 exposure profiles rather than only the maximum PEC values.

From recent experience with a residual herbicide for which the principal input route was drain flow and the predicted period of major exposure was during the winter, the inclusion of the exposure profiles in the risk assessment can raise many regulatory questions. In this case, the concern was associated with the potential risk to algae and aquatic plants. For example, a key question was whether the use of standard laboratory endpoints for algae and *Lemna* (derived under optimum growth conditions) would lead to an overestimation of the risk. This is because growth in the field would be minimal during the main drainage period, which generally occurs between late autumn to early spring. Hence, the profiles can raise the issue of "seasonality." The profiles also can raise questions over the potential duration of exposure (i.e., as indicated by the profile) compared with the duration of exposure in higher-tier ecotoxicology studies. For example, the duration of exposure due to drain flow might be greater than the duration of exposure of a higher-tier algal study. Such questions also need to take account of ecological issues ("real" potential exposure duration under natural conditions in the field) and practical issues (e.g., practical testing constraints on exposure duration). The inherent conservative nature of the risk assessment should also be taken into account.

The TER values that are below regulatory triggers can occur with maximum PEC values from Step 3 R (runoff) scenarios. This can raise the issue of the availability of effective risk mitigation (vegetative filter strips, VFSs) and how this can be incorporated into the risk assessment. Issues surrounding the incorporation of risk mitigation techniques into exposure assessments at Step 4, and associated recommendations, are discussed in the FOCUS *Landscape and Mitigation Report* (FOCUS 2007a, 2007b).

12.6.3   OTHER EXPERIENCES

12.6.3.1   "Pulsed" Exposure

Considering the potential exposure (according to FOCUS) in streams or ditches, in many cases the exposure duration can be less than 1 day and sometimes may not exceed a few hours. Such "pulses" may result from spray drift, runoff, or drain flow. The pulse of exposure in a stream or ditch may be significantly shorter than the exposure period in a standard laboratory acute toxicity test (e.g., a 96-hour fish acute study). This information provides a potential option for refinement (e.g., higher-tier studies with short exposure durations). Pulsed exposure studies have also been conducted in some cases for algae to investigate effects and potential for recovery from relative high-intensity, but short-duration, exposure. These issues are elaborated in Part I of this book.

12.6.3.2   FOCUS Steps 1 and 2 and Metabolites

The outputs from FOCUS Steps 1 and 2 are not generally useful in the decision making for the active substance itself. However, they are useful for providing simplistic and highly conservative PEC values for metabolites (degradates). This enables these metabolites, which are often of much lower toxicity than the active substance, to be understood and "sifted out" of the risk assessment at an early stage.

12.6.3.3   Sediment PEC Values

A FOCUSsw assessment provides PEC values for sediment. However, these values have sometimes not been used in the actual risk assessment. This is because the toxicity study on the standard sensitive sediment-dwelling organism (*Chironomus* sp.) includes the option of spiking either the water column or the sediment. In most cases, the study is designed to include addition of the active substance directly to (i.e., "spike") the water phase, which then gives a combination of exposure from the overlying water and, as partitioning proceeds during the study, from the sediment. The result of this study (usually an NOEC [no observed effect concentration]) can then be expressed as an initial (nominal) concentration in the water column, which can be compared with the maximum PEC value for the water phase.

Note that the workshop report editors identified that this may not always be appropriate, particularly if multiple inputs to an aquatic system are expected, as this is not covered by the ecotoxicology study design exposure regime, in which a single spike is usually used. See note c to Table 3.2 in Chapter 3 for clarification of the appropriate approach that is being recommended for sediment dweller risk assessment (which reflects the practice EFSA now follows when peer reviewing the RMS assessments).

12.6.3.4   Potential for "Overinterpretation" of FOCUS Step 3 Exposure Profiles

As well as maximum and TWA PEC values, time-based profiles of modeled concentrations are also produced at Step 3, as indicated in this chapter. While it is generally understood that the original intention of FOCUSsw was that only the PEC values themselves would be used in risk assessment, higher-tier assessments have now prompted detailed consideration of the profiles themselves. From experience, there is a clear potential for practitioners to overinterpret these profiles. This is discussed in Chapter 4 (Sections 4.4 and 4.5); a more rational and workable approach would be to base risk assessment on a generalized profile that best represents the characteristics of the exposure scenario.

12.7   REGULATORY CONTEXT, REALITY CHECK, AND CONFIDENCE OVERVIEW

James Hingston, Neil Mackay, and Helmut Schaefer

12.7.1   REGULATORY CONTEXT

As discussed in Section 12.4, the FOCUS surface water framework was designed to facilitate aquatic risk assessments at the EU level for the purposes of Annex I inclusion. The following commentary in the FOCUS surface water report provides further background on what the scenarios and framework are intended to represent and (perhaps more importantly) what the framework was not intended to provide:

> The contamination of surface waters resulting from the use of an active substance is represented by ten realistic worst-case scenarios, which were selected on the basis of expert judgment. Collectively, these scenarios represent agriculture across Europe, for the purposes of Steps 1 to 3 assessments at the EU level. However, being designed as "realistic worst case" scenarios, these scenarios do not mimic specific fields, and nor are they necessarily representative of the agriculture at the location or the Member State after which they are named. Also they do not represent national scenarios for the registration of plant protection products in the Member States.

However, since publication of the report and release of the modeling framework, a number of member states have reviewed the FOCUS surface water framework to determine the extent to which such modeling could potentially support national evaluations. Regulatory policies and attitudes toward FOCUS surface water across Europe are continuously evolving, and any summary of national status presented here would likely be outdated by time of publication. Nonetheless, it has been typical for some member states to employ maps presented in the FOCUS surface water report to establish pedoclimatic relevance to a specific regulatory jurisdiction. The maps presented in the FOCUS surface water report do not, however, extend into more recent eastern European member states. Nonetheless, at the time of the framework's completion such maps were generated and can be readily obtained from members of the original FOCUS Surface Water Working Group. In some cases, more intensive evaluations have been carried out employing higher-resolution soil and climate databases. In both cases, it has been typical to identify a subset of scenarios considered to have particular relevance to a given regulatory jurisdiction. Examples include the United Kingdom, where of the 10 scenarios developed within the FOCUS surface water framework, 4 have been identified as having primary pedoclimatic relevance within England and Wales (these are R1, D2, D3, and D4) (see UK Defra Research and Development reports PS2214[1] and PS2220[2]).

12.7.2 REALITY CHECK

The following text, relating to uncertainty, has been abstracted from the executive summary of the FOCUS surface water report. Within this discussion, there is consideration of the question of validation:

> Uncertainty will always be present to some degree in environmental risk assessment. As part of the EU registration process, the use of the FOCUS scenarios provides a mechanism for assessing the PECs in surface water and sediment with an acceptable degree of uncertainty.
>
> The choice of the surface water scenarios, soil descriptions, weather data and parameterization of simulation models has been made in the anticipation that these combinations should result in realistic worst cases for PEC assessments. It should be remembered, however, that the FOCUS surface water scenarios are virtual, in that each is a combination of data from various sources designed to be representative of a regional crop, climate and soil situation, although they have a real field basis. Adjustments of the data to make them useful in a much broader sense have been necessary. As such, none can be experimentally validated.
>
> To further reduce uncertainty, independent quality checks of the scenario files and model shells were performed, and identified problems were removed. An additional check for the plausibility of the scenarios and models is provided by the test model runs made with dummy substances, which have widely differing properties.
>
> Whilst there is still scope for further reductions in uncertainty through the provision of improved soils and weather data at the European level, the FOCUS Surface Water Scenarios Working Group is confident that the use of the standard scenarios provides a suitable method to assess the PECs in surface water and sediment at the first three steps in the EU registration procedure.

A key point raised in the report abstracts is that the scenarios were not intended to "mimic specific fields, and nor are they necessarily representative of the agriculture at the location or the Member State after which they are named." The scenarios are intended to be employed as "virtual, in that each is a combination of data from various sources designed to be representative of a regional crop, climate and soil situation, although they [independently] have a real field basis."

It can be argued that the scenarios cannot be "validated" as such. In the case of certain scenarios, there is, however, a strong link to field research sites — particularly for the representation of drainage scenarios. A summary of such linkages and background information on the field sites and field research is provided next.

D1   Larsson MH, Jarvis NJ. 1999. A dual-porosity model to quantify macropore flow effects on nitrate leaching. J EnvironQual 28:1298–1307.

Larsson MH, Jarvis NJ. 1999. Evaluation of a dual-porosity model to predict field-scale solute transport in a macroporous clay soil. J Hydrol 215:153–171.

Larsbo M, Jarvis NJ. 2005. Simulating solute transport in a structured field soil: uncertainty in parameter identification and predictions. J Environ Qual 34:621–634.

D2   Harris GL, Bailey SW, Mason DJ. 1991. The determination of pesticide losses to water courses in an agricultural clay catchment with variable drainage and land management. Brighton Crop Protection Conference Weeds 3:1271–1277.

Harris GL, Howse KR, Pepper TJ. 1993. Effects of moling and cultivation on soil–water and run off from a drained clay soil. Agric Water Manage 23:161–180.

Harris GL, Nicholls PH, Bailey SW, Howse KR, Mason DJ. 1994. Factors influencing the loss of pesticides in drainage from a cracking clay soil. J Hydrol 159:235–253.

Armstrong A, Aden K, Amraoui N, Diekkrüger B, Jarvis N, Mouvet C, Nicholls P, Wittwer C. 2000. Comparison of the performance of pesticide-leaching models on a cracking clay soil: results using the Brimstone Farm dataset. Agric Water Manage 44(1–3):85–104.

Harris GL, Catt JA, Bromilow RH, Armstrong AC. 2000. Evaluating pesticide leaching models: the Brimstone Farm dataset. Agric Water Manage 44(1–3):75–83.

D3   COST 66 Validation Exercise described in

Vanclooster M, Boesten JJTI. 2000. Application of pesticide simulation models to the Vredepeel dataset: I. Pesticide fate. Agric Water Manage 44(1–3):119–134.

Vanclooster M, Boesten JJTI. 2000. Application of pesticide simulation models to the Vredepeel dataset: I. Water, solute and heat transport. Agric Water Manage 44(1–3):105–117.

D4   Villholth K. 1994. Field and numerical investigation of macropore flow and transport processes [dissertation]. [Copenhagen]: Institute of Hydrodynamics and Hydraulic Engineering, Technical University of Denmark, Lyngby. Series Paper 57.

R1   Gottesbüren B, Aden K, Bärlund I, Brown C, Dust M, Görlitz G, Jarvis N, Rekolainen S, Schäfer H. 2000. Comparison of pesticide leaching models: results using the Weiherbach data set. Agric Water Manage 44(1–3):153–181.

Schierholz I, Schäfer D, Kolle O. 2000. The Weiherbach data set: an experimental data set for pesticide model testing on the field scale. Agric Water Manage 44(1–3):43–61.

R3   Rossi Pisa P, Catione P, Vicari A. 1992. Runoff and watershed studies at the agricultural farm "Terreni di Ozzano" of the University of Bologna. Bologna (Italy): Department of Agronomy, University of Bologna.

Vicari A, Rossi Pisa P, Catione P. 1999. Tillage effects on runoff losses of atrazine, metolachlor, prosulfuron and triasulfuron. 11th EWRS Symposium, Basel, Switzerland, 28 June to 1 July.

Miao Z, Vicari A, Capri E, Ventura F, Padovani L, Trevisan M. 2004. Modeling the effects of tillage management practices on herbicide runoff in northern Italy. J Environ Qual 33:1720–1732.

R4   Leonard RA, Shironhammadi A, Johnson AW, Marti LR. 1988. Pesticide transport in shallow groundwater. Trans ASAE 31:776–788.

Lennartz B, Louchart X, Voltz M, Andrieux P. 1997. Diuron and simazine losses to runoff water in Mediterranean vineyards. J Environ Qual 26:1493–1502.

Voltz M, Lennartz B, Andrieux P, Louchart X, Roger L, Luttringer M. 1997. Transfert de produits phytosanitaires dans un bassin versant cultive mediterraneen: analyse experimentale et implications pour la modelisation. Montpellier (France): Laboratoire de Science du Sol. INRA technical report.

Louchart X, Voltz M, Andrieux P, Moussa R. 2001. Herbicide transport to surface waters at field and watershed scales in a Mediterranean vineyard area. J Environ Qual 30:982–991.

In assessing the performance of the scenarios, it was recognized that total mass loss is perhaps not the most relevant measure for risk assessment. Instead, assessments focused primarily on the representation of maximum peaks. It was concluded that this was the most relevant basis for assessment as the maximum concentration attained is more critical for comparison with acute ecotoxicological endpoints. In many cases, peaks will be short-lived, and chronic exposure will be heavily mitigated by the influence of dilution. However, the implications of longer-term exposure as a consequence of recurrence of peaks may be relevant in many cases, particularly where recovery potential becomes an important facet of the risk assessment. The ability to "validate" or "verify" model performance with respect to recurrence of peaks and longer-term exposure was somewhat limited by availability of relevant field research; consequently, this aspect of model performance has been less vigorously evaluated.

In an ongoing research-and-development (R&D) project funded by Defra/PSD in the United Kingdom, comparisons between the FOCUSsw outputs for pesticide concentrations over time and literature data on observations in the field are being conducted (see UK Defra Research and Development report PS2231[3]). In general, there was insufficient data available in the public domain to be able to undertake a detailed time series analysis for measured exposure via either spray or surface runoff. However, literature data for concentrations of pesticides due to losses via drain flow indicated that the FOCUS surface water scenarios were able to reproduce the overall characteristics of the exposure seen in the field (e.g., there was broad agreement between the FOCUS scenarios and the field exposures with respect to interval between peaks, peak duration, and to a lesser extent the decrease in peak concentrations for successive pulses). Further work in this area is ongoing and would clearly benefit from additional data to be able to compare time series due to spray drift or surface runoff exposures.

12.7.3   MODEL PERFORMANCE: DRAINAGE

More general model scenario performance was assessed by the FOCUS Surface Water Working Group, which considered the following observations:

- Mass losses seem to be largest in well-structured clayey soils and somewhat less from loamy soils. Sands with shallow groundwater are less well investigated but seem to pose a smaller risk than soils exhibiting macropore flow.
- Mass losses clearly depend on compound properties even in the presence of macropore flow.
- For mobile compounds, typically up to 2% to 4% of the applied amount may leach to drains, with 2 extreme values of 8% to 9% reported for weakly sorbed compounds applied on well-structured clay soils (one in autumn in Sweden, one in spring in Italy).

The FOCUS Surface Water Working Group concluded the following:

Simulations for compounds with half-lives ranging from 3 to 30 days and Koc from 10 to 100 (more typical of mobile pesticides actually used in agriculture) indicate simulated annual losses in the range of <0.1% to 3.1% for FOCUS compounds A and B (both with a half-life of 3 days) and <0.1% to 19.3% for FOCUS compounds D and E (both with a half-life of 30 days). Although annual losses are greater than those observed in field studies the range of maximum daily losses are more comparable to field observations. Maximum concentrations depend on both the compound properties and dose rate, but for weakly to moderately sorbed compounds, concentrations from tens to several hundred $\mu g\ L^{-1}$ are commonly reported and were of similar magnitude to the residues in drainflow simulated in the more vulnerable scenarios such as D1, D2 and D6.

12.7.4 MODEL PERFORMANCE: RUNOFF

Comparisons were undertaken with field research in Bologna (similar to the basis for the R3 scenario). Runoff losses ranged between 0.1 and 2% of precipitation. As a consequence of the rainfall pattern, losses of herbicides amounted to a maximum of 0.24, 0.25, 0.05, and 0.003% of the amount applied for atrazine, metolachlor, prosulfuron, and triasulfuron, respectively, and the minimum tillage reduced metolachlor and atrazine losses with respect to conventional tillage (Vicari et al. 1999). PRZM calculations for atrazine in R3 resulted in annual losses of 0.10% and 0.001% for triasulfuron, indicating reasonable general agreement with this single year of experimental data.

In similar research undertaken near Bologna, runoff corresponded to 0.5 and 3.5% of precipitation for normal and minimum tillage, respectively. Maximums of 1.6, 1.1, and 0.07% of the applied amount of metolachlor, atrazine, and terbuthylazine, respectively, were lost via runoff. The annual pesticide losses simulated by PRZM for these chemicals in R3 were 2.0, 1.3, and 0.3% for metolachlor, atrazine, and terbuthylazine, respectively, again indicating reasonable agreement between this scenario and the available experimental data.

Similar research campaigns were carried out in southern France in a vine-growing catchment that is the basis for the R4 scenario. Detailed measurements were reported for seasonal runoff, seasonal pesticide losses, and the concentrations in individual edge-of-field runoff events for normal agronomic applications of diuron and simazine (Lennartz et al. 1997; Louchart et al. 2001). Comparisons between the experimental data for 1995 and 1997 and the results of PRZM simulations for scenario R4 are duplicated in Table 12.14 from Table 6.4.2-1 of the FOCUS surface water report. The results obtained from Scenario R4 using PRZM show good general agreement with the 2 years of experimental data for diuron and simazine, with similar annual losses as well as similar ranges of runoff concentrations.

The FOCUS Surface Water Group concluded the following:

These PRZM simulation results [see Table 12.14] indicate that the model is capable of providing reasonable estimates of the runoff coefficient (fraction of precipitation resulting in runoff) as well as reasonable estimates of cumulative runoff flux. It should be emphasized the FOCUS runoff scenarios provide sound general estimates of runoff and erosion behavior likely to occur given the soil, agronomic and weather data selected for use in each scenario. More detailed, site-specific comparisons of PRZM with experimental runoff events require the use of local soil, agronomic and weather data.

TABLE 12.14

Comparison of Experimental and Simulated Values for Scenario R4

Parameter Being Compared	Values from Field Experiments	Values from PRZM, Scenario R4
Annual runoff (% of annual precipitation)	19 to 22%	24%
Annual diuron loss (% of applied)	0.7 to 0.9 (tilled)	0.7
Annual simazine loss (% of applied)	0.5 to 0.8 (tilled)	0.4
Runoff concentrations of diuron (µg/L, from first 4 events)	1 to 57 (0.5 kg ai/ha) 2 to 100 (2.0 kg ai/ha)	4 to 82 (0.5 kg ai/ha) 11 to 344 (2.0 kg ai/ha)
Runoff concentrations of simazine (µg/L, from first 4 events)	0.3 to 57 (0.28 kg ai/ha) 0.2 to 45 (1.0 kg ai/ha)	0.8 to 57 (0.28 kg ai/ha) 2 to 204 (1.0 kg ai/ha)

12.8   FREQUENTLY ASKED QUESTIONS

12.8.1   VALIDATION IS IMPORTANT TO ENSURE REGULATORY CONFIDENCE IN RISK ASSESSMENT TOOLS; HOW WOULD YOU CHARACTERIZE THE VALIDATION STATUS OF THE FOCUS SW FRAMEWORK?

Prepared by Neil Mackay

The following text has been abstracted from the executive summary of the FOCUS surface water report (FOCUS 2001) and pertains to consideration of uncertainty. Within this discussion, there is consideration of the question of validation:

> Uncertainty will always be present to some degree in environmental risk assessment. As part of the EU registration process, the use of the FOCUS scenarios provides a mechanism for assessing the PECs in surface water and sediment with an acceptable degree of uncertainty.
>
> The choice of the surface water scenarios, soil descriptions, weather data and parameterization of simulation models has been made in the anticipation that these combinations should result in realistic worst cases for PEC assessments. It should be remembered, however, that the FOCUS surface water scenarios are virtual, in that each is a combination of data from various sources designed to be representative of a regional crop, climate and soil situation, although they have a real field basis. Adjustments of the data to make them useful in a much broader sense have been necessary. As such, none can be experimentally validated.
>
> To further reduce uncertainty, independent quality checks of the scenario files and model shells were performed, and identified problems were removed. An additional check for the plausibility of the scenarios and models is provided by the test model runs made with dummy substances, which have widely differing properties.
>
> While there is still scope for further reductions in uncertainty through the provision of improved soils and weather data at the European level, the FOCUS Surface Water Scenarios Working Group is confident that the use of the standard scenarios provides a suitable method to assess the PECs in surface water and sediment at the first three steps in the EU registration procedure.

12.8.2 What Are the Strengths and Weaknesses of the FOCUS SW Modeling Framework?

Prepared by Neil Mackay

A summary of the strengths and weaknesses of the FOCUS SW modeling framework can be found in Section 12.3. Primary strengths are as follows:

- Increased consistency and a common agreed basis for assessment.
- Relative to earlier regulatory schemes is considered more comprehensive.
- Development of a scheme in which the complexity of simulations becomes progressively more complex with each step as demanded by risk characterization (primarily this is a benefit in assessment of metabolites that may only require relatively simple Steps 1 and 2 assessments).
- Ability to simulate a broad range of crop and pedoclimatic combinations within a set of 10 well-researched standard scenarios.
- Scenarios have been developed to directly represent about 33% of agriculture in the European Union at the time of release.
- In terms of protectiveness, the regulatory challenge offered by individual scenarios varies. However, the most vulnerable drainage scenario (D2) has been concluded to represent a 99.3rd percentile of all drained agricultural land in the European Union at the time of release. The most vulnerable runoff scenario (R2) has been concluded to represent a 98th percentile of all agricultural land vulnerable to runoff in the European Union at the time of release.
- Framework includes consideration of primary routes of entry (drift, drainage, runoff).
- Simulation of drainage, runoff, and fate in surface water is via well-established mechanistic regulatory modeling techniques.
- Simulation of drift is via well-established, worst-case standard drift percentages at Steps 1 to 3.
- Ability to represent a range of application regimes (aerial, ground spray, and air blast application as well as granular applications and soil incorporation).
- Ability to generate output in a format sufficient for the purposes of standard ecotoxicological risk assessments.

Important limitations are as follows:

- Software framework does not easily lend itself to representation of mitigation options or even relatively simple refinements in scenarios
- Ability to simulate more complex kinetic formats than SFO is not an option in MACRO and TOXSWA. In PRZM, there is limited capability to simulate biphasic degradation, but this must be handled with care.
- Application timing takes into account rainfall patterns, but PAT strategy may be overly simplistic.
- Ability to accurately represent degradation pathways is limited by model constraints (MACRO can only operate with 1 parent to metabolite link at

a time, while PRZM can simulate parent to 2 metabolites or a series of a primary and secondary metabolite).

- Nonequilibrium sorption mechanisms cannot be represented at present.
- Timescales for simulation have been constrained to a reported period of 12 months for runoff simulations and 16 months for drainage simulations.
- Representation of an upstream catchment is simplistic and highly constrained (chemical contribution, hydrology, etc.).
- Inability to simulate photolysis in TOXSWA.
- Inability to simulate formation of metabolites in surface water systems in TOXSWA.

Further detail is provided in Section 12.3.

12.8.3   CAN A RELATIVELY SIMPLE EXPLANATION BE PROVIDED OF HOW DRAINAGE, RUNOFF, AND DRIFT ARE SIMULATED BY THE FOCUS SW FRAMEWORK?

Prepared by Helmut Schaefer

FOCUS developed a tiered scheme (Figure 12.2), starting with simple extreme worst-case approaches and progressing to more detailed and less worst-case procedures. This was done with the intention to save time and effort for compounds that easily pass the aquatic risk assessment even under worst-case assumptions.

Step 1 is based on extreme worst-case assumptions, and as Step 2 needs the same — relatively low — effort for conducting the PECsw calculations, exposure assessments tend to start actually with Step 2. Step 4 is not part of the FOCUS surface water scenarios report, and no standards are defined for doing Step 4 calculations. However, the FOCUS landscape modeling and mitigation report gives detailed advice, on which refinements can in principle be considered at Step 4.

Step 2 works on a "spreadsheet" type of basis, characterizing the input from drift, runoff, and drainage with percentages of applied amount entering the water body. Only a ditch is implemented as a standard water body.

Loadings are defined as 1 application or a series of individual applications (depending on the intended use), resulting in drift to the water body, followed by a runoff or drainage event occurring 4 days after the last application. Between applications and the runoff or drainage event, degradation in soil is considered. Drift entries are based on the Ganzelmeier/Rautmann tables. Minimum distances assumed to surface water (not as a mitigation measure) are 1 m for low crops (cereals, potatoes, etc.) and 3 m for high crops (orchards, grapes, etc.). Combined runoff or drainage entries as a percentage of soil residue are assumed to depend on region and season (see Table 12.4).

The receiving water body is a ditch 30 cm deep and 1 m wide underlain by sediment 5 cm deep with 5% organic carbon content. Partitioning between water and sediment is based on adsorption coefficients derived in standard batch equilibrium adsorption experiment with agricultural soils. Following a drift event, equilibrium between water and sediment is established within 24 hours. The "runoff or drainage" entry is distributed between water and sediment at the time of loading according to the adsorption coefficient of the compound. Degradation in water and sediment is based on rates derived in water–sediment studies.

Step 3 At Step 3, drainage and runoff are calculated with process-driven simulation models (Figure 12.1), and 3 different water bodies (ditch, pond, and stream) are implemented. Six drainage scenarios and 4 runoff scenarios covering different soil–climate conditions throughout Europe were defined. Runoff is calculated using the PRZM simulation model, which was developed by the USEPA and is used in the US authorization process for runoff estimations. Drainage is estimated with the MACRO model, developed in Sweden by Nick Jarvis. The actual fate in the surface water body is calculated with the TOXSWA model, developed in the Netherlands by Paulien Adriaanse.

Drift as a function of distance is described with regression equations, which are based on the Ganzelmeier/Rautmann tables. Minimum distances to water bodies (not as a mitigation measure) were defined and depend on crop and water body type. Minimum distances can range from 1 m for low crops near a ditch to 6 m for high crops near a pond. The drift percentage is averaged over the width of the water body, starting with the mentioned minimum distance.

Fate in soil needs to be considered in the context of runoff and drainage as entry routes. Degradation in the soil is quantified with rates from either laboratory or field degradation experiments. Sorption to soil is characterized with adsorption coefficients derived in standard batch equilibrium adsorption experiments with agricultural soils.

Runoff from the soil surface is calculated with the PRZM simulation model. Within this model, the estimation of runoff water is based on the so-called "curve number" approach, which considers the soil type, soil moisture, crop type, soil management, and soil permeability. Furthermore, soil erosion is estimated with MUSLE, which considers the runoff volume and peak flow as well as slope, slope length, soil erodibility, soil cover, and soil management. Pesticides enter surface water bodies via runoff water and eroded soil (both processes are often summarized as "pesticide runoff"). The amount of pesticide entering the water body is calculated from the concentration in soil and soil pore water at the day of the runoff event. PRZM calculations are run over several years; however, only a 12-month period is used for the exposure assessment. As some worst-case assumptions were already included, the weather conditions for this period were intended to represent a median case concerning the predicted daily and annual runoff water volume. As hydrological flows vary with season, different weather files were used for different application seasons (see Table 12.9).

For the drainage calculations, the MACRO model is used. Drainage is assumed to be associated with preferential flow (i.e., bypass flow-through macropores, e.g., cracks or wormholes) in the soils. Macropore flow is calculated assuming gravity flow and a power law function for conductivity. Drainage water flow occurs under saturated conditions in the soil layers containing the drainage pipes. The amount of pesticide drained depends on the concentration in this soil layer. MACRO calculations are run over several years; however, only a 16-month period is used for the exposure assessment. Again, the weather conditions for this period were intended to represent a median case concerning the amount of drainage water (see Table 12.10).

Runoff or drainage entries into surface water are driven by storm events. To ensure the realistic worst character of the calculation, the so-called "PAT" was implemented. The objective of the tool is to find an application date in a given time window, which fulfills the following 2 criteria:

1) At least 10 mm of rainfall in the 10 days following application
2) Less than 2 mm of rain each day in a 5-day period, starting 2 days before application, extending to 2 days following the day of application

If no appropriate date can be found, first criteria 1) and then criteria 2) are relaxed. In case of multiple applications, this procedure is done for each individual application.

12.8.4   CAN A RELATIVELY SIMPLE EXPLANATION BE PROVIDED OF HOW TOXSWA SIMULATES CHEMICAL BEHAVIOR?

Prepared by Ettore Capri

TOXSWA describes chemical behavior in a water body at the edge-of-field scale (i.e., a ditch, pond or stream adjacent to a single field) (Adriaanse 1996). Pesticides enter surface water bodies via drainage (output of MACRO becomes input for TOXSWA in FOCUS SW), via runoff water and eroded soil (output of PRZM becomes input for TOXSWA in FOCUS SW), and finally via spray drift. For drainage entries, hourly values of water discharges through drains and the pesticide loads in the discharge during the assessment period are saved to an output file of MACRO (equal divisions of the daily outputs calculated by MACRO), which is then used as input to TOXSWA. For runoff and erosion entries, the erosion loadings and chemical fluxes in runoff and erosion are handled in a similar manner in PRZM. The temporal distribution of the daily runoff and erosion loadings facilitates efficient mathematical solutions of the aquatic concentrations in TOXSWA. The spray drift is calculated as a function of distance of the field from the nearest water body and described with regression equations in the FOCUS drift calculator. The drift loadings are integrated across the width of the water body to provide a mean drift loading for a specific type of water body (ditch, pond, stream). Realistic worst-case environmental combinations have been identified for pesticide inputs from spray drift, runoff, and drainage.

The modeled field water body system in TOXSWA is 2-dimensional and consists of 2 types of subsystem, water layer and sediment. The water and sediment subsystems are coupled by assuming that the concentration in the liquid phase at the sediment surface equals the water phase concentration of the overlying water column. At the bottom of the sediment, there is an inflow or outflow of water with pesticide. In the water layer, substance concentrations vary in horizontal direction but are assumed to be uniform throughout the depth of each compartment, while in sediment concentrations vary in both horizontal and vertical directions.

TOXSWA considers 4 processes: 1) transport, 2) transformation, 3) sorption, and 4) volatilization.

In the water subsystem, the fate processes described are the following:

- Advection and dispersion in the horizontal direction. Diffusion in the horizontal direction has not been implemented as dispersion prevails.
- Exchange with the atmosphere (volatilization) through a diffusive flux across the water-air interface.
- Exchange with the sediment through a combination of an advective and a diffusive flux across the wetted perimeter.
- Transformation is considered as the combined affect of hydrolysis, photolysis, and biodegradation, and it is a function of temperature, but these processes cannot be simulated separately. In practice, however, as degradation input is typically based on water–sediment studies conducted in the dark, the effect of photolysis is excluded. This has the effect of significantly underestimating the overall degradation potential of compounds that are subject to very rapid photolysis. Moreover, transformation is described without distinction between dissolved substance and substance sorbed to suspended solids. It is assumed that the transformation rate of the substance dissolved in the water phase equals the transformation rate of the substance sorbed to suspended solids. Metabolites are not directly considered in TOXSWA but can be represented by performing separate runs and adjusting parent application rates for maximum percentage formed and molecular weight changes. Some suggestions are provided (FOCUS 2001) on how to handle metabolites in TOXSWA according to whether the metabolite is formed in the soil study or the water–sediment study or the same metabolite is formed in the soil study as well as in the water–sediment studies. However, this approach is based on a number of approximations.
- Sorption: In TOXSWA it is assumed that sorption to suspended solids and sorption to sediment are analogous processes to sorption to soil, and that both can be described with the Freundlich equation. Therefore, the nonlinear Freundlich equation is used to describe sorption to suspended solids, but sorption is instantaneous, and time-dependant sorption to suspended solids is not incorporated. Although sorption to macrophytes is described in TOXSWA using a linear isotherm, this feature is not used in the TOXSWA used for the FOCUS surface water scenarios in the SWASH environment.

The system description assumes that the concentration of suspended solids in the water layer is constant. It has also been assumed that pesticides sorbed to suspended solids undergo the same advection and dispersion as dissolved pesticides.

In the sediment subsystem, the substance behavior is described by the following processes:

- Advection and dispersion in the vertical direction.
- Diffusion in the vertical direction.
- Exchange with the water layer through a combination of an advective and a dispersive flux across the wetted perimeter.
- Transformation, described as overall transformation without distinction between substance dissolved and substance sorbed. It is assumed that the transformation rate of the substance dissolved in pore water equals that of the substance sorbed to the solid phase of the sediment.
- Sorption to the solid phase of the sediment. The Freundlich equation is used to describe sorption to sediment, but sorption is instantaneous, and time-dependant sorption to the sediment matrix is not incorporated.

TOXSWA can handle various hydrological conditions simulating water bodies with varying water depths and rates of discharge. All 3 defined water bodies (pond, ditch, and stream) have an adjacent field that contributes drainage or runoff fluxes to the water body. In addition, the ditch and stream receive surface water from an upstream catchment that also contributes drainage or runoff fluxes to the water body, and realistic worst-case inputs from the upstream catchments have been identified.

The FOCUS ditch only occurs in FOCUS drainage scenarios where the land is relatively flat and, in most cases, relatively slowly drained. The ditch is assumed to be 100 m long and 1 m wide, with a rectangular cross section. Its minimum depth is 0.3 m, implying that in all ditches an outflow weir maintains this minimum water level even during periods of very low discharge. It receives drainage fluxes from a 1-ha field adjacent to the ditch and from a 2-ha upstream catchment. Pesticide solute is only present in drainage waters from the 1-ha field adjacent to the ditch. The upstream catchment basin is assumed not to be treated with pesticides; therefore, it is considered that only a third of the area considered in the ditch scenarios is treated with pesticide.

The FOCUS stream occurs in the FOCUS drainage scenarios as well as the FOCUS runoff scenarios. Similar to the FOCUS ditch, the stream is assumed to be 100 m long and 1 m wide, with a rectangular cross section. Its minimum depth is 0.3 m, implying that also in all streams a weir is located that maintains the 0.3-m water level even during periods of very low discharge. On 1 side of the stream, a 1-ha field is located that delivers its drainage or runoff fluxes into the stream. This field is assumed to be treated with pesticides. The stream is also fed by the discharge of an upstream catchment basin of 100 ha, which delivers its constant base flow plus variable drainage or runoff water fluxes to the stream. A surface area of 20% of the upstream catchment basin is assumed to be treated with pesticides, resulting in the dilution of edge-of-field drainage or runoff concentrations by an approximate factor of 5 before it enters the stream. The implications of pesticide contribution from the upstream catchment is simplistically represented via an increase in the drift loading in the TOXSWA input file by a factor of 1.2 (e.g., additional 20% loading).

Pond scenarios represent the simplest arrangement. Each 30×30 m pond receives drainage or runoff waters with associated pesticide in solution from a 4500-m² contributing catchment. No pesticide is present in the base flow that enters the pond. For runoff scenarios, the pond also receives eroded sediment and associated pesticide from a 20-m "corridor" adjacent to the pond.

12.8.5   HOW AND WHEN SHOULD MITIGATION BE REPRESENTED IN THE FOCUS SW FRAMEWORK?

Prepared by James Hingston

Mitigation in the FOCUSsw framework should be represented in a Step 4 assessment only. A Step 4 assessment may be triggered when the aquatic risk assessment fails the appropriate Annex VI triggers based on the exposure assessment from 1 or more of the standard Step 3 scenarios. Exposure assessments at FOCUSsw Steps 1, 2, and 3 should be based on standard default assumptions (e.g., using the generic chemical and fate input parameters, default spray drift input, and standard scenario parameterization at each step) and will be typically based on the likely worst-case GAP (good agricultural practice) proposed.

There is a wide range of options that could be suitable for inclusion in a refined Step 4 assessment. These could involve relatively simple approaches such as the consideration of an increased no-spray buffer zone to reduce inputs via spray drift. Such an approach is likely to be widely accepted at EU and national member-state levels. Additional mitigation of spray drift assuming the use of low-drift application technologies or the reduction of runoff via the use of vegetated filter strips is not well developed at the EU level at present, although these methods may be acceptable in individual member states on a case-by-case basis. Other options for refinement could involve the derivation of alternative refined chemical and fate input parameters based either on a reassessment of existing data or via the generation of additional data to address specific aspects of the fate and behavior of the substance. Finally, Step 4 could involve the development of new location- or region-specific exposure scenarios.

Whichever options for refinement are selected, the results should ideally be presented in a stepwise approach. For example, results of the standard assessment at Step 3 should be provided, followed by the various Step 4 refinement options in separate tiers. In this way, risk assessors can clearly see both the type and magnitude of mitigation required to achieve acceptable exposure concentrations. The level of detail that should be presented for the results of FOCUSsw exposure assessments at any tier will be partly dependent on the complexity of the effects data and the accompanying aquatic risk assessment. When standard first-tier effects data only are available, it may be appropriate simply to present the tabulated results of actual or TWA concentrations. If the effects data include higher-tier information (e.g., mesocosms, consideration of acute-to-chronic ratios, or consideration of recovery potential for example), it may be necessary to provide a representative selection of detailed graphical outputs of the exposure profiles in both water and sediment. This may be important on a case-by-case basis to allow more detailed comparison with the effects database.

Potential approaches to developing Step 4 assessments are discussed in Chapter 9 of the final report of the FOCUS Working Group on Surface Water Scenarios (FOCUS 2001) and the report of the FOCUS Working Group on Landscape and Mitigation Factors in Ecological Risk Assessment. The acceptability of the approaches proposed by earlier drafts of the latter report at the EU level was uncertain following the publication of an EFSA PPR Opinion ("Opinion of the Scientific Panel on Plant Protection Products and Their Residues on a Request from EFSA on the Final Report of the FOCUS Working Group on Landscape and Mitigation Factors in Ecological Risk Assessment" [Question No. EFSA-Q-2006-063] adopted on 13 December 2006 [EFSA 2006]). The final report (FOCUS 2007a, 2007b) was amended in light of this EFSA opinion and has now been noted by the commission's standing committee as guidance to be followed in regulatory submissions and assessments. Experience of applying the guidance as finally noted is still limited.

12.8.6 WHAT OPTIONS CAN BE CONSIDERED TO REFINE EXPOSURE ASSESSMENTS AT STEP 4 (EITHER CHANGING INPUT OR MORE THOROUGH USE OF OUTPUT)?

Prepared by Neil Mackay

At Step 4, it is possible to introduce a number of refinements to the exposure assessment to increase the realism of the risk assessment. At Step 4, these refinement strategies are not prescriptive or standardized. Instead, the strategy is customized to address the issue arising in Step 3. In many cases, there will be a clearly dominant route of entry, and risk refinement strategies may target this route of entry. Typical strategies that are then employed may involve the following:

- Introduction of risk mitigation measures to reduce the significance of the key route of entry (e.g., no-spray zones to address spray drift)
- Refinement of representation of environmental fate to increase the realism or sophistication of the simulations
- Development of a greater understanding of the agricultural and environmental context of exposure (e.g., aquatic proximity assessments considering detailed landscape analyses)

Each aspect is discussed next in brief.

12.8.6.1 Risk Mitigation

Three major routes of entry are represented at Step 3: drainage, runoff, and spray drift. Mitigation of the last route of entry has been well established in regulatory assessments on a member state basis pre-dating Directive 91/414/EEC. There are typically 2 major approaches that can be employed at step 4 to mitigate spray drift either independently or in conjunction: the use of no-spray zones or low-drift nozzle technology. A framework for representation of no-spray zones is provided within SWASH in the form of the FOCUS drift calculator. The algorithms presented there primarily draw on drift curves developed by Rautmann et al. (2001) but are supplemented by data from AgDRIFT 1.11 (SDTF 1999). It should be noted that alternative drift curves have been proposed and are incorporated into the national evaluation scheme in the Netherlands. While the use of no-spray zones is a well-established and readily accepted risk mitigation strategy, options to employ spray-reducing technology as a means of mitigating (or further mitigating) spray drift vary from member state to member state. For example, such schemes can be employed in Germany as a basis for label recommendations, but in the United Kingdom they can only be employed postregistration in local environmental risk assessment plans (LERAPs).

Simple options to mitigate drainage remain rather limited as it is difficult to introduce control measures at the field scale. Instead, other options that have the effect of usage restrictions can be considered on a case-by-case basis in certain member states. One option typically takes the form of restrictions in the timing of application to avoid the active drain flow period. Another possibility in certain regulatory schemes involves restrictions to entirely avoid circumstances (soils or fields) where drainage is practiced. The efficacy of the former option can be readily demonstrated through simulations. The potential to employ the latter option will largely depend on individual member state attitudes toward enforceability.

Like drainage, the assessment of runoff has been formalized through the introduction of the FOCUS surface water modeling scheme. The most common basis for mitigation of runoff has been the introduction of a runoff filter or VFS recommendation. There is a significant and growing body of data regarding the efficacy of such schemes. A very detailed discussion surrounding the formalized representation of the efficacy within step 4 calculations can be found in the report of the FOCUS Working Group on Landscape and Mitigation Factors in Aquatic Ecological Risk Assessment (FOCUS 2007a, 2007b). Note that further commentary can be found in the EFSA PPR review of this FOCUS report (EFSA 2006). For sake of brevity the findings are not reproduced here.

12.8.6.2   Refinement of Representation of Environmental Fate

Derivation of input parameters to support Step 3 calculations are based on guidance notes provided by FOCUS (2001, 2002). Where sufficient good-quality data exist, it is generally recommended to employ geometric mean normalized DegT50 data for soil and arithmetic mean representations of sorption. Additional guidance has recently been published on the derivation of kinetic representations of degradation in soil and a framework to interpret water–sediment studies to derive compartmental modeling endpoints (FOCUS 2006). Under certain circumstances, it may be possible to introduce refinements to simulations at Step 4 to more accurately represent the influence of specific processes. For example,

- Dependencies of behavior on soil characteristics (pH, clay content, etc.); introduction of a process not considered at Step 3 (e.g., aquatic photolysis)
- Aging effects on sorption potential (e.g., nonequilibrium sorption responses)
- Replacement of defaults with data to support customized representation of processes (e.g., wash-off potential)
- Consideration of formulation effects (e.g., wash-off potential, slow-release granular formulations)

Again, such options are not prescriptive in approach. Guidance on the implementation of such options is generally unavailable, and strategies must be developed to recognize the strengths and weaknesses of available data sets. It should be noted that in 1 case (nonequilibrium sorption) guidance is available in the form of the report of the FOCUS Working Group on Degradation Kinetics (FOCUS 2006).

12.8.6.3 Establishing a More Complete Agricultural and Environmental Context

The scenarios that are the basis for Step 3 simulations are necessarily constrained representations of an extremely diverse agricultural system in Europe. Opportunities may exist for refined representation of elements of the agricultural systems that are associated existing or proposed usage. These include

- More accurate representation of irrigation (particularly where chemi-irrigation is involved)
- More accurate representation of crop production cycles or strategies in particular usage environments
- More accurate representation of upstream contributions through analyses of land and chemical use on a catchment basis
- Development of alternative or additional scenarios for crops with limited representation at Step 3
- Establishment of context of exposure (e.g., aquatic proximity assessment) via landscape analyses
- Establishment of temporal context of exposure through development of longer-term simulations (supports event analysis, probabilistic risk assessment, etc.)
- Interpretation of monitoring data sets for existing products

In each case, the strategy employed must be customized to address the issue of concern through a process of problem formulation. The admissibility and appropriateness of individual techniques will be extremely variable depending on quality of supporting environmental fate data, the agricultural and environmental context of existing or proposed use, and policy decisions in individual member states.

12.8.7 In Tier 1 Risk Assessment, There Are Standard Endpoints, Species, and TER Thresholds; How Should These Be Interpreted?

Prepared by Gary Mitchell

This frequently asked question may be approached by considering definitions of 1) a tier 1 risk assessment, 2) a standard tier 1 endpoint, 3) tier 1 species, and 4) tier 1 thresholds (e.g., TER, toxicity–exposure ratio). A tier 1 risk assessment is a basic (unrefined) statement about risk or protection goals that is based on data from a series of standard clean-water laboratory toxicity tests (e.g., an invertebrate [*Daphnia*], fish, algae). A standard tier 1 endpoint is an ecotoxicological screening threshold level or "cutoff limit." Tier 1 species are standard, proscribed species that are considered sensitive or representative of species in the aquatic environment. Tier 1 endpoints (e.g., species LC/EC50 values) are deterministic point estimates. At tier 1, an acute trigger of TER = 100 is utilized to account for possible species variability.

Hence, a tier 1 TER value is an empirical "pass-or-fail" value derived from standard laboratory toxicity tests with proscribed test species as compared to a modeled environmental concentration with consideration of a few defined standard environmental fate characteristics. A tier 1 risk assessment is therefore a "worst-case" determinant for potential risk because it does not account for possible amelioration of toxicity or ecological exposure that may result from physical or chemical properties of the compound and its fate and behavior in the environment.

Interpretation of tier 1 endpoints, species, and TER threshold levels depends to some degree on the practitioner's understanding or definition of these terms. That is, there is no objective means of amending tier 1 TER thresholds to reflect a possible decrease in uncertainty when additional data are made available for the tier 1 assessment. As a consequence in practice, advancement to the next tier (tier 2) is the default requirement when a "fail" trigger (TER < 100) is met at tier 1. There is no other option. Similarly, in the event that a "pass" trigger (TER > 100) is met, then the compound is considered to be of no ecological concern, and no further action (e.g., additional toxicity testing) is required.

12.8.8   WHAT ARE SOME OF THE CRITERIA TO CONSIDER WHEN DECIDING THE MOST APPROPRIATE BASIS FOR INTERPRETING EXPOSURE PROFILES (E.G., MAXIMUM PEC VERSUS TWA PEC)?

Prepared by Steve Norman

Regarding the criteria to consider when deciding the most appropriate basis for interpreting exposure profiles, the first consideration would be whether the risk assessment is for acute or chronic effects.

12.8.8.1   Acute RAs

In general for acute risk assessment, the maximum PEC values should be used. For these assessments, it is generally not appropriate to use a TWA approach because this does not address the risk from the peak in the exposure. In some cases, potential exposure duration in the environment may be significantly shorter than occurs in a 48- or 96-hour continuous-exposure acute test (e.g., as a consequence of rapid hydrolysis or partitioning to sediment). In such cases, toxicity tests may be conducted that include relevant degradation or dissipation processes. In such cases, the study endpoints (in terms of peak concentrations) should be compared with the maximum PEC values (peak concentrations), and the profile of the concentration decline in the study should be comparable to the predicted exposure profile in the field.

12.8.8.2   Chronic RAs

Although relatively short in duration, standard algal growth inhibition studies should be regarded as chronic studies (because cell division, not mortality, is investigated). There is a reasonable argument to compare the endpoints from algal studies with a

TWA PEC because the endpoints from the studies (based on area under the curve and growth rate) are a product of an integrated exposure over the whole test period. Also, growth inhibition is often reversible (unlike mortality), which minimizes the likelihood that the risk is underestimated by not using the maximum PEC.

Note that when alga are being used to represent the risk to all aquatic plants, which will include species with slower growth or different recovery potential, use of a PEC maximum should be maintained (see Chapter 3, Section 3.4).

Often, the problem with chronic risk assessment is that there is a clear mismatch between the exposure duration in a study (typically 21 to 50 days) and the potential duration of significant exposure in the field (which may be markedly shorter due to degradation, partitioning, and flow). These loss processes may result in a shallow or steep decline in the actual PEC over time. Comparison of a maximum PEC (peak concentration) with the NOEC from a chronic study is clearly precautionary, but when this leads to a breach of a TER trigger, refinement of the exposure assumptions is needed. To determine whether a TWA PEC is relevant to the risk assessment, the results of the chronic study should be assessed with care. If the NOEC is based on effects at the LOEC (lowest observed effect concentration) that are observed early in the exposure period (e.g., effects on egg hatch in a fish ELS study), then the use of a TWA PEC may not be appropriate. In such cases, the effect resulted from short exposure duration, which may be more relevant to compare with a maximum PEC.

A TWA PEC is relevant where the effect observed (at the LOEC) resulted from an integration of exposure during a significant proportion of the test duration, such as growth or reproductive success. When using a TWA PEC, the time over which the exposure is integrated must be relevant to the exposure duration in the study that led to a particular effect. Care is needed if there is specific evidence that for a particular compound a short-term exposure may lead to a longer-term effect. However, this specific concern should not detract from the general usefulness of using TWA PEC values in risk assessment. The predicted exposure profiles in FOCUSsw Step 3 may be more complex than a peak followed by a simple decline (e.g., multiple peaks due to drain flow). This in turn complicates the judgment concerning whether the use of a TWA PEC is relevant in a particular case or not.

Note: See Chapter 3, Section 3.4, for further information.

12.8.8.3   Metabolites

Where degradation is a significant loss process supporting the derivation of the TWA PEC, degradates (metabolites) also need to be taken into account in the risk assessment.

12.8.9   MODELS ASSUME INSTANTANEOUS MIXING, BUT IN THE FIELD THERE WILL BE PATCHES WHERE ORGANISMS ARE EXPOSED TO HIGHER-THAN-PREDICTED CONCENTRATIONS AND PATCHES WHERE EXPOSURE IS LOWER THAN PREDICTED; HOW IS THIS ADDRESSED IN THE RISK ASSESSMENT?

Prepared by Gary Mitchell

Patches in the field where aquatic organisms may potentially be exposed to varying PECs are not explicitly addressed by the FOCUS surface water modeling[4] portion of the risk assessment process. However, in practice informed ideas or conclusions on ecological relevance may be drawn (e.g., through expert judgment) in a higher-tier refined risk assessment because during this process 1) maximum PECs are compared to relevant ecotoxicological endpoints, and 2) diverse water bodies and flow rates (e.g., lentic/lotic areas) are accounted for within the FOCUS scenarios (e.g., stream, river, ditch, pond).

Mesocosm studies are the current practice by which a discrete aquatic environment ("patch") may be exposed with controls to a series of theoretical (modeled) test concentrations, including modeled PECs.[5] Mesocosms permit long-term aquatic community studies that generally simulate worst-case field exposure scenarios. Such studies can be used to demonstrate ecological recovery or potential for ecological recovery.

Additional higher-tier evaluations or risk assessments may also include a thorough ecotoxicological or biological evaluation of potentially affected species (e.g., their life history, geographical occurrence, behavior, or likelihood of exposure under the proposed use patterns). For example, sensitive life stages may actively or passively migrate from regions of high xenobiotic concentration to regions of low xenobiotic concentration, or based on life history considerations or climatic factors, these life stages may not be present in the environment when compounds are applied according to GAP.

Any remaining uncertainty related to the possible impact of patches in the environment may alternatively be further addressed probabilistically by using a robust data set, if available. By this approach, the upper end of the tail in a modeled probabilistic curve addresses or represents a "higher-than-predicted" test concentration, and the lower end addresses or represents a "lower-than-predicted" test concentration that may potentially occur as a result of environmental patchiness.

NOTES

1. See http://randd.defra.gov.uk/Default.aspx?Menu=Menu&Module=More&Location=None&ProjectID=12312&FromSearch=Y&Publisher=1&SearchText=ps&SortString=ProjectCode&SortOrder=Asc&Paging=10#Description.
2. See http://randd.defra.gov.uk/Default.aspx?Menu=Menu&Module=More&Location=None&ProjectID=13365&FromSearch=Y&Publisher=1&SearchText=ps&SortString=ProjectCode&SortOrder=Asc&Paging=10#Description.
3. See http://randd.defra.gov.uk/Default.aspx?Menu=Menu&Module=More&Location=None&ProjectID=14889&FromSearch=Y&Publisher=1&SearchText=ps&SortString=ProjectCode&SortOrder=Asc&Paging=10#Description.
4. Of the numerous assumptions made in modeling, the following are expected to cause the maximum degree of inaccuracy in pesticide load estimation: steady-state condition; complete, instantaneous mixing. It is feasible that these 2 assumptions alone or in combination could cause the concentration in a water body to exceed the TER value while the annual loading estimated was not in fact exceeded. In contrast, other factors such as water body volume and timing (seasonality) of the loading events may allow an elevated amount of compound to enter a water body without exceeding the TER. Season-specific loading scenarios will vary due to seasonal differences in, for example, rainfall, the timing of application (winter or spring), and degradation (as related to temperature and sunlight).
5. Immediately after dosing, it is typical to mix the water body in an attempt to distribute the chemical evenly throughout the water column. However, it is likely that some degree of heterogeneity of exposure will remain.

13 Interaction between Fate and Effect Experts

Peter van Vliet, Paulien Adriaanse, Karin Howard, Chris Lythgo, Aiden Moody, Jo O'Leary-Quinn, and Paul Sweeney

CONTENTS

This work group report aims to provide a template for the exchange of aquatic information between ecotoxicologists and environmental fate specialists.

13.1   BACKGROUND

The EU workshop on Linking Aquatic Exposure and Effects in the Registration Procedure of Plant Protection Products (ELINK) was set up with the aim of improving guidance on linking exposure and effects in the aquatic risk assessment for pesticides using the Forum for the Co-ordination of Pesticide Fate Models and Their Use (FOCUS) (surface water). The workshop identified that a template for exchange of information between ecotoxicologists and fate specialists would be useful. The aim of this template is to increase communication and understanding between the 2 groups as well as the scientific quality of the aquatic risk assessment.

13.2   INFORMATION TO BE EXCHANGED

It should be noted that this is a prompt for the types of information to be exchanged. Obviously, it is difficult to produce a definitive list, and there is no substitute for good scientific thinking about the relevant issues. So, we see this as a useful starting position that can be further developed in the light of experience.

Information exchange is especially useful when higher-tier studies (e.g., mesocosm studies) are used. However, information exchange can also be important when concentrations are not maintained within the ecotoxicological study.

The template in Table 13.1 aids information exchange using a fictitious example for the active substance "dasha" and using a fictitious mesocosm to show the types of information needed.

TABLE 13.1
Template for Information Exchange

Information to Be Exchanged

It may be useful for both ecotoxicologists and fate specialists to look at the ecotoxicity study as well as to exchange the following information:

Parameter	Example
1) Details of the good agricultural practice for the pesticide	For example, the pesticide (active substance: dasha) is a fungicide. Details of applications are Crop (field use): Winter wheat Number of applications: 1 Maximum individual dose (kg a.s. (active substance) /ha): 1.0 Maximum total dose (kg a.s./ha): 1.0 Growth stage/month or season: BBCH (Biologische Bundes-anstalt, Bundessortenamt and Chemical industry) 31/May
2) Physical and chemical properties	
Molecular weight	200 g/mol
Saturated vapor pressure	4 * E-04 Pa
Solubility in water	0.2 g/L
3) Fate endpoints	
Active substance: dasha	
Geometric mean lab	
DegT50 soil	15 days at 20 °C and pF (reference moisture condition [field capacity]) 2
kinetics	SFO (single first order)
geometric mean field DT50 soil	10 days
Kinetics:	SFO
Mean Koc	900 L/kg (median $n = 9$)
DT50 water sediment system:	For example, whole system DT50 in the water sediment study was 60 days. Dissipation DT50 from water phase was 4 days.
Details of DT50 values plus other relevant comments	The principle route of dissipation from the water phase was via partitioning to sediment, where it was slowly degraded.
FOCUS input values used	DegT50 water at 20 °C: 1000 days (default worst case) DegT50 sediment at 20 °C: 60 days

FOCUS details
    For example, global maximum PECs
    general information on exposure profiles
	Details of the global maximum predicted environmental concentrations (PECs) for the different pond, stream, and ditch scenarios (e.g., in tables). Highlight key values as appropriate.
	Provide general information on profiles (more detailed information can be provided as needed).
	These are as routinely generated by FOCUS.
	The key routes of entry were considered to be as follows: indicate as appropriate (peak or other considerable concentration), for example:
	Drain flow: Yes
	Runoff: No
	Spray drift: Yes

TABLE 13.1 (CONTINUED)
Template for Information Exchange

Other, for example, Aquatic photolysis Biodegradability	For example, aquatic photolysis was not relevant in the environment. Dasha is not readily biodegradable.
4) Metabolites	For example, no metabolites were formed in the water sediment study. There were no significant soil metabolites entering surface water via runoff or drain flow.
Formation fraction	0.x
Molecular weight	x grams/mole
Geometric mean lab DegT50 soil Kinetics:	
Geometric mean field DT50 soil kinetics:	
mean Koc:	

DT50 water sediment
system:
  Details of DT50
  values plus other
  relevant comments	

FOCUS input values used	

FOCUS details
  For example, global
  maximum PECs
  General information
  on exposure profiles	

Other, for example,
  Aquatic photolysis
  Biodegradability	

Other useful information	
5) Ecotoxicity study information	Note that a similar approach could be used for experimental fate studies in microcosms or mesocosms.
Study type	For example, mesocosm study with 1 application of dasha (as formulation) made in May 2007.
Concentration of the active substance	For example, 5 treatment levels of 0.5, 1.5, 5, 15, and 50 µg a.s./L (spray treatment). Concentrations declined over time. Full details of concentrations over time provided from the mesocosm study report. Graphs describing concentrations with time could also be provided. Details of the size of the mesocosm and the water and sediment depth should be provided.

(continued)

TABLE 13.1 (CONTINUED)
Template for Information Exchange

Sediment details, such as amount of organic carbon	Details of the sediment used should be provided in the mesocosm study report. Information on the amount of organic carbon may be of interest, and this value could be listed for fate here.
pH	pH in the mesocosm study was between pH 6.5 and 7.5 (see study for full details).
Temperature	Temperature was between 10 and 15 °C, with the average temperature 13 °C (see study for full details).

Other useful information, such as
    Light/dark conditions
    Latitude
    Season	

Study end point: The information indicated may be useful.	Relevant mesocosm study end point (e.g., no observed effect concentration/no observed ecologically adverse effect concentration [NOEC/NOEAEC]). Appropriate uncertainty factor. For NOEAEC: Class of effects (see Brock et al. 2000) and duration, if possible specify specific recovery period (e.g., 10 days). Most sensitive organisms affected plus type of effect (e.g. mortality). Duration of exposure to the active substance (e.g., this is described in a graph). Anything else of importance?
ERC[a] (ecotoxicologically relevant concentration)?	Information on the type of concentration, the spatial and temporal scale. See also more detailed information in note[a]
Time to effect?	If available.
RAC (regulatory acceptable concentration)?	Single value or curve?
Other useful information	For example, aquatic macrophyte plants were present in the mesocosm study but were less than 2%.

[a] This is the type of field concentration that is needed as the exposure input for use in the effect tiers. The choice of ERC should be based on ecotoxicological considerations that give the best correlation with the ecotoxicological effects observed. Sources of information that can be used in defining the type of ERC are as follows:

Data on the mode of action of the pesticide.

Which environmental compartment does the organism live in (e.g., water or sediment)?

What is bioavailable to the organism?

Acute and chronic toxicity to standard test species/taxa.

Time to event information indicating the influence of the exposure pattern (e.g., short periods or constant concentration over long periods).

Was the whole test duration necessary to give the measured effect, or would a shorter exposure period have given the same effect?

Identification of the most sensitive life stages as assessed from chronic standard tests.

Further information on ERCs is available in Boesten et al. (2007).

14 Extrapolation Methods in Aquatic Effect Assessment of Time-Variable Exposures to Pesticides

Udo Hommen, Roman Ashauer, Paul van den Brink, Thierry Caquet, Virginie Ducrot, Laurent Lagadic, and Toni Ratte

CONTENTS

This work group report aims to provide an overview of extrapolation methods that can be used to assess ecotoxicological effects of time-variable exposures.

14.1   INTRODUCTION

In the past, ecotoxicological studies on the effects of plant protection products were mostly conducted under (quasi) constant exposure conditions (e.g., semistatic tests) or pulsed exposure simulating drift entries (e.g., in mesocosm studies). Now, refined exposure models like FOCUS (Forum for the Co-ordination of Pesticide Fate Models and Their Use) Steps 3 or 4 simulations (FOCUS 2001, 2007a, 2007b) predict complex exposure scenarios due to the combination of variable drift, run-off, and drainage entries in different types of ecosystems (ditch, stream, pond). Thus, the question arises how the results of the typical standard and higher-tier tests can be used for risk assessment considering this diversity of possible exposure scenarios. Therefore, the objective of the ELINK work group for extrapolation methods was to identify and evaluate tools to extrapolate effects between exposure scenarios.

Different types of extrapolation might be necessary to extrapolate from effects observed in ecotoxicological tests to effects in the field due to the use of a plant protection product:

1) Extrapolation between exposure patterns: This type of extrapolation from exposure profile used in the test to the exposure profile encountered in the field situation within the biological level of organization is always needed (e.g., from an acute fish laboratory test to survival of fish exposed in the field or from a community tested in a mesocosm to the community in the field). This extrapolation is the main focus of this report.

2) Extrapolation from one biological level of organization to another: Many ecotoxicological tests are designed to measure organism-level endpoints like survival, growth, or reproduction; while the protection aims are related to populations, community structure, and ecosystem function. The combination of physiological and ecological modeling might be applied to extrapolate between hierarchical levels. This type of extrapolation is not the subject of ELINK, but ecological models might also be used

for extrapolation between exposure scenarios on higher levels than the organism.

3) Other extrapolations (e.g., species-to-species extrapolation) are not the focus of this work group and are not discussed here.

Regarding the type of substances, the focus is on organic plant protection products. Other chemical stressors like metals are not considered.

During the first ELINK workshop (Bari, Italy, March 2007), 4 case studies were presented that exemplified typical situations for which extrapolation between exposure patterns is needed. The ELINKmethrin case study, for example (see Appendix 3), represents a typical data set of a rapidly dissipating substance resulting in extrapolation problems from standard acute and chronic tests to field situation with up to 5 applications. Similar problems arise when extrapolating from a species sensitivity distribution (SSD) based on 48 survival data or from a mesocosm study with 3 weekly applications to the 5 sharp exposure peaks that can be expected in the field. In general, the group identified the following 4 test endpoints at which extrapolation to other exposure patterns is most often required:

- Survival of fish and invertebrates (acute toxicity tests)
- Growth and reproduction of fish and invertebrates (chronic tests)
- (Population) growth of algae or macrophytes (algae or *Lemna* growth inhibition tests)
- Effects on and recovery of populations (algae, macrophytes, invertebrates) in micro- or mesocosm studies

14.2   OVERVIEW OF AVAILABLE TOOLS

Extrapolating pesticide effects on an organism from 1 (tested) exposure scenario to another can be done in 2 ways: using descriptive models in which effects are related directly to external concentrations or using mechanistic, process-oriented models that describe explicitly toxicokinetics (TK; uptake and elimination) and toxicodynamics (TD; damage and recovery in the organism) to predict the toxic effects on the organism level. In most cases in ecotoxicology, we do not want to stop on the level of the organism since our interest is on populations, communities, or ecosystems. Thus, if the objective is also to extrapolate on a higher biological level than the organism, population, food web, or ecosystem models might be necessary.

Therefore, 5 main types of models for extrapolation can be differentiated:

- Direct-link models that relate directly the effect to the external concentration, such as by comparing time-weighted averages (TWAs) of predicted exposure and toxicity experiments (e.g., no observed effect concentration [NOEC] expressed as TWA)
- Toxicokinetics and toxicodynamics (TK/TD) models that consider TK and TD as separate steps to mechanistically explain and predict effects at the organism level

- Population models that describe the population dynamics of 1 species depending on the exposure to a toxicant
- Food web and ecosystem models that describe dynamics of a set of populations linked via predator–prey or competition relationships and — in ecosystem models — influenced by abiotic factors
- Empirical models that use observed responses for different exposure patterns to estimate effects of untested patterns by statistical or information theory-based methods

These main groups are described in the following sections with the focus on the general principle and how exposure is related to effects. For more details, references to reviews or publications on specific model applications are given. In Section 14.3, the tools are evaluated with respect to applicability, data requirements in addition to the standard tests, and benefits of the generated output. For some of the tools, application examples are provided in Section 14.4.

14.2.1   Direct-Link Toxicity Models

The direct-link toxicity models are descriptive models that do not differentiate between TK and TD. They provide a direct link between exposure and effects; thus, they are called "direct-link models" (Ashauer et al. 2006a). These models were the first attempts not only to describe toxicity as a function of toxicant concentration in the medium but also to consider the duration of exposure.

In the review of Ashauer et al. (2006b), some of these models are described and discussed in more detail: Haber's law, power term models, exponential mortality based on logistic function, the probit plane model, survival analysis, and cumulative episodic exposure. Maybe the best-known 1-step model is "Haber's law," which assumes that toxicity depends on the product of concentration and time. This is the basis for the use of the TWA, for which exposure concentration is integrated over time and then divided by the duration of the test. Thus, different exposure patterns but with the same TWA are assumed to have the same effects. The TWA is often used in pesticide risk assessment to account for time-varying exposure in the evaluation of toxicity tests (e.g., in semistatic *Daphnia* reproduction tests, Organisation for Economic Co-operation and Development [OECD] Guideline 211, European and Mediterranean Plant Protection Organization [EPPO] 2003a, 2003b). A more detailed analysis of TWA can be found in the work of Ashauer et al. (2007a, 2007b, 2007c). Criteria for when the TWA approach can be used in the aquatic risk assessment of pesticides can be found in Section 3.4 of Chapter 3.

Boesten et al. (2007) presented a tiered approach for linking exposure and effects that also uses the TWA concept. The basic of this "ERC approach" is the definition of the ecotoxicologically relevant concentration (ERC) for the given problem. This ERC is a type of concentration that gives the best estimation of the risk: It should consider the environmental compartment, the mode of action and bioavailability of the toxicant, the exposure pattern, and the test duration necessary to cause an effect. An example for ERCs might be maximum PECs (predicted environmental concentrations) or TWAs in the water column, in pore water of the upper layer of the sediment, or in the sediment itself. Thus, complex exposure dynamics are simplified to some characteristics like peaks, curves, or TWAs.

Boesten et al. (2007) suggested a tiered approach for the use of the ERC in risk assessment for time-varying concentrations:

1) Comparing single RAC (regulatory acceptable concentration*) and PECs (i.e., minimum RAC of the relevant time window compared to maximum PEC).
2) Comparing graphically RAC and PEC curves (describing their time courses): If the RAC curve is always above the PEC curve, then there is no consequential acceptable low risk.
3) Comparing corresponding TWAs of RAC and PEC (i.e., using the TER approach).

14.2.2 TOXICOKINETICS/TOXICODYNAMICS MODELS

The TK/TD models describe the processes that link exposure to effects in an organism. In this report, the terms are used with the following meaning:

- The TK describes the fate of a toxicant in an organism depending on the processes of absorption (uptake, depending on the external concentration), distribution within the organism, metabolism, and elimination (ADME).
- The TD describes how the internal concentration of a toxicant affects the organism considering injury and recovery as well as their link to the effect endpoint in the organism (i.e., survival).

The predicted endpoint of the TK submodel is the concentration at the target site. The simplest description of the TK is the 1-compartment first-order kinetics model, which is also the most commonly used in aquatic ecotoxicology (e.g., Moriarty 1975; Kooijman et al. 2004). It describes the time course of the internal (whole-body) concentration of the toxicant C_{int} depending on the external concentration C_{ext} using uptake and elimination rate constants (k_{in}, k_{out}), and the external concentration (Equation 14.1).

$$dC_{int}/dt = k_{in} \times C_{ext} - k_{out} * C_{int} \tag{14.1}$$

Uptake and elimination rates are measurable in experiments, and estimation methods are available (e.g., in bioconcentration studies, references in Ashauer et al. 2006a). Since the TK describes the time course of the toxicant concentration in the organism, this approach can also be used to simulate whether repeated pulsed exposures lead to a buildup of internal concentrations or whether the time between pulses

---

* The RAC is the measured endpoint in an assessment, e.g., the EC50 or NOEC from a single species test, divided by the appropriate assessment factor, e.g., 10 or 100. The RAC can also be derived from a higher tier study, dividing the NOEC or NOEAEC (no observed ecologically adverse effect concentration) of a mesocosm study by an assessment factor considering the remaining uncertainty. Examples are given in Brock et al. (2006) and Boesten et al. (2007).

is long enough for complete elimination (depuration) of the compound. The time to eliminate 95% of the compound is approximately $t_{95} = 3/k_{out}$. This depuration time is informative for risk assessment of fluctuating or pulsed exposures and provides the basis for any effect simulation in TK/TD models.

A great amount of work has also been done on 2- and 3-compartment models of pharmacokinetics of insecticides in insects (references in Lagadic et al. 1993, 1994, and more recently in Greenwood et al. 2007). These approaches used for developing and optimizing insecticides should be further explored in the context of ecotoxicological risk assessment and for invertebrates in general. Besides simulating the concentration at the target site more precisely, more complex TK might have their benefits in species-to-species extrapolation. Multicompartment TK are also called physiologically based pharmacokinetic (PBPK) models.

The time course of effects is described by the TD. All models described here base their TD on the TK. Effects may prolong even if the toxicant is "completely" depurated (Ashauer et al. 2007a, 2007b, 2007c). Completely on the scale of the whole organism is usually approximated as 95% or 99% depuration; thus, the lingering effect on the scale of the whole organism may be caused by the remaining toxicant molecules, physiological damage, or both.

The models described in this chapter are hazard models, that is, they assume that death, although depending on the toxicant concentration in the organism or the damage, is at least partly stochastic. Hence, the hazard rate $h(t)$ is the probability of death at a given point in time. The integral of the hazard rate, the cumulative hazard $H(t)$, appears in the equation that defines the survival probability $S(t)$, which is the probability that an organism survives until time t.

$$S(t) = \exp[-H(t)] \tag{14.2}$$

A more detailed review was given by Ashauer et al. (2006a), who used the term "2-step models" according to Lee et al. (2002) because TK and TD are considered as separate steps in TK/TD models compared to the direct-link models.

14.2.2.1 Effect (Hazard Rate) Is Proportional to Internal Concentration

The review of Ashauer et al. (2006a) referred to these models as critical body residue (CBR) models. The simple hazard model (SHM) represents the hazard modeling approach of this concept. It is the simplest TK/TD model because it directly relates the hazard rate to the internal concentration $C_{int}(t)$:

$$h(t) = k \times C_{int}(t) \tag{14.3}$$

Assuming that the effect is proportional to internal concentration means that the time course of effect will mimic the time course of the body residue. That implies instantaneous and complete recovery of the organism; that is, any reduction in the internal concentration immediately translates in a proportional reduction of the effect (i.e., hazard rate). If there were anything less than instantaneous and complete recovery of the organism, then there would be a delay in the expression of the effect. Furthermore, this model assumes that there is no threshold (i.e., even the smallest amount of body residue causes some effect). Since this model assumes instantaneous and complete recovery of the organism, it should only be used for compounds with a reversible mode of action, such as baseline toxicity or uncoupling of oxidative phosphorylation.

The simple hazard model has been used to describe effects of time-varying exposure by Widianarko and Van Straalen (1996) and Jagers op Akkerhuis et al. (1999).

14.2.2.2 Effect (Hazard Rate) Is Proportional to Internal Concentration above a Threshold

In the review of Ashauer et al. (2006a), this model is referred to as the threshold hazard model (THM). The hazard rate is proportional only to the amount of toxicant above a certain internal concentration:

$$h(t) = k \times \max[C_{int}(t) - C_{threshold}, 0] \tag{14.4}$$

Like the SHM, this model also assumes that the time course of effects will mimic the time course of the internal concentration above a threshold, implying instantaneous and complete recovery of the fitness of the organism. Accordingly, instantaneous and complete recovery of the fitness of the organism means that this model should only be used for compounds with an instantly and completely reversible mode of action, such as baseline toxicity or uncoupling of oxidative phosphorylation.

The DEBtox (dynamic energy budget) model (Kooijman and Bedaux 1996) is based on the same TD assumptions as the THM, but it differs from the THM in 2 main points:

- The TK are expressed in units of external concentrations, which has the advantage that only 1 TK parameter, the "elimination rate constant," needs to be estimated. Thus, the NEC (no effect concentration) in the original DEBtox (Kooijman and Bedaux 1996) is an external concentration. The other estimated parameters in the original DEBtox model are the elimination rate constant, a killing rate that links the difference between NEC and actual concentration [$C_{int}(t)$] to the hazard rate $h(t)$, and a background hazard, which accounts for effects that may occur in the control.
- The DEBtox model is integrated within the DEB (dynamic energy budgets) theory. This means that there are additional and different relationships between internal concentrations and effects on energy-related endpoints. These relationships can be used to attribute toxic effects to different energy allocation-related processes within the organism (activity, maintenance, growth, sexual maturation, reproduction). These relationships may be different (also mathematically) from the relationship for survival but could prove very useful for sublethal effects assessment. Effects on growth and fertility are assessed under a similar hypothesis, plus supplementary assumptions for the type of physiological perturbation encountered at the individual level (e.g., effects on fertility might be due to mortality of eggs during oogenesis, a decrease in assimilation of energy, an increase of maintenance costs for the defense against the toxicant, an increase of the costs

of growth that competes with fertility, or an increase of the costs of reproduction). This results in a set of 3 and 5 different equations for growth and
fertility, respectively (these equations are not given here; see Kooijman and
Bedaux 1996). Note that these effect scenarios are assumed to be independent and exclusive, which might not always be true.

There are many modifications and extensions of the standard DEBtox model. Two
of them are

- Péry et al. (2001) described a DEBtox modification to account for time-varying exposure in toxicity tests. This modification allows taking into account
  the decrease of the external concentration of toxicant, due to (a)biotic degradation, bioaccumulation, or metabolism. It might be useful to compare degradation kinetics from 1 exposure condition to the other, providing exposure
  consists of a single pulse. Yet, this opportunity remains to be tested.
- The standard DEBtox assumes constant environmental conditions, including constant exposure concentration, to ease the estimation of model parameters, which might be a limit for its application in the field. Pieters et al.
  (2006) applied a modified DEBtox to experiments with *Daphnia magna*
  exposed to a single pulse of fenvalerate under varying food conditions. This
  helps to account for the impact of the trophic state of the receiving water
  body on the kinetics and effects of the tested pesticide; this might be relevant when extrapolating between exposure scenarios.

14.2.2.3 Effect (Hazard Rate) Is Proportional to Damage

The idea that effects are proportional to damage is based on the models by Ankley
et al. (1995) and Lee et al. (2002). The damage assessment model (DAM) by Lee et al.
makes no a priori assumption about the reversibility of the fitness of the organisms:
Repair and recovery processes are explicitly modeled. In the DAM, the cumulative
hazard $H(t)$ is proportional to the damage $D(t)$, which leads to increasing survival
probabilities for fluctuating or pulsed exposure. By definition, survival probabilities
(probability to survive until time t) cannot increase; hence, the original DAM is not
applicable to fluctuating or pulsed exposures.

Using a similar idea as that behind the DAM (i.e., that the effect is proportional
to "damage"), we can formulate a model as follows:

$$dD/dt = k_k \times C_{int}(t) - k_r \times D(t) \tag{14.5}$$

where k_k is a killing rate constant, k_r is the repair/recovery rate constant, and $D(t)$ is the
damage at any point in time. The damage describes the reduced fitness of the organism.
Because of its generic definition, it is applicable to various mechanisms of action. To
avoid confusion with the original DAM, this model is called the damage hazard model
(DHM). The link to survival is made by the standard approach used in hazard models:

$$h(t) = k_p \times D(t) \tag{14.6}$$

$$S(t) = \exp(-H(t)) \tag{14.7}$$

where k_p is a proportionality constant, $h(t)$ is the hazard rate, and $S(t)$ is the survival probability (probability that an organism survives until time t). This model permits the time course of effects to differ from the time course of the internal concentration; that is, the fitness of the organisms may still be reduced (= damage increased) after most of the toxicant is depurated. Furthermore, this 2nd kinetic step can explain delayed toxic effects. Yet, this model still assumes that even small amounts of toxicants will have some effect (i.e., the model contains no threshold). This model has not been used yet.

14.2.2.4 Effect (Hazard Rate) Is Proportional to Damage above a Threshold

Based on the DHM and the threshold concept from the THM (or DEBtox), a new, combined model has been developed by Ashauer et al. (2007a) specifically for the simulation of effects from fluctuating or pulsed exposure to pesticides (threshold damage model, TDM).

The TDM also uses Equations 14.1, 14.2, and 14.5 but differs in the link between the hazard rate $h(t)$ and damage in that it includes a threshold. For clarity, the whole TDM is given here:

$$dC_{int}/dt = k_{in} \times C_{ext}(t) - k_{out} \times C_{int}(t) \tag{14.1}$$

$$dD/dt = k_k \times C_{int}(t) - k_r \times D(t) \tag{14.5}$$

$$h(t) = \max[D(t) - C_{threshold}, 0] \tag{14.8}$$

$$S(t) = \exp(-H(t)) \tag{14.2}$$

where k_k is a killing rate constant, and k_r is the recovery rate constant. The damage in the TDM is defined as the reduction of the fitness of the organism.

The TK model parameters were derived from measured internal concentrations in uptake–elimination experiments (Ashauer et al. 2006b), and the TD parameters were derived from toxicity tests with sequential pulsed exposure. After calibrating the model, it was tested by extrapolating survival in additional toxicity tests with fluctuating and sequential exposure to 2 pesticides with differing mechanisms of action, pentachlorophenol and chlorpyrifos (Ashauer et al. 2007b). Later, the model was also parameterized and tested for carbaryl (Ashauer et al. 2007c) and extended for mixtures and sequential exposure to multiple compounds (Ashauer et al. 2007a). The test organism in these studies was the freshwater invertebrate *Gammarus pulex*.

Similar to calculating total depuration times as the time required to eliminate 95% of the toxicant, it is also possible to calculate total recovery times for the organisms. Because TK (elimination) and TD (organism recovery) occur in parallel, the calculation of total recovery times requires running the model for the exposure under question. For example, following exposure to a 1-day pulse that kills 50% of the population, the calculated total recovery times of *G. pulex* are 3, 15, and 25 days for pentachlorophenol, carbaryl, and chlorpyrifos, respectively. The total recovery time could be used to define the length of the exposure concentration profile that has to be considered as a single exposure event, that is, that window within which effects from subsequent exposures are not independent of each other.

The TDM is mathematically very similar to the receptor kinetics model (RKM) proposed by Jager and Kooijman (2005). Even though this model was proposed by members of the DEBtox group, it is a different model from the standard DEBtox model. The receptor model accounts for the possible reversibility of effects of organophosphorous pesticides on acetylcholine (AChE) in guppies, fathead minnows, and springtails. It differs from the TDM in the assumption whether saturation of target sites will affect the toxicant–target site interaction or not (receptor model: yes; TDM: no). Instead of damage, the receptor model describes the time course of the fraction of occupied target sites. The effect is then proportional to the fraction of inhibited target sites that exceeds a certain critical level (i.e., a threshold). As long as the fraction of occupied target sites is very small, so that it can be neglected, the TD of the receptor model and the TDM should be very similar.

$$dF/dt = k_{kf} \times C_{int}(t) - k_{rf} \times F(t) - k_{kf} \times C_{int}(t) \times F(t) \tag{14.9}$$

The 2 first terms on the right-hand side of Equation 14.8 are identical to Equation 14.5 with the difference that k_{kf} and k_{rf} are specific rate constants for the occupation and recovery of target sites, respectively. The third term accounts for those receptors that are already occupied. The link to survival is then made similarly to Equation 14.8:

$$h(t) = k_f \times \max[F(t) - C_{threshold}, 0] \tag{14.10}$$

where k_f is a proportionality constant.

Growth is taken into account in the RKM but not in the TDM. This might be of importance when modeling effects of repeated pulses over a period that is long enough to allow the chosen test species to grow significantly (and thus modify the number of available targets). But, this implies that survival data alone would generally not be sufficient to accurately determine the value of the parameters in the case of the RKM (see Jager and Kooijman 2005).

14.2.2.5 Summary of the Relationships between the TK/TD Models

So far, TK/TD models do not consider distribution and metabolism of the toxicant within the organism. Thus, the description of the TK is usually restricted to the processes of uptake and elimination only, and the models differ mainly in their assumptions on the TD (see Table 14.1): Is the hazard proportional to internal concentration or to the damage? Is there a threshold for the effect? How are repair and recovery processes considered? Consequently, the TD concepts differ in the range of toxic mechanisms for which they are valid. Hence, the explicit statement of the underlying assumptions, as it is attempted here, may aid in the selection of the appropriate model for a compound with a given mode of action.

TABLE 14.1
Overview of TK/TD Models

Model	Toxicokinetics: Uptake and Elimination	Hazard Proportional to CI	Hazard Proportional to Damage	Threshold Concentration for Effect	Target Interaction/ Reversibility of Binding	Recovery/Repair of Effects
SHM (simple hazard model)	x	x	—	—	—	Instantaneous and complete
DEBtox (dynamic energy budget)	TK related to C_{ext}, only elimination considered	x	—	x	—	Instantaneous and complete
THM (threshold hazard model)	x	x	—	x	—	Instantaneous and complete
DAM (damage assessment model)	x	—	x	—	—	x
DHM (damage hazard model)	x	—	x	—	—	x
TDM (threshold damage model)	x	—	x	x	—	x
RKM (receptor kinetics model)	x	—	x	x	x	x

14.2.3 Population Models

For the risk assessment of pesticides, population models allow extrapolating from organism-level effects to population-level effects. They can be used, for example, to extrapolate from inhibition of reproduction to effects on population growth and size. In addition, since recovery of a population within a certain time frame might be acceptable in the regulatory context, population models are the only tools available for extrapolation (Forbes et al. 2009).

There are many examples of models focusing on the dynamics of single populations in theoretical ecology, conservation biology, population management, and ecotoxicology. Reviews on population models in ecological risk assessment can be found, for example, in the work of Pastorok et al. (2001) and Munns et al. (2007). In population models for ecological risk assessment of chemicals, the exposure has often been simplified to constant exposure (by changing some model parameters, e.g., reducing the reproduction rate) or by 1 pulse exposure (by reducing the abundance at 1 point in time).

In only a few cases, population models are coupled directly to TK/TD models (e.g. Billoir et al. 2007, Ducrot et al. 2007). In these studies, population outputs were linked to dynamic individual effect models (DEBtox), but external concentrations of pollutants (metals) were treated as constant (i.e., simplified exposure scenario). Coupling TK/TD models to account for dynamic exposure scenarios (e.g., TDM) to population models should permit accounting for both time-varying exposure concentration and effects.

Population models are usually differentiated by the type of the modeled entity and the consideration of spatial variability. Here, we follow the terminology of the SETAC (Society of Environmental Toxicology and Chemistry) workshop on population-level risk assessment (Munns et al. 2008): unstructured models, biologically structured models, individual-based models (IBMs), metapopulation models, and spatially explicit models.

14.2.3.1 Unstructured (Scalar) Models

The total abundance of the population is modeled here as a single number using a differential or difference equation model. Thus, the models have only 1 state variable describing the population, the total population abundance. A classical example is the logistic model (Verhulst equation; Verhulst 1845), in which the change of the population size depends on the actual abundance, the population growth rate, and the carrying capacity:

$$dN/dt = Nr\,(1 - N/K) \tag{14.10}$$

A deterministic analysis of population dynamics accounting for toxicant effects with unstructured models was first proposed by Hallam et al. (1983). Barnthouse (2004) used this type of model to analyze recovery times depending on the magnitude of effects and population growth rates.

Unstructured models may incorporate factors such as demographic and environmental stochasticity, which are among the main source of variation in population dynamics. Demographic stochasticity is a significant factor in small populations, while environmental stochasticity is important for both small and large populations (Lande 1993). In Section 14.4.3, an example is given of how to use this simple model to calculate effects and recovery for different exposure scenarios.

The logistic model can also be expanded to consider the dependency of the population growth rates from variable environmental factors like temperature, light, or nutrient or food levels. Yet, unstructured population models are seldom used in an ecotoxicological context except when several models of this type are combined to an ecosystem model.

14.2.3.2 Biologically Structured Models

Biologically structured models are relevant in ecotoxicology because they can account for sensitivity differences in classes of individuals depending on their age, size, or development stage when assessing effects of toxicants on population dynamics. Effects on individual fecundity, growth, and survival are integrated in a single endpoint at the population level. Among possible population-level endpoints, the population growth rate λ is generally used in ecotoxicological studies (see Forbes and Calow 1999).

Various methods have been used to assess effects on toxicants on λ, with most common approaches based on Euler–Lotka and Leslie models (Leslie 1945, 1948). Both models share a common hypothesis, mainly a constant environment with unlimited resources, leading to the exponential growth and stable age structure of the population. They allow the calculation of the value of λ based on age-specific survival and fertility rates. They mainly differ in their mathematical formulations (linear vs. matricial algebra) and type of criteria for the classification of individuals.

An overview of matrix models was given by Caswell (2001). Matrix models allow the classification of individuals based not only on their age (as in the Euler–Lotka model) but also on other biological characteristics such as size, physiological state, development stage, and so on that may be relevant from an ecotoxicological point of view. Therefore, response to stress for the various life stages (or size class, etc.) can be directly incorporated into the model projections of the value of λ.

Whatever the chosen model, a minimal required data set consists of class-specific survival and fertility rates. Standard chronic toxicity or life-cycle test data (e.g., the *Daphnia* reproduction test) are sufficient to implement Euler–Lotka models and 2-stage matrix models (i.e., models that only distinguish juveniles and adults). Complementary data (i.e., life-cycle trait values, density dependence data) are mandatory to feed more complex demographic models. Appropriate data can be obtained by

1) Experimenting. For instance, life table response experiments (LTREs) were designed specifically to feed matrix models (Caswell 2001).
2) Modeling. TK/TD models are useful in this context. See, for instance, the work of Billoir et al. (2007) and Ducrot et al. (2007), who coupled DEBtox models for individual effects (modeled for various toxicants and invertebrate species) with matrix models.

Elasticity analysis allows testing the sensitivity of population growth rate to the variation of each of the life history parameter values (De Kroon et al. 1986, 2000) by measuring the change in the value of λ that occurs when the value of 1 of the life-cycle traits is modified (in a given range) by the modeler. Elasticity analyses can be used to identify the stages of the life history that should be the focus of management efforts or conversely to identify life history strategies that could be most vulnerable to a given stressor for a given species (Gleason and Nacci 2001).

Strong hypotheses of demographic models, such as constant environment or exponential growth of the population, can be relaxed. This is of particular interest when extrapolating effects from 1 population to the other (for the same species and toxicant). More complex demographic models (and corresponding software) are available to consider density dependence, environmental or population stochasticity, and toxic effects (e.g., Spencer and Ferson 1998). See, for instance, the work of Grant (1998), who incorporated various types of density dependence into a basic Leslie model using data on the toxicity of dieldrin to the copepod *Eurytemora affinis*.

14.2.3.3 Individual-Based Models

In individual- (or agent-) based models, each organism instead of the whole population or groups of similar individuals is modeled separately. This approach can result in a very high number of state variables because the state of the model is defined by the properties of all the modeled individuals. On the other hand, the model itself might be not very complicated. The first IBMs were developed in the 1970s (e.g., Kaiser 1976, 1979) and became very popular in ecology in the 1990s (DeAngelis and Gross 1992; Grimm 1999). Their advantage is that in principle every aspect of the biology, including individual behavior, can be used in the model. In addition, IBMs are per se stochastic models (otherwise, the approach makes no sense), which allow us to model effects of individual differences on population-level endpoints such as extinction.

The IBMs have often been used for the analysis of general ecological questions and for population viability analysis. Until now, applications of IBMs to analyze or predict chemical effects seem to be rare (e.g., Hommen et al. 1993; Klok and de Roos 1996; Jaworska et al. 1997; Koh et al. 1997; Preuss et al. 2009) and to assume usually constant exposure. A recent example of a spatially explicit IBM to model effects of pesticides on aquatic waterlouse *Asellus aquaticus* is given in Section 14.4.3 (MASTEP [metapopulation model for assessing spatial and temporal effects of pesticides]; van den Brink et al. 2007).

14.2.3.4 Metapopulation Models

While the first 3 population model types focus on 1 population in which all organisms can be (at least theoretically) interacting with each other, metapopulation models divide a population into several patches linked only via dispersal. These patches might be real islands in the ocean or a lake, but they can also be, for example, ponds or small areas of a specific terrestrial habitat in a landscape. The first metapopulation models described the dynamic of the number of occupied patches (Levins 1969, 1970), but more detailed models account also for the abundance in each of the patches (e.g., Hanski 1999). Therefore, metapopulation models can be comprised of a set of scalar, structured, or individual models in which the populations are linked via immigration and emigration rates.

Sherratt and Jepson (1993) were among the first to propose the use of metapopulation models (either stochastic or deterministic) to simulate the long-term impact of pesticides on terrestrial invertebrates, but only from a theoretical point of view. Although this model was not specifically developed for aquatic organisms, there is evidence that aquatic arthropods may exist as metapopulations at the landscape scale (Jeffries 1994; Briers and Warren 2000); therefore, the use of such a model for the assessment of the effects of chemicals on these organisms seems feasible. Spromberg et al. (1998) developed a single-species metapopulation model that is discrete and deterministic and incorporates dose–response curves and biotic growth rates. They simulated the effects of both persistent toxicants (e.g., metals, radionuclides) using a constant exposure concentration and degradable compounds (e.g., organophosphates) using a constant degradation rate. However, they did not make any comparison with real data.

14.2.3.5 Spatially Explicit Models

The spatially explicit class of population models is not exclusive to classes 1 to 4. However, in most cases spatially explicit models are IBMs in which individuals move in a virtual landscape or metapopulation models in which the location and dimension of the patches and the migration of the organisms between patches is considered. A special kind of spatially explicit models is grid-based models. Here, the landscape is modeled like a chessboard, with each cell having specific abiotic and biotic properties (e.g., habitat quality and abundance). If the change of the cells' properties is based on rules instead of formulas, these models are usually called "cellular automata" (e.g., Balzter et al. 1998 and the references therein).

In Section 14.4.4, a spatially explicit IBM of *Asellus aquaticus* is presented (MASTEP, metapopulation model for assessing spatial and temporal effects of pesticides; van den Brink et al. 2007).

14.2.4 Food Web and Ecosystem Models

Historically, aquatic ecosystem models were first developed in the 1970s for lakes to simulate and predict eutrophication (e.g., Park et al. 1974; Thomann et al. 1975; Scavia 1980). Aquatic ecosystem modeling is frequently based on the implementation of compartment models that exchange fluxes of nutrient or energy. The compartments may correspond to a species or a trophic level or functional group ("trophospecies") or to nutrients. Alternatives to compartment models, such as artificial neural networks for quantitative simulation (Lek et al. 1996 and references therein) or loop analysis for qualitative analysis (Hulot et al. 2000; Justus 2006), exist but so far it seems that they have not been used for ecological risk assessment of chemicals.

Food webs are a useful framework to assess the magnitude and importance of trophic relationships in an ecosystem. Food web linkages ultimately determine the fate and flux of every population in an ecosystem, and food web characterization is usually required as an initial step in understanding an ecosystem. In this context, food web design is a central task because the assumptions made when identifying the elements that constitute the food web and the connection between these elements may have a great influence on the outcomes of the model. Therefore, handling of species aggregation in food web models is still a controversial issue (Allesina et al. 2005).

Food web models may be used to assess the exposure of various types of organisms to environmental contaminants, especially highly persistent and bioaccumulable organic compounds such as PCBs (polychlorinated biphenyls) or DDT. They usually forecast steady-state equilibrium concentrations based on the characteristics of the compounds (e.g., octanol–water partition coefficient) and of the organisms (e.g., lipid content).

Reviews of food web and ecosystem models used for ecological risk assessment may be found in Koelmans et al. (2001) and Pastorok et al. (2002).

Three examples of dynamic food web models applied to pesticides are CASM (comprehensive aquatic systems model), AQUATOX (simulation model for aquatic systems for performing ecological risk assessment), and C-COSM (a model of the fate of nutrients and pesticides and their effects on biota in microcosms). The CASM (Bartell et al. 1992, 1999, 2000; Naito et al. 2002) is a bioenergetic ecosystem model that simulates the daily production dynamics of populations, including predator–prey interactions, in relation to daily changes in light intensity, water temperature, and nutrient availability. The dynamics of each population is described by physiological parameters that control growth in relation to daily changes in light intensity, water temperature, available nutrients, respiration, feeding, and mortality. CASM incorporates a toxic effects submodel that extrapolates from the single-species toxicity data, such as LC50s (median lethal concentrations) and EC50s (median effective concentrations that affects 50% of the test organisms), to determine the potential effects of chemicals on biomass production in an aquatic ecosystem. The direct effect of a chemical on each model population is calculated by estimating the change in the bioenergetic rates for the expected water concentration of the chemical. Example risk calculations for 10 chemicals, including some pesticides, can be found in the work of Naito et al. (2003). In its original form, CASM assumes constant exposure conditions, but it has been expanded to deal with time-varying exposure (Naito et al. 2002). However, all the examples published so far relate to constant-exposure scenarios.

AQUATOX (Park and Clough 2004; Park et al. 2004; http://www.epa.gov/water science/models/aquatox/) is another bioenergetics model that simulates the dynamics of various abiotic (e.g., dissolved oxygen concentration) and biotic state variables through time in relation to changes in some driving variables (e.g., water temperature, light, nutrient, or toxicant loadings). By combining modeling modules with databases for properties of different chemicals, ecosystem types, and taxa, it addresses potential impacts of chemical stressors on phytoplankton, periphyton, submersed aquatic vegetation, zooplankton, zoobenthos, and fish populations in rivers, lakes, or ponds.

AQUATOX considers several fate processes (turbulent diffusion between epilimnion and hypolimnion; partitioning of a compound between water, sediment, and biota; microbial degradation; photolysis; hydrolysis; volatilization; uptake by algae and macrophytes; uptake across the gills of animals; biotransformation; and depuration by organisms). Therefore, exposure of organisms is described on a time-varying concentration basis. Acute and chronic toxic effects from the concentration of a toxicant in a given organism are assessed using the CBR approach. Thus, in principle AQUATOX is able to simulate effects resulting from variable exposure.

C-COSM (Traas et al. 2004) is a dynamic model to predict abiotic fate of nutrients and pesticides and their effects on biota in microcosms. It builds on existing models (Traas et al. 1996; Traas, Janse, et al. 1998; Traas, van den Brink, et al. 1998). Nutrient cycling and nutrient uptake by primary producers was incorporated, taken from existing eutrophication models for shallow waters (Janse 1997). The food web is divided in functional groups and in turn a priori subdivided on the basis of taxonomical or size differences. The model is based on the law of mass conservation, and the derivatives are essentially mass balance equations for the rate of change of biomass, nutrients, and a toxicant. State variables are expressed in mass density. The model consists of 4 modules: an abiotic module for dead organic matter (detritus) and the fate of nutrients and toxicants, containing 3 state variables for organic matter, 5 for the toxicant cycle, 8 for the phosphorus cycle, and 7 for the nitrogen cycle; a primary producer's module containing 4 state variables for dry weight of algae, periphyton, and macrophytes, 4 for adsorbed pesticide, and 2 for nutrients taken up by macrophytes; a consumer's module containing 9 state variables for dry weight of planktonic grazers and macroinvertebrates; and a mass balance module containing 4 state variables to check mass balance for each cycle (Traas et al. 2004). The mortality term in the derivative for each functional group consists of natural mortality and additional mortality due to the toxicant, estimated on the basis of their laboratory sensitivity (LC50).

14.2.5 Empirical Models

It is also possible to use information on effects observed for different exposure patterns in experiments or during monitoring programs in the field for extrapolation by using statistical approaches or techniques like case–base reasoning. An example for the latter is PERPEST (Predicting Ecological Risks of Pestiudes), an empirical model of pesticide effects based on a database of micro- and mesocosm studies (van den Brink et al. 2002; van den Brink, Brown et al. 2006). In its present state, the PERPEST database is based on effects of pesticides on various endpoints (e.g., community metabolism, phytoplankton, and macroinvertebrates) as observed in micro- and mesocosm studies and classified according to their magnitude and duration. The PERPEST model searches for analogous situations in the database based on the concentration and relevant (toxicity) characteristics of the compound. This allows the model to predict effects of pesticides for which no effects on a semifield scale have been published and concentrations that not have been evaluated in micro- or mesocosm studies. The PERPEST model results in a prediction showing the probability of classes of effects (no, slight, or clear effects plus an optional indication of recovery) on the various grouped endpoints. In 2007, the database was refined with respect to inclusion of recent experiments and fungicides. In this way, more varied exposure patterns are included in the database. This provides the possibility to use, for example, the information on the many pyrethroid studies with different numbers of and intervals between pulses to predict the effects of exposure scenarios of a new pyrethroid.

14.3 EVALUATION OF THE TOOLS

The tools described were discussed with respect to their pros and cons as tools for extrapolation effects between exposure patterns not only in ELINK Work Group 5 but also in different work groups during the 2nd ELINK workshop in Wageningen, the Netherlands.

The tools differ in their data needs in addition to that obtained in standard tests and in higher-tier tests. The tools provide a variety of results, reaching from ecologically relevant concentrations (ERCs) extracted from exposure profiles and effect studies over estimates of the time needed by an organism to recover from a toxicant pulse exposure to predictions of recovery rates in a landscape or of ecosystem-level effects. An overview of the discussed criteria for each type of tool is given in Table 14.2.

The largest advantage of the direct-link models, especially the ERC approach, is their simplicity. The ERC approach does not need additional data, it is not limited to the level of the organism like the mechanistic TK/TD models, and it can also be directly applied to situations when extrapolation is needed from population experiments or mesocosm studies to situations in the field with variable exposure patterns. However, the approach might result in too conservative risk estimations (e.g., when there is only a short, small overlap of the exposure and effect profile). Furthermore, the underlying assumptions, that is, Haber's law ($c \times t$ = constant effect), may not be appropriate for the species and compound under scrutiny. Nevertheless, the approach may serve as a first step in the extrapolation task.

Direct-link models are often used in population or food web models to link effects to the (external) toxicant concentration instead of explicitly considering TK and TD. For example, the growth rate might be modeled to depend directly on the actual external concentration via a sigmoid dose–response function obtained in a toxicity test under semistatic conditions. However, when models such as direct-link models are not based on process descriptions, the extrapolation to scenarios that are very different from those with which the models were calibrated is difficult to justify. Ashauer et al. (2006a) concluded that direct-link models should not be the first choice to model effects of pulsed exposure, and for some of the concepts, application in dynamic models is even impossible.

For first-order TK models, the parameters k_{in} and k_{out} can be measured in a bioaccumulation study or fitted from survival data over time as is done, for example, in DEBtox (Kooijman and Bedaux 1996). However, fitting the elimination rate from survival data as done in DEBtox is prone to error because this method inherently assumes that the time course of effects mimics the time course of internal concentrations without delay (i.e., it assumes instantaneous and complete TD recovery). If this assumption is violated, as it is at least for specifically acting compounds such as carbamates or organophosphates, then the elimination rate constant will be underestimated. Hence, it is recommended that because of their fundamental importance for risk assessment of pulsed exposures or fluctuating concentrations the TK parameters should be measured (i.e., estimated from measured internal concentrations).

The TD models described in Section 14.2 differ in the data sets that are required for parameterization. Increasing model complexity (i.e., increasing numbers of model parameters) requires more detailed data sets. Also, they may require measurements in

TABLE 14.2
Evaluation of Tools to Extrapolate Effects between Exposure Patterns

Approach	Example	Output	Data Requirements	Applicability for ELINK	Weaknesses	Recommended Use
Direct link (ERC approach)	ERC	TER by comparing peaks or TWAs or evaluation of the overlap of exposure and effect profiles	No additional data; can be used for standard and higher-tier data	Now (no specific software, no additional data required)	Might result in too conservative estimations. Assumptions (Haber's law, i.e., $c \times t$ = constant effect) may not hold; depends on species and compound	First step in extrapolation task for standard or higher-tier tests
TK/TD models for acute (lethal) effects on invertebrates or fish	TDM	Threshold and effect values for untested exposure patterns (e.g., critical time for toxicological recovery between pulse events)	Species and substance specific (elimination and update rate constants, survival data at about 3 exemplary pulse patterns)	Now (but see data requirements) Examples available for *Gammarus* and *Daphnia*	Not applicable without additional experiments. Further experiments acceptable in Tier 1. Restricted to organism-level effects. No examples for fish yet	If ERC indicates potential acute risk on invertebrates and fish, especially for extrapolating effects to number, magnitude, and interval of pulses (e.g., calculation of critical interval for recovery between pulses)
TK/TD models for long-term sublethal effects	DEBtox	Threshold and effect values for untested exposure patterns (single pulse)	Species and substance specific (survival plus growth or reproduction data for at least 3 successive observations during the exposure period)	Near future; coupling to population models of fish or invertebrates	See remarks made above for TDM. Need for further development to account for time-varying concentrations of pesticides in case of repeated pulses and recovery kinetics of the individuals between pulses	For extrapolation of inhibition of reproduction or growth from standard test exposure scenarios to (predicted) field exposure patterns

(continued)

TABLE 14.2 (CONTINUED)

Evaluation of Tools to Extrapolate Effects between Exposure Patterns

Approach	Example	Output	Data Requirements	Applicability for ELINK	Weaknesses	Recommended Use
Simple population models	Logistic model	Population dynamics for different exposure scenarios	Population growth rates, if needed dependencies from for example, temperature, light, nutrients	Now	Life cycle and TK/TD usually ignored but might be included by use of biologically structured models and coupling to TK/TD models	Estimation of intrinsic population recovery for phyto- and (some) zooplankton and other invertebrate species, especially of sharp pulse scenarios
Complex population models	MASTEP	Prediction of population-level effects and recovery, including recolonization	Detailed information of species ecology. If coupled to TK/TD, see their requirements also. If spatially explicit, data on spatial distribution of exposure and presence of species as well as mobility and dispersal	Species specific, available for *Asellus* yet, similar models for some other species under development	High data requirements; complex models may be more difficult to communicate; model cannot so easily be replicated	Higher-tier tool for estimation of recovery, including recolonization for "more complex" life cycles (e.g., spatiotemporal extrapolation from mesocosm studies)
Food web models	AQUATOX	Dynamics of exposure and effects in a food web, including indirect effects	Identification of most important taxa of the food web (e.g., key species, aggregation of species to model compartments). Bioenergetics and ecotoxicological parameters per compartment	Now (software available, including parameter database)	Simplified TK/TD, simplified models for populations, not spatially explicit	Analysis of general patterns (direct and indirect effects). Analysis of effects and internal recovery in mesocosm studies and maybe extrapolation to other scenarios
Empirical models	PERPEST	Classified effects	Mesocosm study database, substance properties	Future (exposure pattern not described in sufficient detail yet, is under development)	Predictions totally dependent on data availability; not applicable for new classes of compounds	Estimation of ecosystem effects using information available on similar situations

addition to survival data from standard toxicity tests. For example, information about TD can be inferred from survival data of sequential pulsed exposure tests (Ashauer et al. 2007b, 2007c). Some information that may be required for risk assessment of repeated pulsed exposures is not contained in standard toxicity tests; hence, these data are insufficient to appropriately address the problem. For example, the question whether there may be a buildup of internal concentrations may only be answered if internal concentrations were measured to parameterize the TK model. Also, the question whether subsequent pulses may lead to increased mortality because organisms may not have had time to fully recover cannot be answered based on standard toxicity tests that consist of a single, constant exposure only. A benefit of hazard modeling is that the background hazard rate (i.e., the mortality rate in the control treatment of a toxicity test) can be easily derived and integrated with any hazard model.

In general, TK/TD models provide a more mechanistic basis for extrapolation than the more or less empirical direct-link models, which may not be applied outside their calibration range. Furthermore, TK/TD models can be integrated in a variety of population models, either classical or IBMs.

Most TK/TD models and examples available yet focus on effects on survival, but any model that is calibrated on sublethal effect data may be used to extrapolate sublethal effects to different exposure scenarios with similar limitations as for lethal endpoints. Most TK/TD models may require some modification to be applicable to nonlethal endpoints, whereas the principal modeling concepts remain the same.

The DEB theory may be particularly suited to simulate sublethal effects that result from changes in the energy allocation within the organism because energy usage and budgets within organisms are explicitly simulated. The main advantage of the standard DEBtox model in the context of ELINK is that there is a software tool that allows parameter estimation from standard toxicity tests (DEBtox version 1.0; Kooijman & Bedaux 1996; http://www.bio.vu.nl/thb/deb/deblab/debtox/). The DEBtox model can be used to analyze toxicity test data, providing several successive measurements of the endpoints are made, by fitting the model to the full data set (effects depending on toxicant concentration and time). Nevertheless, it would be wrong to assume that extracting the most information from existing data will automatically enable us to answer new questions such as those of effects extrapolation to fluctuating or sequential pulsed exposure patterns. Some necessary information may be missing in both the models and the data. Increasingly complex questions, such as those posed by risk assessment of fluctuating or sequential pulsed exposure, may require more complex models and data sets than current standard toxicity tests. For example, the basic DEBtox model for survival may not be applicable to a large number of pesticides because of the inherent assumptions of the model formulation. The range of applicability has never been evaluated yet, and there have been no attempts to evaluate fitted elimination rate constants with measured elimination rate constants.

In the DAM, the cumulative hazard $H(t)$ is proportional to the damage $D(t)$, which leads to increasing survival probabilities for fluctuating or pulsed exposure. By definition, survival probabilities (probability of an organism to survive until time t) cannot increase; hence, the original DAM is not applicable to fluctuating or pulsed exposures, and it must not be used in the context of ELINK.

The generic definition of the TDM, and the fact that no restricting assumptions about the speed of recovery are made, makes the approach applicable to a wide range of toxic mechanisms and classes of compounds. Ashauer et al. (2006a) concluded in their review that the THM (the modification of the DEBtox they discussed in their review) and the modified DAM are the best approaches for modeling effects of time-varying exposure to aquatic organisms. However, they also pointed to the fact that neither the THM nor the modified DAM has been tested yet on data from pulsed exposure experiments. Subsequently, key aspects of both the THM (based on the standard DEBtox model) and the DAM have been integrated in the TDM. This model has been specifically developed and tested for the purpose of predicting toxic effects on aquatic organisms, under and extrapolating to, sequential exposure pulses, and fluctuating concentrations.

The TDM has been tested on survival experiments with *G. pulex* exposed to 3 pesticides (chlorpyrifos, carbaryl, and pentachlorophenol; see Section 14.4.2). The extrapolation capabilities of the TDM have been evaluated by calibrating the model with 1 set of experiments and then predicting the outcome of additional, independent experiments with different exposure patterns (Ashauer et al. 2007a, 2007b, 2007c). For compounds with slow organism recovery, such as chlorpyrifos, it has been observed that subsequent pulses can cause increased toxicity, even though there was enough time for depuration (Ashauer et al. 2007a, 2007b).

The strengths of the process-based TK/TD models are that they can be parameterized with laboratory experiments and then be used to extrapolate to different exposure patterns. Particular advantages of the TDM are that it 1) is applicable to different modes of action; 2) yields total organism recovery times; and 3) includes process knowledge such as TK and TD. Its application as an extrapolation tool has been evaluated for a range of exposure patterns and 3 different pesticides. The total recovery times (Ashauer et al. 2007c) are particularly useful for risk assessment of repeated pulses because they inform whether the organisms have had enough time to recover in between pulses. Hence, it facilitates an evidence-based decision on the time window of the exposure concentration profile that needs to be assessed as 1 consecutive event. For example, those subsequent peaks that occur before the organisms have recovered from previous pulses (i.e., subsequent pulse falls within total recovery time) need to be considered together, whereas those that follow with sufficient delay (i.e., later than the total recovery time) may be considered independent events.

Population models may vary in their complexity, including very simple models like the logistic growth model up to models simulating life cycles of interacting individuals in a landscape. Thus, 1 of the most difficult tasks for the useful application of a population model in the risk assessment process is the decision on the appropriate type of model. During the SETAC LEMTOX workshop, "Ecological Models in Support of Regulatory Risk Assessments of Pesticides," the need for "similar solutions for similar problems" was formulated, which led to the request to develop a kind of guidance document on good modeling practices covering model development, analysis, documentation, and evaluation (Forbes et al. 2009). It is hoped that this guidance will be a follow-up activity of the workshop.

For the purpose of ELINK, the benefit of population models was seen mainly in integration of individual-level endpoints to population-level endpoints (e.g., by coupling TK/TD models for sublethal effects) with the population model and in the prediction of recovery potential for different exposure patterns. In fact, population models may be the only available tool for spatiotemporal extrapolation of recovery rates (e.g., MASTEP; Section 14.4.4). Depending on the species to be modeled, the availability of data, and the need to consider recolonization, the model may be quite simple or more complex.

One of the advantages of food web and ecosystem models is their ability to predict indirect effects of toxicants (e.g., effects due to changes in predation or competition) and the interactive effects of contaminants and other stressors. They give the opportunity to extrapolate information from the effects observed on isolated species (e.g., results of acute and chronic toxicity tests) to a set of interconnected populations. Among their limits is the need for an accurate description of many (eco)physiological processes that require many data, the possible bias associated with aggregation of species into compartments, and the treatment of uncertainties (Koelmans et al. 2001). The effect of parameter variance, time horizon of predictions, and alternative model algorithms on model predictions is discussed by Poethke et al. (1994). Their conclusions that "model predictions on the effects of toxicants in pelagic food webs have to be taken with extreme precaution" may still be valid. If dynamic food web models are coupled with TK models, they allow estimating food chain transfer of toxicants.

At present, the indirect effect of pesticides in aquatic ecosystems are mainly assessed via micro- and mesocosm studies. However, these test systems usually do not include fish and often no or only a limited number (species and individuals) of macrophytes. Therefore, models might be the only tool for analyzing possible consequences in food webs, including the experimentally more difficult to handle groups. However, at present the focus of ecological modeling in pesticide risk assessment is on the extrapolation of effects and recovery of populations.

To our knowledge, empirical models are not yet applicable to the extrapolation between exposure levels. PERPEST (van den Brink et al. 2002; van den Brink, Brown et al. 2006) may be a promising tool for the future if the database is expanded and the exposure patterns are described in more detail. This work is currently under way as a part of a doctoral project (van den Brink, personal communication).

For all the mechanistic models (TK/TD, population, and food web models), techniques are available to explore the sensitivity of the model against input parameter values to identify the parameters that should be measured as precisely as possible. In addition, Monte Carlo simulations allow the quantification of the uncertainty of the model output depending on the variability or uncertainties of the model parameters.

14.4   APPLICATION EXAMPLES

14.4.1   ERC CONCEPT

An example for the application of the ERC concept was given by Boesten et al. (2007) using data for the herbicide linuron. The effect assessment was based on a mesocosm study with 3 monthly applications. The highest nominal concentration with no effects on ecosystem structure and only slight and transient effects on oxygen metabolism was used to determine the RAC (5 µg/L). The relative time window for the comparison of exposure and effect was considered to be 7 days (length of standard *Lemna* test); thus, a 7-day RAC curve was extracted from the fate data

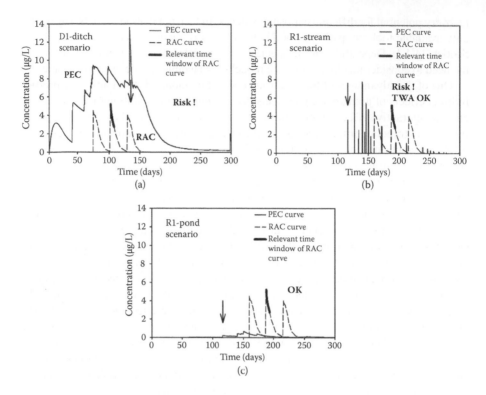

FIGURE 14.1 Example of the use of the ERC approach. (Adapted after Boesten et al. 2007. From Boesten, Köpp, Adriaanse, Brock, Forbes. *Conceptual model for improving the link between exposure and effects in the aquatic risk assessment of pesticides.* Ecotoxicol Environ Saf 66:291–308. Copyright Elsevier 2007. Reprinted with permission.)

for linuron in the 5 µg/L mesocosm (Figure 14.1). For the exposure assessment, 3 FOCUS step 3 scenarios were used.

In tier 1, it was checked whether the maximum of the PEC was always below the minimum of the RAC curve in the relevant time window. This was only the case for scenario C. Thus, on tier 2, the RAC and PEC curves were compared for scenarios A and B. In both scenarios, the PEC curve was not always below the RAC curve, which triggers tier 3 comparison of TWAs. Because the PEC curve was always above the RAC curve in scenario A, TWA would not help, but in scenario B the TWA PEC was below the TWA RAC for any time window exceeding 1 day. Thus, risk was considered acceptable for scenarios B and C but not for A. A refined risk assessment for scenario A using recovery of effects for the determination of the RAC can be found in the work of Boesten et al. (2007).

14.4.2 THRESHOLD DAMAGE MODEL

The TDM as described in Ashauer et al. (2007b, 2007b) will serve as an example for the use of TK/TD models for extrapolating between different pulse scenarios because it is the only TK/TD model yet that has been applied to pulsed exposure of pesticides.

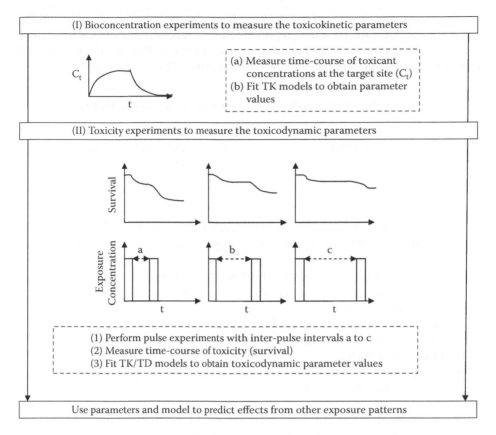

FIGURE 14.2   General methodology to parameterize TK/TD models, for example, the TDM.

The general methodology to parameterize the TDM or similar TK/TD models is outlined in Figure 14.2. First, it is necessary to perform bioconcentration experiments in which the time course of the internal concentration of the compound is measured. Organisms are exposed, then transferred to fresh media, and the experiment is continued for some time to allow better estimation of the parameters of the TK submodel. Measurements of total radioactivity of radiolabeled compounds may suffice in many cases. A 2nd set of experiments is designed with several exposure pulses with intervals a to c in between. The shortest interval a should be long enough to allow for complete depuration, whereas the longest c should be chosen such that substantial TD recovery has occurred. Survival over time is monitored, and then the TDM is fitted to the survival data, while the internal concentration is simulated with fixed parameters to obtain the TD parameters.

The model may be implemented in any software that allows optimization (parameter estimation) of a nonlinear model that requires numerical integration. Fairly user-friendly, suitable software packages exist, for example, ModelMaker 4 (http://www.modelkinetix.com/) or OpenModel (http://www.nottingham.ac.uk/environmental-modelling/OpenModel.htm).

Once the parameters of the model have been obtained, the model can be run for any exposure pattern, for example, fluctuating or repeated-pulsed exposures, and the total time required for the organism can be calculated. Detailed discussions and examples can be found in the work of Ashauer et al. (2006b, 2007b, 2007c).

14.4.3 SIMPLE POPULATION MODELS: DIFFERENTIAL EQUATION MODELS

Barnthouse (2004) has suggested the use of the logistic model of population growth (Verhulst 1845) to quantify the effect of frequency and intervals of disturbance on the recovery rate of aquatic populations. This approach is exemplified here to quantify recovery times of rotifers for different pulse exposure scenarios based on the following assumptions:

- The pesticide is a substance with very fast dissipation (DT50 [period required for 50% disappearance] = 1 day). Thus, only the peak concentrations matter.
- Only acute lethal effects on the rotifers are considered. The dose–response relation is described via a sigmoid (logistic) function characterized by the LC50 and the slope.
- Spatial variability is ignored (homogeneous mixture of the pesticide in the water column, no immigration of rotifers from no or less-exposed sections of the water body).
- The population dynamics of the rotifers can be simplified via the logistic function (Equation 14.10).

For the example given here, rotifer growth rates were extracted from a mesocosm study by log-linear regression. In the study, the test item was applied 4 times in weekly intervals with 7 different treatment levels (T1 to T7). Exponential growth in the treated mesocosms started usually 1 week after the last application and was faster in the higher ($r = 0.4$ day^{-1}) than at the lower treatment levels ($r = 0.19$ day^{-1}; Figure 14.3).

Intrinsic growth rates and dose–response relations from the mesocosm study were used to simulate recovery rates for different exposure scenarios (Figure 14.4). For the simulated treatment level (T3), the model predicts recovery (reaching 95% of the carrying capacity) around 60 days after the first application. In the mesocosms, the controls reached their maximum on day 28 before abundances decreased again. Growth in treatment level 3 (T3) was significantly delayed, and the abundance of control was reached around 8 weeks after the first application. Thus, despite the fact that the logistic model could of course not simulate the decrease of rotifer abundances in the last weeks of the study, it may provide estimations of the recovery potential for different exposure scenarios.

The former model needs only data on the population growth rate and the dose–response relation. The model can easily be refined to consider that growth may depend on several environmental conditions. For example, Driever et al. (2005) modeled the population growth of the duckweed *Lemna minor* with a differential equation model that included dependencies of the growth rate on temperature, nutrient concentration, and crowding.

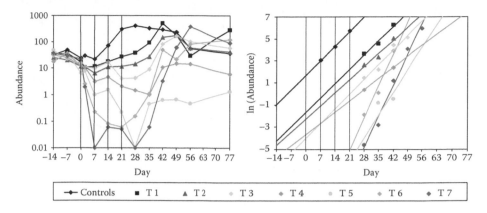

FIGURE 14.3   Rotifer abundance per treatment in controls and treatment levels T1 to T7 in a mesocosm study (right) and linear regressions for the time period with approximately exponential growth (left).

In a similar way, Weber et al. (2007) modeled the growth of algae depending on temperature, light, external and internal phosphorus concentration, and (external) toxicant concentration. The objective was to allow extrapolating from the standard algae tests to different exposure patterns (e.g., multiple peaks). Thus, the default values for the 11 model parameters were based on literature values and additional laboratory studies, while for describing the toxicity, the results of a standard algae growth inhibition tests are sufficient. First results for simulating population growth in a standard test are shown in Figure 14.5.

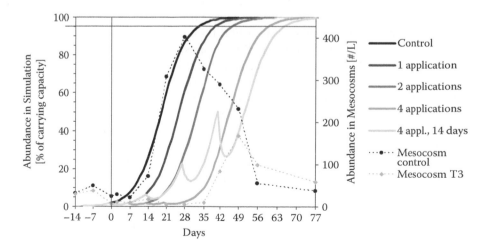

FIGURE 14.4   Simulated abundance of rotifers for control conditions and 4 exposure scenarios and corresponding experimental data from a mesocosm study.

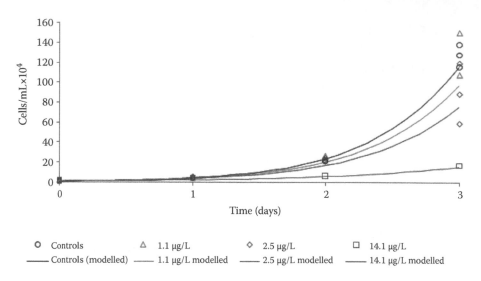

○ Controls   △ 1.1 µg/L   ◇ 2.5 µg/L   □ 14.1 µg/L
——— Controls (modelled)   ——— 1.1 µg/L modelled   ——— 2.5 µg/L modelled   ——— 14.1 µg/L modelled

FIGURE 14.5   Population growth of *Desmodesmus subspicatus* as observed in laboratory tests and predicted by the model of Weber et al. (2007).

14.4.4 COMPLEX POPULATION MODEL: MASTEP

The population model (MASTEP, metapopulation model for assessing spatial and temporal effects of pesticides; van den Brink et al. 2007) was developed to describe effects on and recovery of the waterlouse *Asellus aquaticus* after exposure to a fast-acting, nonpersistent insecticide as a result of spray drift in a pond, ditch, or stream according to the FOCUS scenarios. The primary objective of MASTEP is the analysis and prediction of recovery due to recolonization and is based on movement patterns of individuals (and water) in a structured landscape. The model is able to use the spatial and temporal distribution of the exposure in different treatment conditions based on FOCUS calculation as an input. A dose–response relation derived from a hypothetical mesocosm study was used to link the exposure with the effects via a sigmoid direct-link model. TK and TD have not been included yet. The modeled landscape was represented as a lattice of 1 by 1 m cells and included exposed cells close to a treated field and cells not exposed along untreated fields. The model includes processes of mortality of *A. aquaticus*, life history, random walk between cells, density dependence of population regulation, and in the case of the stream scenario, medium-distance drift of *A. aquaticus* due to flow (Figure 14.6). All parameter estimates were based on expert judgment and the results of a thorough review of published information on the ecology of *A. aquaticus*. An individual base approach was used to model the life cycle and movement of *Asellus*.

In the treated part of the water body, the ditch scenario proved to be the worst-case situation due to the absence of drift of *A. aquaticus*. Effects in the pond scenario were smaller because the pond was exposed from 1 side, allowing migration from the less-contaminated other side. The results of the stream scenario showed the

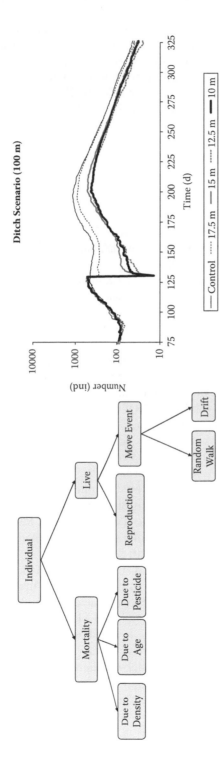

FIGURE 14.6 Conceptual model of MASTEP describing the activities of individual *A. aquaticus* (left) and simulated population dynamics after insecticide application in a FOCUS ditch scenario (treatments were different buffer zones).

importance of including drift for the population recovery in the 100-m stretch of the stream that was treated. It should be noted, however, that the inclusion of drift had a negligible impact on numbers in the stream as a whole (600 m).

At present, MASTEP has been used to simulate effects of 1 insecticide entry and focused on the spatial patterns over time. MASTEP can be used for extrapolation between different exposure patterns as long as the dose–response relationship derived from the mesocosm experiment is valid (i.e., peaks are valuated). It would also be possible to couple the model with a TK/TD model.

14.4.5 FOOD WEB MODEL APPLICATIONS

Naito et al. (2003) compared biomass reduction predicted by the Lake Suwa version of the CASM model (CASM_SUWA) with the results of various experiments and with NOECs obtained from mesocosm tests. In this study, the concentrations of chemicals in water were assumed to be constant, although CASM can handle time-varying concentrations in risk estimations.

The qualitative comparison of the model results for plant protection products with those of the reported mesocosm tests indicated that some observations made in mesocosm studies supported at least in part the direct and indirect effects predicted from simulation using the model. This was the case for atrazine, diazinon, and simetryn.

For 7 toxicants (LAS [Linear Alkylbenzene Sulfonates], phenol, copper, PCP, atrazine, chlorpyrifos, and diazinon) the BR20 (concentration resulting in a biomass reduction by 20%) derived from the model was of the same order of magnitude or lower than mesocosm-derived NOECs, with BR20–NOEC ratio values ranging from 0.013 to 2.

Although AQUATOX has been designed to facilitate new applications and scenario development, very few comparisons between AQUATOX outputs and experimental data have been performed so far (Park et al. 2004), and most of them have not been published in peer-reviewed journals. The AQUATOX reference manual provides an example of application of the model for the ecological risk assessment of the insecticide chlorpyrifos in an experimental pond enclosure, but no comparison with original pond-measured data is made (Park et al. 2004). Rashleigh (2007) used AQUATOX to simulate the effects of a point source contamination of a reservoir by dieldrin but again without comparison with real data.

Sourisseau et al. (2008) applied AQUATOX to simulate functioning of artificial streams used for ecological risk assessment and performed a multivariate sensitivity analysis of the model. They showed that the calibrated model was able to adequately describe the dynamics of most of the simulated biological compartments of the streams. Using data from other streams, it was shown that between-stream natural variability was a source of discrepancy between observed and simulated data. Sensitivity analysis showed that the model was highly sensitive to the parameters related to the temperature limitation, maximum rate of photosynthesis of producers, and consumption by consumers. This strongly suggests that particular attention should be devoted to the estimation of these parameters if this model has to be used for ecological risk assessment of toxicants in aquatic ecosystems.

14.5 CONCLUSIONS AND RECOMMENDATIONS

Various tools or models for extrapolation of effects and recovery between different exposure patterns were discussed by ELINK Work Group 5, leading to the following conclusions and recommendations:

- Direct-link models as used in the ERC approach (Boesten et al. 2007) can be directly applied to standard and higher-tier experiments. A tiered approach is used if point estimations of exposure and effects, TWAs, or curves are compared. However, the approach might lead to a conservative estimation of risk. Furthermore, the underlying assumptions, that is, Haber's law ($c \times t$ = constant effect), may not be appropriate for the species and compound under scrutiny. In these cases, other mechanistic tools may allow a refinement of risk estimations.

- TK/TD models appear as a promising type of models since their focus is on the understanding of processes. Their predictions include effects at the individual level or the toxicological recovery of individuals. However, additional data are required like uptake and elimination rate constants from bioaccumulation studies and, depending on the model, survival experiments with multiple-pulsed exposures. TK/TD models are recommended if, for example, the ERC approach indicates a potential acute risk on invertebrates and fish. Several models exist, but until now only the TDM has been successfully applied to analyze the effects of multiple peaks on invertebrates. To our knowledge, applications of TK/TD models for sublethal effects (e.g., on reproduction or growth) under variable environmental toxicant concentrations have not been published yet.

- Simple population models, such as differential equation or biologically structured models, may be useful to extrapolate intrinsic population recovery rates (i.e., recolonization can be assumed not to be important). They are relatively easy to parameterize but on the other hand may be inappropriate for complex life-cycle types or when effects on behavior are important. Thus, they will likely be used only for relatively simple organisms exhibiting short generation times, such as algae, some macrophytes (e.g., *Lemna*), zooplankton, and many other invertebrate species. Usually, the effects are directly linked to the external concentration. However, internal toxicant concentrations can be accounted for if the population models are coupled to a TK/TD model.

- Complex population models, such as IBMs, overcome disadvantages of the simple population models but usually cost much higher effort in parameterization. These models can be easily made spatially explicit, allowing predictions of external recovery by recolonization from uncontaminated water bodies. Theoretically, all relevant processes can be included into the model, including movement, dispersal, and other behavior. Again, coupling with TK/TD models is straightforward and allows consideration of individual exposure due to the activity of the separately modeled individuals.

The main area of application is seen in the spatiotemporal extrapolation from mesocosm or other higher-tier studies.

- Food web and ecosystem models are considered valuable tools for analyzing complex ecological interactions due to predation and competition, which may lead to indirect effects of toxicants. However, for the time being, the assessment of indirect effects is considered to be mainly based on micro- and mesocosm studies. Food web models can serve as additional tools for analyzing these complex studies rather than predicting effects on their own. Nevertheless, in combination with micro- and mesocosm studies they allow extrapolations between different exposure scenarios, including nontested ones.

- Empirical models are able to extrapolate effects between different exposure scenarios but need a large database about effects observed for different exposure patterns. Such a database is not available now because mesocosm studies are usually designed to test different treatment levels but not different exposure scenarios (i.e., number of and interval between peaks), and monitoring studies usually lack a sufficiently detailed description of the exposure situation.

15 Ecological Characterization of Water Bodies

CONTENTS

This work group report aims to provide information on the ecological characterization of edge-of-field surface waters to aid the design and evaluation of higher-tier tests.

15.1   ECOLOGICAL CHARACTERIZATION OF EDGE-OF-FIELD SURFACE WATERS: AN INTRODUCTION

Anne Alix

When performing higher-tier risk assessments for plant protection products, it is, in almost all cases, necessary to (re)evaluate the available data, whether submitted in European dossiers or in dossiers in support of member state authorization procedures, in the context of their representativeness of the real world (Liess et al. 2005). This is because higher-tier data, especially model ecosystem studies, aim to simulate ecological phenomena in a more realistic way. In addition, the occurrence of effects as observed in model ecosystem studies needs to be presented in a way that can be understood and accepted by decision makers. In building their risk assessment, evaluators often consult literature databases or contact specialists to gain confidence in the level of ecological realism that the higher-tier studies provide. This is, more often than not, done on a case-by-case basis as there is currently neither a common or recommended source of information nor any agreed mechanism of how it should be used. Having an agreed approach, in combination with appropriate sources of information, would help risk assessors better interpret and extrapolate results of ecotoxicological experiments and models to the real world. Such an information resource would also be helpful in checking that no important issue was missed in the risk assessment. Finally, performing risk assessments on the basis of a clear ecological description of the environment to be protected will be of help to evaluators in communicating with, and explaining their decisions to, other stakeholders.

To address the issues outlined, it is essential to characterize, from an ecological perspective, those typical water bodies in the agricultural landscape that we aim to protect. This chapter presents examples of how this can be done by presenting case studies for streams in Germany and Spain, ditches in the Netherlands, and all water body types with an emphasis on ponds in England and Wales. We realize that these case studies do not give a complete picture for all types of water bodies that occur in the agricultural landscape of different European member states. In addition, we are aware of other studies, already published or in progress, that also aim to describe the ecological characteristics of typical water bodies in the agricultural landscape (e.g., studies in Europe on the ecology of ponds, streams, and ditches; Schäfer et al. 2007; Davies et al. 2008). We believe, however, that the case studies presented here may be of help in future discussions of how ecological scenarios should be linked to the exposure scenarios of FOCUS (Forum for the Co-ordination of Pesticide Fate Models and Their Use) to improve the ecological realism of the risk assessment for plant protection products within the context of 91/414/EC.

In our case studies, we did not aim to directly link our scenarios with "reference conditions" as defined within the Water Framework Directive (WFD) as these reference conditions rather correspond to more or less-pristine water bodies and to environmental disturbance and stress in general and less so to water bodies in the agricultural landscape, potentially exposed to plant protection products in particular. Nevertheless, to obtain a realistic worst-case ecological scenario for these water bodies in the agricultural landscape, we selected sites characterized by a relatively high biodiversity and devoid of long-term chemical stress, indicating functionality and a capacity for resilience to perturbation. All data collected were filtered to exclude monitoring sites potentially impacted from urban or industrial situations.

The data collected for streams located in northern and central Germany were generated through 3 biomonitoring projects that investigated the conditions for exposure to pesticides and related effects. These streams are located in small lowland and low mountain range areas, in landscapes subject to moderate or intensive agricultural use. The use of agricultural land in the surroundings has been described with regard to the actual exposure to pesticides and to the influence of land use adjacent to or nearby these water bodies. In all circumstances, pesticide concentrations measured in water samples had to be lower than the threshold level of effect for each active substance. The streams were of similar width but shallower than the typical stream used in FOCUS surface water scenarios. The possible consequences of these differences in terms of exposure levels are discussed further. No detailed data for macrophytes, periphyton, and plankton could be gathered within the time course of the ELINK (EU Workshop on Linking Aquatic Exposure and Effects in the Registration Procedure of Plant Protection Products) project. The detailed ecological characterization of these German streams is presented in Section 15.2.

Dutch ditches represent a good example of heavily managed water bodies. The information used was selected from a huge database generated for drainage ditches in all parts of the Netherlands. The selected ditches are macrophyte dominated, as biodiversity, including macroinvertebrates, is in general higher in macrophyte-rich systems than in macrophyte-free systems. The characteristics of macrophyte-dominated ditches in the Netherlands particularly differ from the FOCUS ditch scenarios in that they have higher amounts of organic matter in the form of macrophytes and organic debris in the top layer of the sediment. The lower range (10th percentile) of the dimensions of these ditches more or less corresponds with the ditch dimensions adopted in the FOCUS scenarios. Drainage ditches can be characterized as dynamic in space and time due to mechanical cleaning and dredging activities. This management regime favors plant and animal species with pioneering properties. The Dutch ditch case study has its focus on macroinvertebrates and macrophytes and their potential sensitivity to pesticide stress. In addition, the potential vulnerability of invertebrates to pesticide stress is evaluated (e.g., by comparing the species composition between ditches adjacent to agricultural fields and natural habitats). In drainage ditches adjacent to agricultural fields, a lower proportion of macroinvertebrate individuals with a univoltine life cycle was observed, while the reverse was the case for individuals with a bivoltine and multivoltine life cycle. However, a similar relationship could not be demonstrated when comparing the relative distribution of macroinvertebrate taxa over different voltinism categories between ditches adjacent to agricultural fields and those adjacent to natural habitats. No detailed data could be gathered on fish, periphyton, and plankton within the time course of the ELINK project. The detailed ecological characterization of Dutch ditches is presented in Section 15.3.

In Spain, data collected were from 2 types of water bodies associated with arable land: 1) permanent rivers and 2) temporal (ephemeral) rivers. The permanent rivers selected are located in the region of the Júcar Basin, in the east of Spain. The water flow of these rivers is influenced by the Mediterranean seasonal climatic variability. Júcar Basin is an agricultural area where 46% of the surface is covered by agriculture. Differences in species richness and relative abundance were observed in areas impacted by agriculture compared with nonimpacted areas. The ephemeral rivers (ramblas) are described for the region of the southeast. This area is characterized by intensive horticulture and fruit tree cultivation. These temporal water bodies present biological communities well adapted to changing conditions. The dominant groups are algae, macrophytes, and macroinvertebrates, with resilient species and organisms being often more resistant to a variety of disturbances. The examples for Spanish water bodies are described in detail in Section 15.4.

Data collected for the United Kingdom were extracted from an important database that describes the typical water bodies encountered in each of 13 landscape classes identified in England, Wales, and Scotland. For the purpose of this work, landscape class 6, was selected as it accounts for a considerable percentage of cereal-growing land where associated water bodies are potentially vulnerable to pesticide entry. This landscape class is characterized by gently sloping, slowly permeable clays and heavy loams with subsoil drainage. While a range of water body types is associated with landscape class 6, for this exercise we focused on ponds, for which there is a detailed description of morphology and bankside features, such as distance from the bank top to an arable crop, abundance of aquatic invertebrates and vegetation, bank scrub, tree density, and adjacent land use. For macroinvertebrates, number of species and relative abundance among invertebrate groups are available, which could also be related to other water body types occurring in the same landscape class. For macrophytes, species richness and types could also be compared for all water body types in the same landscape. The detailed ecological characterization of these UK ponds is presented in Section 15.5.

The ecological characterization that was done for the selected water bodies includes a physical description together with inventories for fauna and flora. The main parameters are summarized in Table 15.1. Although not every parameter is reported for each of the case studies, it is possible to compare 2 or more different European water bodies against all but 3 parameters.

Water column width and depth, water flow, or slope are key parameters to describe the hydrology of systems. Quantitative information on these key parameters may be important as well to explain the presence or absence of certain species and ecological traits. In addition, width, depth, and water flow also influence the level of exposure of species in the aquatic systems. Consequently, the information gathered can be compared with the selected properties of the FOCUS surface water scenarios.

Other parameters, such as organic matter content of the sediment layer, temperature, and pH, may also influence both community composition and environmental fate of plant protection products within a system. However, quantitative data were not systematically available.

An ecological characterization was provided in all case studies but not always in the same detail and considering all relevant trophic or taxonomic groups. However, in all case studies information was made available for macroinvertebrates. Even considering macroinvertebrates alone, these databases remain highly valuable as a record of assemblages associated with typical water bodies.

For the same group of organisms, the level of taxonomic detail in which the taxa are reported also varied between the case studies, and clearly quantitative data are more valuable than qualitative. While it is acknowledged that the collected data cannot be used directly in risk assessment, it gives an indication of what type of information might be of importance when assessing the impact of pesticide stress. In constructing ecological scenarios, it is deemed important to describe rules regarding what information is recorded and processed.

Besides characterizing the occurrence of taxa, species traits like voltinism of the observed species or taxa provide valuable information, particularly to evaluate the potential for recovery. On this topic, possible missing data may be found in published databases (e.g., Tachet et al. 2000; Usseglio-Polatera et al. 2000; Statzner et al. 2004), taking into consideration the possible extrapolation from 1 European region to another. In the case studies, the dispersal capacity of pesticide-sensitive species

TABLE 15.1

Information Collected on Abiotic and Biotic Factors for a Range of Typical European Water Bodies

Topic	Parameter	Dutch Ditch	German Stream	Spanish Stream	Spanish Ramblas	UK Ponds
Hydrology	Surface area	X				X
	Width	X	X	X	X	X
	Depth	X	X	X	X	X
	Velocity		X	X	X	
	Seasonal aspects				X	
	Conductivity	X		X	X	X
	Temperature			X		
	pH	X	X	X		X
	Nutrients	X	X			X
	Depth organic debris layer	X				
	Organic matter	X		X		
	Suspended solids					X
	Dry bulk density	X				X
	Sediment layer	X				X
	Biotic factors—Richness of macrophytes	X				X
	Proportions of emergent, submerged, and floating-leaved plants					X
Ecology	Frequency of main taxa for macrophytes	X			[a]	
	Peak biomass for macrophytes	X				
	Plant species					X
	Richness of algae					
	Richness of invertebrates	X				X
	Frequency of main taxa for arthropods	X	X	X	[a]	
	Frequency of main taxa for nonarthropods	X		X		
	Relative distribution of macroinvertebrates over taxonomic groups	X	X	X		X
	Relative distribution of macroinvertebrates over voltinism categories	X	X	X		
	Ecological traits of the main macroinvertebrates encountered			X	[a]	
	Ecological indices for invertebrates			X		
	Abundance of fish species		X			
	Relative frequency of abundance for fish species		X	X		

[a] Qualitative information (taxa listed).

received little attention. However, it is recognized that besides the dispersal capacity of the species of concern, also the ecological infrastructure, such as the connectivity between suitable freshwater habitats, plays an important role.

Despite the limitations of the case studies presented in this chapter, the data collected may help risk assessors in identifying which species, taxa, or species traits should be considered when conducting and interpreting higher-tier tests to assess the risks of plant protection products in typical water bodies of the agricultural landscape. On the basis of the ecological information in the case studies, some recommendations with regard to taxa and species traits that should be taken into account in the risk assessment have been proposed. These recommendations may help in selecting taxa to be present in model ecosystem experiments or for additional laboratory testing or modeling exercises.

As life-cycle traits related to voltinism affect the capacity of a species to recover from chemical (among others) stress, the data collected for Dutch ditches were further interpreted with regard to the consequences of voltinism on the sensitivity of taxa of an assemblage. The aim was to compare the relative sensitivity to insecticides among aquatic insect species that differed in the number of generations per year. The analysis of species sensitivity distributions (SSDs) showed no significant difference in sensitivity between species based on the voltinism criteria. As a consequence, the threshold level of effects obtained from a model ecosystem study predominantly characterized by the presence of sensitive species with short generation times can be extrapolated to aquatic communities that include species with longer life cycles. If the risk assessment involves recovery aspects, however, it is recommended that species traits and dispersal properties of the sensitive taxa of concern are considered.

One important point that emerges from the data set collected in the United Kingdom is the systematic method of data collection, which allows comparisons between different water body types and landscape class. Such information could be of great value to extrapolate conclusions from a risk assessment based on higher-tier data (interpreted by supportive data from ponds in landscape class 6) to other landscapes or water body types. It is important to establish the ecological representativeness of water bodies in landscapes that differ from those in which data have been generated.

15.2 ECOLOGICAL CHARACTERIZATION OF SMALL STREAMS IN NORTHERN AND CENTRAL GERMANY

Jörn Wogram

15.2.1 INTRODUCTION

The FOCUS surface water concept provides specific exposure scenarios for different types of water bodies and climatic conditions for higher-tier risk assessment according to Council Directive 91/414 EEC (*EC* 2002). By contrast, there are no corresponding scenarios available on the effects side. The reason for this gap is that, in principle, the aquatic higher-tier effects assessment is based on additional ecotoxicological testing rather than on effect modeling. In practice, the most common approaches are 1) the testing of additional species in single-species lab tests, 2) single-species tests performed under more realistic exposure conditions, or 3) tests in model ecosystems (Campbell et al. 1998; Giddings et al. 2001). However, it might be questioned regarding how well such higher-tier approaches represent the ecological characteristics of the wide spectrum of types of water bodies in the European Union. For instance, the composition of coenoses in natural water bodies may differ significantly from those in microcosm and mesocosm tests or from species sets used for SSDs. In terms of abiotic characteristics affecting the environmental fate and behavior of pesticides, the same applies to a comparison of exposure models, model ecosystems, and natural water bodies. Thus, the remaining task with regard to a focused risk assessment for the aquatic environment is to define suitable ecological scenarios to serve as a reference when designing or evaluating exposure models, effect models, and model ecosystems in the context of a higher-tier risk assessment. One precondition to achieving this goal is a suitable description of the coenoses to be expected in the different types of water bodies as well as of the abiotic factors affecting the environmental fate and behavior of pesticides.

15.2.2 IN SEARCH OF A REFERENCE

In the context of the WFD (*Directive 2000/60/EC* 2000), reference conditions have been defined for abiotic conditions as well as for the composition of biotic assemblages (coenoses) in different types of water bodies. The WFD might suggest using these descriptions as a reference also in the risk assessment of pesticides. However, these reference aquatic habitats reflect pristine, undisturbed conditions, including exceptionally vulnerable species and low-nutrient contents, in contrast to conditions prevailing in the agricultural landscape, potentially resulting in an overprotective risk assessment.[*] This report aims at a description of the conditions typical of the agricultural environment.

---

[*] Nevertheless, potential conflicts with the aims of the WFD should be avoided. The obligation set by the WFD is to achieve the "good ecological status" of waterbodies by 2015, where "good" is defined by a certain maximum deviation from reference conditions. Consequently, a risk assessment under council directive 91/414 should be protective also in terms of the quality criteria for the "good ecological status". In order to prevent any discrepancies, the description of typical coenoses compositions given in this report should be compared to the indicators defining the good ecological status for the different types of waterbodies in Germany (e.g., Meier et al, 2006 with respect to invertebrates). However, this is beyond the focus of this report.

As regards the biological conditions, it was agreed that the ecological scenarios should reflect the "realistic best case" to be expected in this type of landscape. This realistic best case is defined as the ecological conditions to be found within a landscape of moderate agricultural use and no or at most low exposure to pesticides. Coenoses found under such conditions are more diverse and more vulnerable than under conditions of more intense agricultural use (Schäfer et al. 2007), reflecting a realistic worst-case scenario in terms of ecological sensitivity.

This report is meant as a contribution to a compilation of ecological scenarios for the pesticide risk assessment in the European Union. It describes ecological characteristics of small streams in northern and central Germany with a specific focus on factors relevant for the fate and the ecological effects of pesticides. Data considered for this report derive from the following projects:

Project 1. The pesticide exposure and effects-monitoring project by the Federal Environment Agency performed by the working group of Matthias Liess and Ralf Schulz at the Technical University of Braunschweig (published in Liess et al. 2001 and Wogram 2001)

Project 2. The effects-monitoring project funded by the Industrieverband Agrar e.V. (German pesticide industry association) with data from the same region (Pantel 2003)

Project 3. The fish stress ecology and pesticide biomonitoring study by the Technical University of Braunschweig (diploma thesis, Wogram 1996; partly published by Sturm et al. 1999)

All of the 3 projects were performed in small lowland and low mountain range streams in landscapes of moderate-to-intensive agricultural use in northern and central Germany. It needs to be determined in future projects regarding what extent these conditions are representative of other parts of Germany or for special cases (e.g., streams in peaty areas). Besides a description of the variation of ecological conditions among streams (aiming at a statistical evaluation at maximum *n*), 1 typical stream is described in more detail to provide a more robust example (see Section 15.2.5).

15.2.3 ABIOTIC CHARACTERISTICS

15.2.3.1 Dimensions

Width, depth, and current velocity were measured extensively in project 3 (Wogram 1996) in 15 small permanent lowland streams in a 30-km radius around Braunschweig, Niedersachsen (Table 15.2). All parameters were measured at low discharge level in June and July 1995. Those factors are also described in project 1 but with less reliability for width and depth as these factors were only roughly estimated.

In terms of width, the streams investigated broadly reflect the FOCUSsw model stream (Table 15.2). However, the mean width–depth ratio was much higher than in the FOCUSsw model stream, a ratio of 3.3 reflecting the minimum rather than the mean ratio found in reality. As indicated by the data from project 3, a more realistic mean ratio would be 10. Ohlinger and Zenker found a similar result (median ratio of approximately 10) in 39 small streams (width <1 m) in a wine-growing area in southern Germany (Schulz et al. 2007).

TABLE 15.2
Characteristics of 15 Small Lowland Streams in an Agricultural Landscape in Northern Germany (Data from Project 3, Wogram 1996) and a Comparison with the FOCUSsw Model Scenario

		1995 Braunschweig Study (15 streams)			
Parameter	FOCUS Stream	Median	Arithmetic Mean	Maximum	Minimum
Mean depth (deepest point in the transverse section) (cm)	30	12.9	15.4	39.4	2.4
Width (cm)	100	102.5	116.5	222	60
Width–depth ratio	3.3	9.8	11.7	32.5	4.2
Current velocity (m/s)		0.17	0.16	0.31	0.07

CONSEQUENCES FOR RISK ASSESSMENT

As indicated by a much higher mean width–depth ratio than assumed in FOCUSsw, the (maximum) predicted environmental concentration in surface water (PECsw) calculated on the basis of the FOCUSsw model stream might underestimate the real exposure of edge-of-field headwater streams to pesticides by a factor of up to 10 or even higher. The consequences of this finding for the overall level of protection should not be ignored in future risk assessment.

Furthermore, it should be assessed whether the typical values for current velocity (median $\cong 0.2$, sometimes below 0.1 m/s) are in line with the assumptions in the FOCUSsw stream exposure models (beyond the focus of this report).

15.2.3.2   Physical and Chemical Characteristics

It was agreed in Working Group (WG) 6 at ELINK Workshop II that the following physical and chemical factors should be considered in the ecological scenarios: pH, temperature, organic matter in the sediment, sorption, and degradation in macrophytes.

For pH, sediment composition, and macrophyte abundance, corresponding data are available from projects 1 (Table 15.3) and 2 (Pantel 2003). For temperature, no representative data are available. However, it can be assumed that in this type of water body with a high surface–volume ratio, water temperatures are highly correlated to air temperature.

pH: Projects 1 (Table 15.3) and 2 (Pantel 2003) consistently reported that pH values typically ranged from 7.0 to 8.5, thus indicating that acidic or extremely alkaline conditions (as typical for ditches in peaty areas or occasionally for ditches rich in nutrients, respectively) are very unlikely for small lowland or low mountain range streams in agricultural landscapes in Germany.

Sediment composition: Data from project 1 show that in small lowland or low mountain range streams in agricultural landscapes, sediment is dominated by sand,

TABLE 15.3
Characteristics of Small Lowland ($n = 10$) and Low Mountain Range ($n = 4$) Streams in Agricultural Landscapes in Northern and Central Germany (Data from Project 1, 1996 to 2000, April to September)

Parameter	Median	Arithmetic Mean	SD	Minimum	Maximum
Width (mean) (cm)[a]	83.3	81.5	39.4	20	150
Depth (maximum) (cm)[a]	11.2	15.5	10.8	5	35
Current velocity (m s^{-1})	0.2	0.2	0.1	0.001	0.37
Submerged plants (coverage of sediment) (%)[a]	25.1	28.4	27.9	0	86
Mud or organic silt (%)[a]	12.5	16.4	14.6	0	59
Loam (<0.1 mm) (%)[a]	25.0	28.0	25.0	0	69
Sand (0.1 to 2 mm) (%)[a]	24.5	31.8	29.3	0	92
Gravel (2 to 60 mm) (%)[a]	5.9	14.6	14.7	0	40
Stones (>60 mm) (%)[a]	2.7	9.1	13.3	0	40
pH (mean)	8.0	7.9	0.4	7.2	8.4

[a] Rough estimates.

loam, and organic silt, while gravel typically is not dominant in this type of water body. Besides the relatively low current velocity, soil erosion from arable fields and maintenance measures (dredging) probably promote fine material to the disadvantage of gravel and stones.

Macrophyte abundance: As shown by data from project 1, the proportion of the sediment area covered by submerged macrophytes varies widely between single water bodies (0 to 85%). A coverage degree of 25% can be considered typical.

CONSEQUENCES FOR RISK ASSESSMENT

In the risk assessment for small headwater streams in Germany (except streams in peaty areas, which were not covered by the discussed studies), pH values of 7.0 to 8.5 should be considered.

The influence of sorption to and degradation in macrophytes on the overall fate and behavior of pesticides in streams is probably limited as the degree of coverage by submerged vegetation is relatively low and absent in some streams.

15.2.4 BIOLOGICAL CHARACTERIZATION

Sufficient data are available for abundance of fish (project 3) and macroinvertebrates (projects 1 and 2), while no detailed data are available for aquatic macrophytes, periphyton, and zooplankton (e.g., copepods and rotifers).

15.2.4.1 Fish

In 15 of the streams in the surrounding area of Braunschweig, fish fauna was investigated by means of electrofishing in summer of 1995 (project 3, Wogram 1996). Those streams were permanent lowland streams of moderate-to-extensive structural degradation with the characteristics shown in Table 15.3.

Fish occurred in all streams with an absolute maximum depth of more than 10 cm (measured at low water level within a 100-m stream section), even in those with a median depth of less than 3 cm. In streams with a mean depth of less than 10 cm at low water level, only the 2 stickleback species *Gasterosteus aculeatus* and *Pungitius pungitius* were abundant. In streams with a mean depth of 10 to 40 cm at low water level, species of the families Anguillidae, Cyprinidae, Esocidae, Gasterosteidae, Percidae, and Salmonidae occurred.

As regards the frequency of occurrence (Figure 15.1) as well as relative abundance (Figure 15.2), the stickleback species *Gasterosteus aculeatus* and *Pungitius pungitius* were found to be the most common species. Other species found in more than 1 stream and in more than 1 individual, respectively, were the cyprinid species loach (*Barbatula barbatula*) and gudgeon (*Gobio gobio*), eel (*Anguilla anguilla*), and juvenile individuals of northern pike (*Esox lucius*).

Some other species, such as the cyprinid species *Tinca tinca, Rutilus rutilus, Pseudorasbora parva, Blikka bjoerkna,* and *Alburnus alburnus,* were only found in single individuals. In the context of a later project in 1998 (Wogram et al. 2000), additional species were found in some of the same streams with more than 1 individual per stream and effort: minnow (*Phoxinus phoxinus*), roach (*Rutilus rutilus*), and juvenile individuals of perch (*Perca fuviatilis*) and brown trout (*Salmo trutta* f. *fario*; unpublished data, not shown).

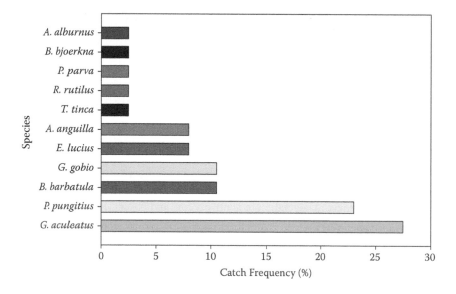

FIGURE 15.1   Catch frequency of fish species in samples (*n* = 39) from 15 small lowland streams in an agricultural landscape of northern Germany. (From project 3, Wogram 1996.)

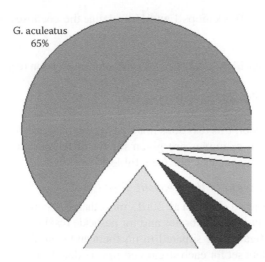

FIGURE 15.2   Relative abundance of fish species in 15 small lowland streams of an agricultural landscape in northern Germany. (From project 3, Wogram 1996.)

CONSEQUENCES FOR RISK ASSESSMENT

A potential exposure of fish to pesticides should always be considered in risk assessment for small lowland streams as fish typically occur even in streams much smaller than the FOCUSsw model stream and even in summer at low water level.

In particular, the following fish families should be considered in risk assessment for small lowland streams: Anguillidae, Cyprinidae, Esocidae, Gasterosteidae, Percidae, and Salmonidae. In low mountain range streams (not investigated in project 3), additional species may be expected (e.g., *Cottus gobio*, Cottidae).

15.2.4.2   Macroinvertebrates

According to Annex VI of Council Directive 91/414 EEC, the aim of an environmental risk assessment is to protect "populations of non-target species from long-term repercussions" caused by the use of pesticides. Based on the assumption that the value of a species does not depend on its ecological relevance, the description of the coenoses in this report does not focus on the question "Which taxa are dominant?" but rather "Which taxa are found frequently?"

15.2.4.2.1   Do We Need to Know All Species?

Based on the assumption that 1) species sensitivities toward toxicants are more similar within than between phylogenetic taxa (e.g., Wogram & Liess 2001; Maltby et al. 2005) and 2) the ecological vulnerability of a species depends on its specific ecological characteristics (e.g., Liess and von der Ohe 2005), it was agreed within

WG 6 at the ELINK Workshops I and II to define the coenoses on the basis of the following:

- taxa of a rather high level of taxonomical identification (e.g., genus, family, order)
- typical and relevant traits of these taxa

Following this concept, this text does not aim at an exhaustive description of species. Nevertheless, typical examples are given for the different taxa to ensure transparency in the variety of species traits (see Table 15.6).

15.2.4.2.2 Description of the Realistic Best Case

In the context of project 1 (Liess et al. 2001), more than 100 sampling sites in 30 streams were sampled for macroinvertebrates and for pesticide residues in sediment and water. The streams investigated are situated in northern and central Germany, respectively (Table 15.4). The data set for each stream comprises data from a period of 2 to 4 years at 2 to 8 biological samplings (surber sampler, Macan 1958) per year and an event-triggered sampling for pesticide residues in the period from April to July. However, a pesticide-sampling regime sufficient to detect also the absence of pesticides in the water at a sufficient level of certainty was only carried out in 14 of the 30 streams (2 event-triggered methods used in parallel over a period of at least 2 years; not shown).

For this report, data from project 1 were reanalyzed. The total range of sampling sites was scanned for criteria indicating the realistic worst case (see Section 15.2.2). Criteria for the selection were the type of land surrounding the water body (at least 50% agricultural use: arable land or grassland) and a toxicologically nonsignificant pesticide exposure. The following indicators were used for the verification of low exposure to pesticides:

- Organic farming only in the surrounding of the streams (also considering stream sections situated upstream).
- Grassland only in the surrounding of the streams (no tilth).
- In cases of conventional agriculture in the surrounding area, no pesticide residues found or pesticides found only below ecotoxicological threshold level. This criterion applies only to intensively monitored sites.

As a result of the scanning process, 14 sites (Table 15.4) on 13 streams were selected for the description of the realistic best case.

Taxa numbers: On average, 37 different macroinvertebrate taxa were found per sampling site. The majority of taxa comprised arthropods, particularly insects (Figure 15.3), with the order Trichoptera providing the highest number of taxa (8.3 taxa per site, Table 15.5). Other orders with considerably high numbers of taxa were Coleoptera, Diptera, and Ephemeroptera.

Abundance: In terms of relative abundance, the order Amphipoda, which was mainly represented by only 1 species, namely *Gammarus pulex,* was the most important taxon (Figure 15.3 and Table 15.5). Other codominant taxa (>10%) were Trichoptera, Diptera, and Eulamellibranchia. However, as already stated, the dominance status of a taxon should not be considered a driving factor in the risk assessment.

TABLE 15.4

Sampling Sites Selected for the Characterization of the Realistic Best Case in Small Lowland and Low Mountain Range Streams in Agricultural Landscapes (Data from Northern and Central Germany, Project 1, Liess et al. 2001)

Site Number (Project Internal)	Name of Stream	Location	Landscape	Type of Agricultural Land Use (%)	Criterion for Selection
4	Rolfsbüttler Bach	Rolfsbüttel (north of Braunschweig, NDS)	Südheide (lowland)	Tilth (60), grassland (40)	No or low pesticide exposure detected
8	Vollbüttler Mühlenriede	Ribbesbüttel (north of Braunschweig, NDS)	Südheide (lowland)	Tilth (75), grassland (25)	No or low pesticide exposure detected
59	Wabe	Erkerode (east of Braunschweig, NDS)	Elm (low mountain range)	Grassland	Grassland
102	Sennebach	Sillium (southwest of Braunschweig, NDS)	Hildesheimer Börde (loess plain)	Grassland	No or low pesticide exposure detected
121	n.a.	Wulksfelde (north of Hamburg, SH)	Oberes Alstertal (lowland)	Tilth	Organic farming
123	n.a.	Wohldorf (north of Hamburg, HH)	Oberes Alstertal (lowland)	Tilth (30), grassland (70)	Organic farming
126	Flottriede (up)	Hämelerwald (west of Braunschweig, NDS)	Lowland	Tilth (80), grassland (20)	Organic farming
127	Flottriede (down)	Hämelerwald (west of Braunschweig, NDS)	Lowland	Tilth (40), grassland (60)	Organic farming
134	Blumenhagener Bach	Lauenau (west of Hannover, NDS)	Deister (low mountain range)	Grassland	Grassland
140	Rodenberger Aue	Rodenberg (Süntel, NDS)	Weserbergland (low mountain range)	Grassland	Grassland
142	Hollenbach	Zersen (Süntel, NDS)	Weserbergland (low mountain range)	Grassland	Grassland
144	Mühlengraben	Elfenborn (north of Rinteln, NRW)	Rintelner Becken (lowland)	Tilth (50), grassland (50)	Organic farming
148	n.a.	Warburg-Nörde (north of Kassel, H)	Kasseler Becken (lowland)	Tilth (66), grassland (33)	Organic farming
152	Hambach	Oberhambach (south of Darmstadt, H)	Bergstrasse (low mountain range)	Grassland	Grassland

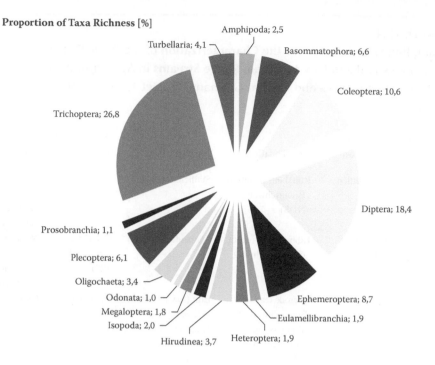

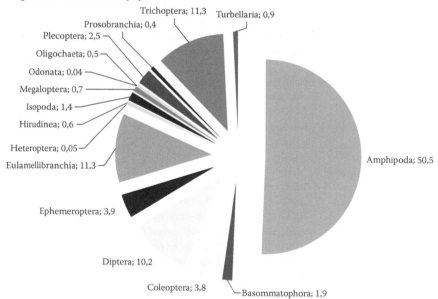

FIGURE 15.3   Relative distribution over main taxonomic groups of aquatic macroinvertebrate taxa (upper panel) and abundance (lower panel) in streams in agricultural landscapes in northern and central Germany (reanalyzed data from project 1).

TABLE 15.5

Distribution of Voltinism Categories Over Taxa and Abundance in Different Orders of Aquatic Macroinvertebrates in 14 Sites (13 Streams) in Northern and Central Germany (Reanalysis of Data from Project 1)

Order	Number of Taxa (Mean ± SD)	Frequency of Abundance (%)	Proportion of Taxa Numbers (Mean ± SD) (%)	Proportion of Abundance (Mean ± SD) (%)	Voltinism	Proportion of Taxa Numbers (Mean ± SD) (%)	Frequency of Abundance (%)
Amphipoda	1.1 (±0.3)	100	2.5 (±1.2)	50.5 (±25.1)	Pluri	0	0
					Uni	100 (±0)[a]	100[a]
					Semi	0	0
Basommatophora	2.5 (±1.6)	86	6.6 (±6.0)	1.9 (±4.0)	Pluri	4.2 (±14.4)	7
					Uni	95.8 (±14.4)	86
					Semi	0	0
Coleoptera	3.6 (±2.6)	100	10.6 (±4.3)	3.8 (±5.5)	Pluri	0	0
					Uni	60.4 (±22.2)	93
					Semi	39.6 (±22.2)	93
Diptera	3.3 (±0.6)	100	18.4 (±4.5)	10.2 (±8.2)	Pluri	40.7 (±12.5)	100
					Uni	59.3 (±12.5)	100
					Semi	0	0
Ephemeroptera	2.2 (±1.4)	79	8.7 (±6.4)	3.9 (±4.2)	Pluri	44.8 (±30.9)	79
					Uni	45.0 (±33.0)	57
					Semi	10.2 (±12.7)	43
Eulamellibranchia (Bivalvia)	1.3 (±0.7)	71	1.9 (±1.4)	11.3 (±20.8)	Pluri	0	0
					Uni	100	71
					Semi	n.d.[b]	n.d.[b]

(continued)

TABLE 15.5 (CONTINUED)

Distribution of Voltinism Categories Over Taxa and Abundance in Different Orders of Aquatic Macroinvertebrates in 14 Sites (13 Streams) in Northern and Central Germany (Reanalysis of Data from Project 1)

Order	Number of Taxa (Mean ± SD)	Frequency of Abundance (%)	Proportion of Taxa Numbers (Mean ± SD) (%)	Proportion of Abundance (Mean ± SD) (%)	Voltinism	Proportion of Taxa Numbers (Mean ± SD) (%)	Frequency of Abundance (%)
Heteroptera	1.8 (±1.4)	57	1.9 (±2.1)	0.05 (±0.09)	Pluri	0	0
					Uni	100	57
					Semi	0	0
Hirudinea	2.2 (±1.1)	79	3.7 (±2.5)	0.6 (±0.7)	Pluri	0	0
					Uni	100	79
					Semi	0	0
Isopoda	1.5 (±0.5)	57	2.0 (±2.1)	1.4 (±3.4)	Pluri	0	0
					Uni	100[a]	57[a]
					Semi	0	0
Megaloptera	1.0 (±0)	79	1.8 (±1.1)	0.7 (±1.4)	Pluri	0	0
					Uni	0	0
					Semi	100	79
Odonata	1.5 (±0.6)	29	1.0 (±1.8)	0.04 (±0.12)	Pluri	0	0
					Uni	66.7 (±47.1)	21
					Semi	33.3 (±47.1)	14
Oligochaeta	1.2 (±0.4)	93	3.4 (±1.6)	0.5 (±0.7)	Pluri	70.5 (±32.0)	86
					Uni	29.5 (±32.0)	50
					Semi	0	0

Plecoptera	2.5 (±2.3)	100	6.1 (±5.7)	2.5 (±2.9)	Pluri	0	0
					Uni	95.8 (±9.6)	64
					Semi	4.2 (±9.6)	21
Prosobranchia	1.0 (±0)	50	1.1 (±1.2)	0.4 (±1.2)	Pluri	0	0
					Uni	100	50
					Semi	0	0
Trichoptera	8.3 (±5.8)	100	26.8 (±5.5)	11.4 (±8.1)	Pluri	3.0 (±5.8)	29
					Uni	97.0 (±5.8)	100
					Semi	0	0
Turbellaria	2.0 (±1.0)	93	4.1 (±2.4)	0.9 (±1.2)	Pluri	100	93
					Uni	0	0
					Semi	0	0

[a] Species are uni- to bivoltine depending on ecological conditions.

[b] Sampling method was not suitable for the detection of large semivoltine mussel species (*Anodonta* spp., *Unio* spp.).

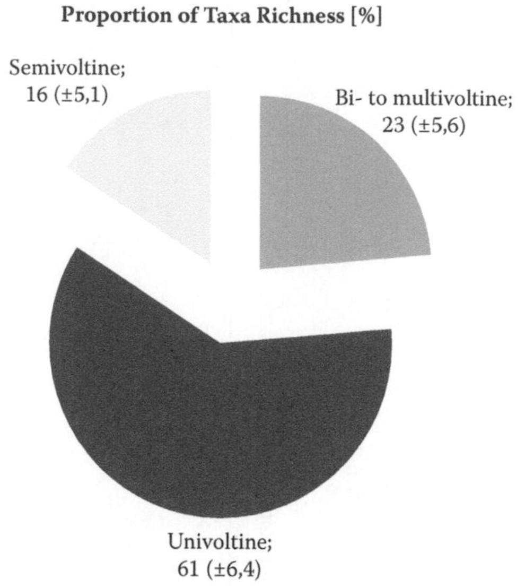

Proportion of Taxa Richness [%]

FIGURE 15.4   Relative distribution of voltinism categories over taxa in aquatic macroinvertebrates in streams in agricultural landscapes in northern and central Germany (reanalyzed data from project 1).

Frequency: An interesting result is that each of the orders investigated had a relatively high level of frequency, with the lowest frequency for Odonata with 29% of the sites, whilst most of the orders occurred in all or nearly all of the sites (Table 15.5).

Voltinism: Generally, most taxa found in the streams had a univoltine life cycle (Figure 15.4). However, also bi- to plurivoltine as well as semivoltine species were present. Relative abundance of different voltinism classes revealed a similar pattern (not shown).

To enable a suitable assessment of the risk to macroinvertebrates in streams, taxa richness and frequency of voltinism categories should also be described separately for the main taxa (Table 15.5). One of the most relevant results is that in most of the insect orders the majority of species have a univoltine life cycle, and within the insect orders, the univoltine species feature high levels of frequencies. Nevertheless, semivoltine taxa are also common in some orders (Coleoptera, Ephemeroptera, Megaloptera, Plecoptera, Odonata) with frequencies of 93, 43, 79, 21, and 14%, respectively (Table 15.5).

For reasons of transparency (and potentially also for the selection of suitable focal species), examples of typical semi-, uni-, and bi- to plurivoltine species from a case study are given in Table 15.6.

TABLE 15.6
Distribution of Voltinism Categories Over Taxa and Abundance in Different Orders of Aquatic Macroinvertebrates in the Vollbüttler Mühlenriede, Northern Germany (Analysis of Data from the Liess et al. Working Group at the University of Braunschweig, Years 1995 to 1999, Unpublished)

Order	Voltinism	Number of Taxa	Taxa (%)	Individuals (%)	Species (Example)
Amphipoda	Univoltine	1	100	100	*Gammarus pulex*[a]
	Bi- to plurivoltine		0	0	
	Semivoltine		0	0	
Basommatophora	Univoltine	4	100	100	*Lymnaea stagnalis*
	Bi- to plurivoltine		0	0	
	Semivoltine		0	0	
Coleoptera	Univoltine	9	100	100	*Platambus maculatus*
	Bi- to plurivoltine		0	0	
	Semivoltine	3	21	1	*Elmis aenea*
Diptera	Univoltine	7	50	24	Tabanidae unid.
	Bi- to plurivoltine	4	29	75	Chironomidae unid.
	Semivoltine		0	0	
Ephemeroptera	Univoltine	3	38	3	*Baetis vernus*
	Bi- to plurivoltine	4	57	6	*Baetis rhodani*
	Semivoltine	1	13	88	*Ephemera danica*
Eulamellibranchia	Univoltine	2	100	100	*Pisidium* sp.
	Bi- to plurivoltine		0	0	
	Semivoltine		0[b]	0[b]	
Heteroptera	Univoltine	4	100	100	*Nepa cinerea*
	Bi- to plurivoltine		0	0	
	Semivoltine		0	0	
Hirudinea	Univoltine	4	100	100	*Glossiphonia complanata*
	Bi- to plurivoltine		0	0	
	Semivoltine		0	0	
Isopoda	Univoltine	2	100	100	*Asellus aquaticus* [a]
	Bi- to plurivoltine		0	0	
	Semivoltine		0	0	
Megaloptera	Univoltine	1	100	100	*Sialis lutaria*
	Bi- to plurivoltine		0	0	
	Semivoltine		0	0	
Odonata	Univoltine	2	67	38	*Pyrrhosoma nymphula*
	Bi- to plurivoltine			0	
	Semivoltine	1	33	62	*Calopteryx splendens*
Oligochaeta	Univoltine	1	100	100	Tubificidae unid.
	Bi- to plurivoltine		0	0	
	Semivoltine		0	0	

(*continued*)

TABLE 15.6 (CONTINUED)
Distribution of Voltinism Categories Over Taxa and Abundance in Different Orders of Aquatic Macroinvertebrates in the Vollbüttler Mühlenriede, Northern Germany (Analysis of Data from the Liess et al. Working Group at the University of Braunschweig, Years 1995 to 1999, Unpublished)

Order	Voltinism	Number of Taxa	Taxa (%)	Individuals (%)	Species (Example)
Plecoptera	Univoltine	1	100	100	*Nemoura cinerea*
	Bi- to plurivoltine		0	0	
	Semivoltine		0	0	
Trichoptera	Univoltine	20	95	98	*Anabolia nervosa*
	Bi- to plurivoltine	1	5	2	*Mystacides longicornis*
	Semivoltine		0	0	
Turbellaria	Univoltine		0	0	*Polycelis nigra/tenuis*
	Bi- to plurivoltine	2	100	100	
	Semivoltine		0	0	

[a] Species are uni- to bivoltine depending on ecological conditions (referred to as multivoltine in Section 15.3).

[b] Sampling method was not suitable for the detection of large semivoltine mussel species (*Anodonta* spp., *Unio* spp.).

CONSEQUENCES FOR RISK ASSESSMENT

As indicated by the data on typical lowland and low mountain range streams, the following macroinvertebrate orders should generally be considered in risk assessment of edge-of-field streams in Germany: Amphipoda, Basommatophora, Coleoptera, Diptera, Ephemeroptera, Eulamellibranchia, Heteroptera, Hirudinea, Isopoda, Megaloptera, Odonata, Oligochaeta, Plecoptera, Prosobranchia, Trichoptera, and Turbellaria.

In terms of biological traits, it should be considered in the risk assessment that a univoltine life cycle is most common in most of the taxa, and that even semivoltine taxa are quite frequent. Emphasis should be placed on the potential risk to semivoltine insect species as these have been shown to be exceptionally vulnerable to pesticide stress (Liess et al. 2001; Beketov & Liess in review).

It should be noted that the sampling method used in project 1 was not suitable for the detection of large semivoltine mussel species (*Anodonta* spp., *Unio* spp.), which are known to be generally typical for small streams. Thus, the relevance of this potentially vulnerable taxon for the risk assessment of edge-of-field streams remains uncertain. The same applies to some other taxa that were not investigated in project 1 (e.g., microcrustaceans and mites).

15.2.5   CASE STUDY: TYPICAL LOWLAND STREAM IN AN AGRICULTURAL LANDSCAPE IN NORTHERN GERMANY

15.2.5.1   Introduction

To provide a transparent example for the type of water body described in this report, 1 of the streams investigated in each of the projects (1, 2, and 3, see earlier) is described in more detail: The Vollbüttler Mühlenriede, a small siliceous stream with a mean depth of 28 cm and a mean width of 222 cm (at low water level in summer) situated some 20 km north of Braunschweig, is embedded in a landscape of moderate agricultural use (Figure 15.5).

Crops cultivated in the surrounding area of the stream are cereals, potato, maize, some asparagus, and rape, but there is also some land used as meadow and pasture. The stream suffers from structural degradation (Figure 15.6) and is maintained by weed mowing (riparian vegetation: every year; emerse and submerse vegetation: approximately every 2 years). As shown in project 1 (Wogram 2001), the exposure to pesticides is low. In the years 1998 and 1999, only some contamination events with concentrations below effect threshold levels were found for some herbicidal and fungicidal substances (not shown). As assessed by the relative abundance of sensitive and vulnerable macroinvertebrate species, there is no indication of an alteration of the community caused by the exposure to pesticides (Wogram 2001).

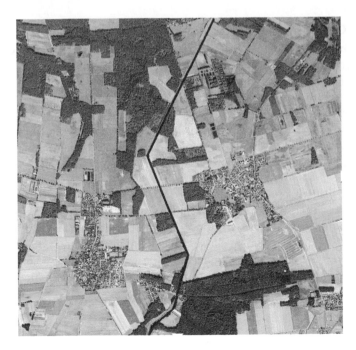

FIGURE 15.5   Air photograph of the Vollbüttler Mühlenriede (black line).

FIGURE 15.6   The Vollbüttler Mühlenriede in winter 2000. As a consequence of low runoff potential and buffer strips along adjacent fields, exposure to pesticide is generally low in this stream. (From Wogram 2001.)

15.2.5.2 Composition of the Macroinvertebrate Coenosis

In the Vollbüttler Mühlenriede, 77 different macroinvertebrate taxa were found. The majority of taxa comprised arthropods, particularly insects, with the order Trichoptera providing the highest number of taxa (20 taxa; Figure 15.7). Other orders with considerable numbers of taxa were Coleoptera, Diptera, and Ephemeroptera.

In terms of relative abundance, the order Amphipoda (represented by only 1 species, *Gammarus pulex*) was the most important taxon. Other dominant taxa were Trichoptera, Diptera, and Ephemeroptera.

15.2.5.3 Voltinism

Generally, most taxa found in the stream were univoltine species. However, also bi- to plurivoltine as well as semivoltine species were present (Figure 15.8). Relative abundance of different voltinism classes revealed a similar pattern. In addition, voltinism categories are described separately for the main taxa (Table 15.6). For reasons of transparency, examples of typical semi-, uni-, and bi- to plurivoltine species are also given in Table 15.6.

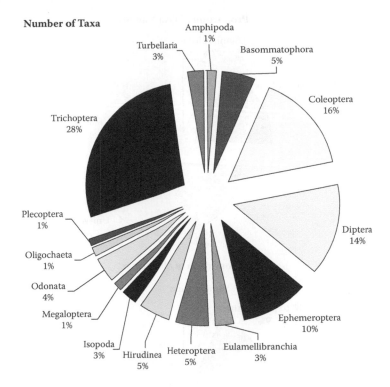

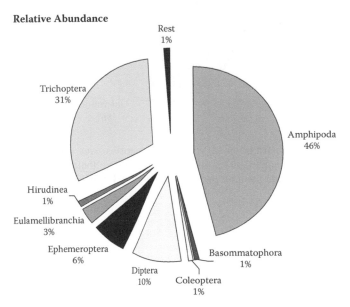

FIGURE 15.7 Relative distribution over main taxonomic groups of aquatic macroinvertebrate taxa (upper panel) and abundance (lower panel) in the Vollbüttler Mühlenriede, northern Germany (analysis of data from project 1, years 1995 to 1999).

Proportion of Taxa Numbers

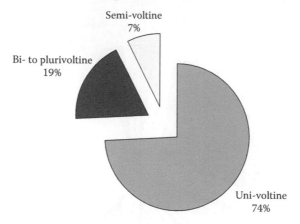

Proportion of Abundance

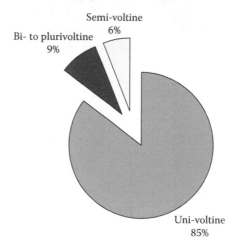

FIGURE 15.8 Relative distribution of voltinism categories over taxa (upper panel) and abundance (lower panel) in aquatic macroinvertebrates in the Vollbüttler Mühlenriede, northern Germany (analysis of data from project 1, years 1995 to 1999).

15.3 ECOLOGICAL CHARACTERIZATION OF DRAINAGE DITCHES IN THE NETHERLANDS TO EVALUATE PESTICIDE STRESS

Theo Brock, Gertie Arts, Dick Belgers, and Caroline Van Rhenen-Kersten

15.3.1 INTRODUCTION

An important aim of the registration procedure of plant protection products is to protect water organisms from possible exposure to these chemicals. To protect aquatic biodiversity from pesticide stress, it is necessary to have insight in the abiotic and biotic properties of the water bodies that can be found in the agricultural landscape. In the Netherlands, the most abundant water bodies in the agricultural landscape comprise drainage ditches. In these shallow waterways, aquatic macrophytes may form the bulk of the biomass. Species richness, densities, and biomass of macroinvertebrates are reported to be considerably higher in biotopes dominated by macrophytes than in comparable macrophyte-free sites (see, e.g., Beckett et al. 1992; Sagova et al. 1993; Brock and Van der Velde 1996). Consequently, a relatively high macroinvertebrate biodiversity is found in macrophyte-dominated ditches (Brock et al. 1980; Scheffer et al. 1984; Nijboer et al. 2003). The aim of this section is to give a quantitative and qualitative description of relatively biodiverse, macrophyte-dominated drainage ditches that can be found in the agricultural landscape of the Netherlands.

15.3.2 DENSITIES AND PHYSICAL CHARACTERISTICS OF DITCHES IN THE NETHERLANDS

The density and physical characteristics of ditches vary considerably in different landscape types of the Netherlands. Table 15.7 presents an overview of the surface area of different types of water courses (in m^2/km^2) in different areas of the Netherlands. Ditches that do not carry water in dry periods are relatively abundant in the provinces of Noord-Brabant, Gelderland, Overijssel, and Drenthe, while ditches that usually do not fall dry are more abundant in the lower parts of the Netherlands. Water-carrying ditches with a width less than 6 m are relatively abundant in the Dutch agricultural landscape. Wider ditches (>6 m) are relatively abundant in the bog area of Drenthe, the sea clay area of Westland, the fen area in Friesland, and the Beemster polder (Table 15.7).

15.3.3 IMPACT OF DITCH CLEANING AND DREDGING

An important characteristic of drainage ditches is that they are heavily managed to secure water transport. This management involves the partial or complete removal of vegetation mass by mechanical harvesting (ditch cleaning) and the removal of the upper mud layer to maintain water depth (ditch dredging). Twisk et al. (2003) reported that in the Netherlands the following machines are used for mechanical cleaning and dredging:

- *Ditch scoop.* A ditch scoop scrapes the vegetation out of the ditch together with a certain amount of mud. The scoop is either completely closed or has bars with large open spaces in between. Its range is about 3 m.

TABLE 15.7

Estimated Surface Area of Different Types of Water Courses (Mainly Ditches) (m²/km²) for Different Areas in the Netherlands[a]

	Trench or Dry Ditch	Width <3 m	Width 3 to 6 m	Width >6 m
Sand area Noord-Brabant	18752	5455	522	5240
Sand area Gelderland	11707	3713	1251	2727
Boulder clay area Overijssel	12640	4913	1067	739
Boulder clay area Gelderland	13894	4345	0	3602
Boulder clay area Drenthe	15385	9168	0	2555
River area Gelderland West	2098	2374	4361	3283
River area Gelderland Oost	15215	2623	707	2653
Sea clay area Groningen	2558	21408	5850	8841
Sea clay area Zuid-Holland	574	12535	4203	35
Sea clay area Noord-Holland	372	16690	7128	6580
Sea clay area Westland	22	23700	8991	19369
Fen area Friesland	1594	22123	7209	12107
Fen area Zuid-Holland	3409	64775	1850	6342
Noordoostpolder	1262	11339	1053	6052
Flevoland	7336	2075	0	4715
Beemster	492	27205	20525	10598
Bog area Drenthe	2645	735	2403	34185
Strandwallen Noord-Holland	1528	19330	6683	3504
Limburg (Krijt en Loss)	1398	4038	0	963

[a] Source top 10; Van der Gaast and Van Brakel 1997.

- *Mowing basket.* A mowing basket cuts plants off above the roots and then scoops the vegetation out of the ditch. The basket consists of bars with large open spaces in between. The range can be as much as 8 m.
- *Mowing drum.* A mowing drum removes plants from the ditch by grinding them into small pieces between the 2 rotating drums and spraying the remains over the adjacent fields.
- *Suction pipe.* A suction pipe removes mud from the ditch by means of a hose that extracts the mud and sprays it over the adjacent fields.
- *Pull-shovel.* A pull-shovel scrapes the mud from the ditch bottom and deposits it on a heap in the adjacent fields. A tractor subsequently spreads the mud over the fields.
- *Punched pull-shovel.* This machine works like the pull-shovel but is perforated to allow the water (and organisms) to run back into the ditch.

The abundance and species richness of aquatic organisms in ditches are reported to decrease with increasing frequency of ditch cleaning from 0 to 6 times per year (STOWA 1993). Now, however, mechanical harvesting of macrophytes usually takes place once a year in peat ditches (period September to October) and 2 times a year in clay and sand ditches (periods June to August and September to October). Dredging of ditches may be performed at intervals of 5 to 10 years (e.g., Peeters and Scheffer 2004). Mechanical ditch cleaning or dredging always has a temporary (negative) effect on the presence and abundance of aquatic macrophytes and macrofauna. This impact is dependent on the machines used, the frequency of cleaning and dredging, and the time of the year of mechanical management. For example, Twisk et al. (2003) reported that the use of the mowing drum resulted in lower species numbers of macrophytes than the use of the mowing basket or the ditch scoop. Arts et al. (1988) reported that removal of *Hottonia* and *Ranunculus* vegetation early in the growing season may result in a dominance of fast-growing macrophytes like *Elodea* and *Ceratophyllum*. Furthermore, Hesen et al. (1994) reported that the removal of dense submerged macrophyte vegetation in summer may result in an increase of floating *Lemna* vegetation. The macroinvertebrate community associated with dense *Lemna* vegetation is much lower than that in communities dominated by submerged macrophytes (e.g., Brock et al. 1980).

Ditch cleaning in early summer is reported to have negative effects on fish and amphibians ("Aangepast maaibeheer" 1998). Peeters and Scheffer (2004) mentioned significant lower macrofauna richness in ditches 8 days postcleaning by means of the mowing basket and ditch scoop. After dredging, full recovery of the macroinvertebrate community in ditches of the Hollands Noorderkwartier was observed after 24 months (Peeters and Scheffer 2004). Also, Twisk et al. (2000) observed that the presence of larvae of Trichoptera and amphibians could be lower during the first years after dredging. The recolonization potential of macroinvertebrates may be increased if the whole ditch system is not cleaned or dredged within a short time period, but patches of submerged vegetation are maintained that can act as refugia.

From the data presented, it appears that the normal operating range of environmental and ecological conditions (including species richness) can be characterized as dynamic in space and time due to the mechanical cleaning and dredging regime. This management regime favors aquatic plant and animal species with pioneering properties.

15.3.4 ECOLOGICAL CHARACTERIZATION OF DITCHES WITH A RELATIVELY HIGH BIODIVERSITY

Nijboer et al. (2003) presented an overview of the vegetation and macroinvertebrates that can be found in drainage ditches of the Netherlands. On basis of this information, they distinguished 13 different community types for Dutch ditches, of which community type MP (= cluster 1 in the classification of Nijboer et al.) has the highest macroinvertebrate biodiversity (mean number of taxa per sample was 67 with a range from 32 to 126). In total, 376 samples of macroinvertebrates were analyzed in this cluster–community type, containing a total of 676 different taxa and 741 231 different individuals of macroinvertebrates. For this reason, the MP-type ditches serve best to derive an "ecological scenario" that can be used for regulatory purposes within the context of the registration of plant protection products. The MP-type ditch of the Dutch agricultural landscape is moderately eutrophic to eutrophic and characterized by the presence of well-developed macrophyte vegetation. The MP ditches can be

TABLE 15.8

Some Important Abiotic and Biotic Properties of the MP Type of Ditches (Cluster 1) in the Netherlands[a]

	Median (10th to 90th percentile)
Water depth (m)	0.6 (0.3 to 1.0)
Width of ditch (m)	5 (2 to 10)
Electronic conductivity (mS/m)	40 (20 to 130)
pH	7.9 (7.3 to 8.5)
Depth organic debris layer (cm)	5 (1 to 35)
Total phosphorus (mg P/L)	0.5 (0.2 to 1.6)
Total nitrogen (mg N/L)	4.8 (2.3 to 8.8)
Ammonium (mg/L)	0.3 (0.0 to 1.6)
Nitrate (mg/L)	0.5 (0 to 4)
	Mean (range)
Macroinvertebrate species richness	67 (32 to 126)
Macrophyte species richness	(5 to 29)

[a] After Nijboer et al. 2003.

found in areas with a soil of clay (42%), peat (35%), and sand (23%). The water is characterized by relatively low chloride content (<500 mg/L). Note that in the coastal area of the Netherlands, the chloride content of surface water may be considerably higher due to seepage of saltwater. Some of the other important abiotic properties of the MP ditches are reported in Table 15.8.

In Table 15.9, detailed information is provided on bulk density, organic matter content, and volume fraction of liquid in sediment layers of experimental ditches that differ in age. In recently excavated or thoroughly "cleaned" ditches, the organic matter content of the upper sediment layer is much lower than in older

TABLE 15.9

Some Characteristics of Different Sediment Layers in Young and Old Experimental Ditches of the Sinderhoeve Complex (Renkum, the Netherlands)

Young Ditches (2 Years after Construction)				Old Ditches (5 Years after Construction)			
Sediment Layer (cm)	Organic Matter (%)	Dry Bulk Density (kg·dm⁻³)	Volume Fraction of Liquid	Sediment Layer (cm)	Organic Matter (%)	Dry Bulk Density (kg·dm⁻³)	Volume Fraction of Liquid
0 to 1	4.3	0.65	0.68	0 to 1	26	0.1	0.9
1 to 3	1.6	1.46	0.40	1 to 2	19	0.2	0.8
3 to 6	1.8	1.56	0.36	2 to 4	6	0.7	0.7
Below 6	1.9	1.54	0.36	4 to 10	2	1.6	0.4

macrophyte-dominated ditches, while the reverse is the case for the bulk density and volume fraction of liquid of the upper sediment layer.

The frequently observed macrophytes in the MP ditches are reported in Table 15.10. Note that this table only gives information on the frequency of occurrence of macrophytes. A frequently observed macrophyte does not need to be dominant or to be present with a relatively high biomass. The frequently observed macrophytes

---

TABLE 15.10

Frequently Observed Macrophytes in Vegetation-Rich, Eutrophic Ditches in the Netherlands (of the Type MP According to the Classification of Nijboer et al. 2003)[a]

	Frequency (%)
Lemna minor/Lemna gibba	68
Spirodela polyrhiza	52
Elodea nuttallii	50
Ceratophyllum demersum	47
Filamentous algae	44
Lemna trisulca	36
Callitriche sp.	30
Hydrocharis morsus-ranae	30
Glyceria fluitans	22
Potamogeton pectinatus	18
Potamogeton pusillus	18
Sagittaria sagittifolia	17
Butomus umbellatus	17
Wolffia arrhiza	16
Eleocharis palustris	15
Nuphar lutea	14
Equisetum fluviatile	14
Berula erecta	14
Alisma plantago-aquatica	13
Nymphaea alba	9
Azolla filiculoides	8
Potamogeton natans	8
Ranunculus circinatus	8
Chara sp.	8
Myriophyllum spicatum	7
Elodea canadensis	7
Hottonia palustris	7
Potamogeton crispus	6
Potamogeton lucens	6
Riccia fluitans	5

[a] The total number of ditch sites investigated was 237 (=100%). Only macrophytes are listed that could be found on at least 5% of the sampling localities. Data from Nijboer et al. (2003).

TABLE 15.11

Peak Biomass (in Grams Dry or Ash-Free Dry Weight per m²) of Macrophytes Observed in Macrophyte-Dominated Dutch Ditches

Dominant Macrophytes	Biomass (g/m²)	Location Ditch	Reference
Elodea nuttallii, Chara sp., *Myriophyllum spicatum, Sagittaria sagittifolia*	153	Sinderhoeve Renkum (Gld)	Van Geest et al. 1999
Lemna gibba	165	's-Gravenpolder (Z)	Brock 1988
Lemna minor	55 to 178	Sinderhoeve Renkum (Gld)	Arts et al. submitted
Potamogeton natans	165	Terschelling (Fr)	Van Vierssen 1982
Myriophyllum spicatum, Elodea nuttallii, Sagittaria sagittifolia	241	Sinderhoeve, Renkum (Gld)	Van Wijngaarden et al. 2006
Chara sp.	255	Sinderhoeve, Renkum (Gld)	Van den Brink et al. 1996
Potamogeton compressus	274	Vlijmen (N-Br)	Brock 1988
Potamogeton gramineaus	375	Nieuwkuijk (N-Br)	Brock 1988
Ceratophyllum demersum	593	Ooy (Gld)	Brock 1988
Eleocharis acicularis	676	Vlijmen (N-Br)	Brock 1988
Potamogeton trichoides	775	Vlijmen (N-Br)	Brock 1988
Pilularia globulifera	956	Nieuwkuijk (N-Br)	Brock 1988
Stratiotes aloides	1156	Nieuwkuijk (N-Br)	Brock 1988
Elodea nuttallii	1193	Vlijmen (N-Br)	Brock 1988

comprise species floating on or near the water surface (e.g., *Lemna minor/gibba, Spirodela polyrhiza*); nonrooting submerged macrophytes (e.g., *Ceratophyllum demersum*, filamentous algae, *Lemna trisulca*); rooted submerged macrophytes (e.g., *Elodea nuttallii, Callitriche, Potamogeton pectinatus, Potamogeton pusillus*); rooted floating-leaved macrophytes (e.g., *Nuphar lutea, Nymphaea alba, Potamogeton natans*); and rooted emergent macrophytes (e.g., *Sagittaria sagittifolia, Butomus umbellatus, Eleocharis palustris, Equisetum fluviatile*).

In the work of Nijboer et al. (2003), no information is provided on biomass of macrophytes in ditches. Literature data on the peak biomass of macrophytes in Dutch ditches are given in Table 15.11. Dependent on the dominant macrophytes present and their growth forms, reported peak biomass values in macrophyte-dominated ditches range from 153 to 1193 g (ash-free) dry weight per square meter. Peak biomass values of macrophytes in Dutch drainage ditches are usually found in the period July to August. In spring, biomass values for macrophytes of 50 to 250 g dry weight/m² may be observed in ditches. Because of the dense macrophyte growth in Dutch ditches, these systems are "cleaned" regularly by removing the vegetation to maintain the drainage function of the water courses. The macrophytes and macroinvertebrates typical for Dutch drainage ditches are adapted to this regular cleaning regime.

Frequently observed aquatic arthropods and nonarthropods in the macroinvertebrate community of MP ditches are reported in Tables 15.12 and 15.13, respectively. Note that these tables only give information on the frequency of occurrence of macroinvertebrates. A frequently observed macroinvertebrate does not need to be abundant in the majority of sampling localities. The most frequently reported aquatic arthropods in MP ditches (Table 15.12) comprise the isopod *Asellus aquaticus, Arrenurus* species (Hydracarina), the ephemeropteran *Cloeon dipterum,* the heteropteran *Sigara striata,* and several chronomids (e.g., *Glyptotendipes* sp., *Procladius* sp., *Cricotopus* gr. *sylvestris*) and coleopterans (e.g., *Haliplus ruficollis, Noterus crassicornis, Graptodytes pictus*).

The most frequently reported nonarthropods (Table 15.13) comprise snails (e.g., *Anisus vortex, Bithynia tentaculata, Planorbis planorbarius*), leaches (*Erpobdella octoculata, Glossiphonia heteroclita*), and oligochaetes (e.g., *Stylaria lacustris*).

15.3.5 VOLTINISM OF MACROINVERTEBRATES IN DUTCH DITCHES

The report of Nijboer et al. (2003) does not give a detailed description of the abundance and life-cycle characteristics of aquatic macroinvertebrates in ditches. For this reason, a more detailed description is given in this section for the macroinvertebrate community in 1) 19 ditches with clay sediment and adjacent to agricultural fields and 2) 18 ditches with clay sediment adjacent to forested or nature conservation areas (see Figures 15.9 and 15.10).

In the agricultural clay ditches and in the protected area clay ditches, respectively, 254 and 252 different macroinvertebrate taxa were found. In both types of ditches, the majority of taxa comprised arthropods, particularly Diptera (including Chironomidae), Coleoptera, Hydracarina, and Heteroptera. The most abundant nonarthropod taxa comprised snails and oligochaetes (upper panels in Figures 15.9 and 15.10).

In the samples from the 19 agricultural clay ditches and in those from the 18 nature area clay ditches, respectively, 12 763 and 11 129 individuals of macroinvertebrates were counted. In both types of ditches, the majority of individuals comprised aquatic arthropods. In the agricultural clay ditches, the relative contribution of Diptera (including Chironomidae) was relatively high (32%), while in the nature area ditches both Diptera (24%) and Gastrapoda (25%) showed a high relative contribution in the total number of individuals (lower panels in Figures 15.9 and 15.10). In both types of ditches, the most abundant macrocrustaceans were Isopoda (particularly *Asellus aquaticus*).

Table 15.14 presents the relative distribution of macroinvertebrates sampled in agricultural clay ditches and in the nature area clay ditches over different voltinism categories. A univoltine organism completes its life cycle once a year. A bivoltine organism has 2 generations per year, and a multivoltine organism has more than 2 generations per year. A semivoltine organism needs more than a year to complete its life cycle. Note that in different types of ecosystems and geographical areas, the voltinism of the same species may be different. For example, in the productive Dutch drainage ditches species like the amphipod *Gammarus pulex,* the isopod *Asellus aquaticus,* and the insect *Cloeon dipterum* usually are multivoltine, while in less-productive aquatic ecosystems in Scandinavia these species may be univoltine.

In appears from the data presented in Table 15.14 that in ditches all categories of voltinism are abundant, except semivoltine organisms. Furthermore a relatively large

TABLE 15.12

Frequently Observed Aquatic Arthropods in the Macroinvertebrate Community of Vegetation-Rich, Eutrophic Ditches in the Netherlands (of the Type MP According to the Classification of Nijboer et al. 2003)[a]

Species/Taxon	Frequency (%)	Taxonomic Group
Asellus aquaticus	92	Crustacea
Arrenurus globator	80	Hydracarina
Cloeon dipterum	79	Ephemeroptera
Sigara striata	66	Heteroptera
Glyptotendipes sp.	60	Chironomidae
Procladius sp.	60	Chironomidae
Cricotopus gr. sylvestris	59	Chironomidae
Caenis robusta	59	Ephemeroptera
Arrenurus crassicaudatus	58	Hydracarina
Haliplus ruficollis	58	Coleoptera
Limnesia undula	57	Hydracarina
Noterus crassicornis	56	Coleoptera
Graptodytes pictus	55	Coleoptera
Arrenurus sinuator	54	Hydracarina
Hyphydrus ovatus	52	Coleoptera
Hygrotus inaequalis	52	Coleoptera
Ilyocoris cimicoides	52	Heteroptera
Ischnura elegans	49	Odonata
Triaenodes bicolor	48	Trichoptera
Piona conglobata	47	Hydracarina
Parachironomus gr. arcuatus	46	Chironomidae
Sigara falleni	44	Heteroptera
Notonecta glauca	43	Heteroptera
Plea minutissima	42	Heteroptera
Laccophilus minutus	42	Coleoptera
Sialis lutaria	41	Megaloptera
Argyroneta aquatica	40	Aranea
Hydroporus palustris	40	Coleoptera
Cataclysta lemnata	40	Lepidoptera
Clinotanypus nervosus	39	Chironomidae
Endochironomus albipennis	39	Chironomidae
Endochironomus tendens	39	Chironomidae
Arrenurus latus	39	Hydracarina
Hygrotus versicolor	37	Coleoptera
Peltodytes caesus	37	Coleoptera
Noterus clavicornis	36	Coleoptera
Gammarus pulex	36	Crustacea
Polypedilium gr. nubeculosum	35	Chironomidae
Helophorus brevipalpis	34	Coleoptera

TABLE 15.12 (CONTINUED)
Frequently Observed Aquatic Arthropods in the Macroinvertebrate Community of Vegetation-Rich, Eutrophic Ditches in the Netherlands (of the Type MP According to the Classification of Nijboer et al. 2003)[a]

Species/Taxon	Frequency (%)	Taxonomic Group
Tanypus kraatzi	33	Chironomidae
Caenis horaria	32	Ephemeroptera
Atripsodes aterrinus	31	Trichoptera
Laccophilus hyalinus	30	Coleoptera
Hydrodroma despiciens	30	Hydracarina
Haliplus immaculatus	30	Coleoptera
Limnesia fulgida	29	Hydracarina
Anacaena limbata	28	Coleoptera
Piona alpicola	28	Hydracarina
Tanytarsus sp.	27	Chironomidae
Haliplus fluviatilis	27	Coleoptera
Gammarus tigrinus	27	Crustacea
Proasellus meridianus	27	Crustacea
Piona coccinea	27	Hydracarina
Corixa punctata	26	Heteroptera
Xenopelopia sp.	25	Chironomidae
Psectrocladius varius	25	Chironomidae
Cymatia coleoptrata	25	Heteroptera
Microtendipes chloris agg.	24	Chronomidae
Hesperocorixa limnaei	24	Heteroptera
Enochrus testaceus	23	Coleoptera
Unionicola crassipes	23	Hydracarina
Limnesia maculata	23	Hydracarina
Piona nodata	22	Hydracarina
Ocetus furva	22	Trichoptera
Haliplus lineatocollis	21	Coleoptera
Laccobius minutus	21	Coleoptera
Eylais extendens	20	Hydracarina
Holocentropus picicornis	20	Trichoptera
Piona pusilla	20	Hydracarina
Piona imminuta	18	Hydracarina
Haliplus heydeni	17	Coleoptera
Heleocharis obscurus	17	Coleoptera
Microvelia reticulata	17	Coleoptera
Hydracarina cruenta	17	Hydracarina
Neumania deltoides	17	Hydracarina
Pionopsis lutescens	17	Hydracarina
Acricotopus lucens	16	Chironomodae

(continued)

TABLE 15.12 (CONTINUED)
Frequently Observed Aquatic Arthropods in the Macroinvertebrate Community of Vegetation-Rich, Eutrophic Ditches in the Netherlands (of the Type MP According to the Classification of Nijboer et al. 2003)[a]

Species/Taxon	Frequency (%)	Taxonomic Group
Gerris lacustris	16	Heteroptera
Arrenurus buccinator	16	Hydracarina
Psectrocladius gr. *sord/limb.*	15	Chironomodae
Mideopsis orbicularis	15	Chironomodae
Cryptochironomus sp.	15	Chironomidae
Endochironomus gr. *dispar*	15	Chironomidae
Helochares lividus	15	Coleoptera
Hydrobius fuscipes	15	Coleoptera
Arrenurus securiformis	15	Hydracarina
Neumania vernalis	15	Hydracarina
Ablabesmyia longistyla	14	Chironomidae
Laccobius bipunctatus	14	Coleoptera
Hydrachna conjecta	14	Hydracarina
Arrenurus fimbriatus	14	Hydracarina
Corynoneura scutellata agg.	13	Chironomidae
Spercheus emarginatus	13	Coleoptera
Podura aquatica	13	Collembola
Nepa cinerea	13	Heteroptera
Elylais hamata	13	Hydracarina
Hygrobatis longipalpis	13	Hydracarina
Tiphys ornatus	13	Hydracarina
Cyrnus flavidus	13	Trichoptera
Paratanytarsus sp.	12	Chironomidae
Dicrotendipes nervosus	12	Chironomidae
Anacaena globulus	12	Coleoptera
Chaoborus flavicans	12	Diptera
Agabus undulatus	11	Coleoptera
Enochrus melanocephalus	11	Coleoptera
Hesperocorixa sahlbergi	11	Heteroptera
Hydraclina globosa	11	Hydracarina
Cladotanytarsus sp.	10	Chironomidae
Proasellus coxalis	10	Crustacea
Brachypoda versicolor	10	Hydracarina
Erythromma najas	10	Odonata

[a] The total number of samples investigated was 376 (= 100%). Only arthropods are listed that could be found in at least 10% of the sampling localities. Data from Nijboer et al. (2003).

TABLE 15.13
Frequently Observed Aquatic Nonarthropod Macroinvertebrates of Vegetation-Rich, Eutrophic Ditches in the Netherlands (of the Type MP According to the Classification of Nijboer et al. 2003)[a]

Species/Taxon	Frequency (%)	Taxonomic Group
Anisus vortex	88	Gastropoda
Bithynia tentaculata	88	Gastropoda
Planorbis planorbarius	79	Gastropoda
Physa fontinalis	76	Gastropoda
Valvata piscinalis	74	Gastropoda
Lymnaea stragnalis	71	Gastropoda
Erpobdella octoculata	71	Hirudinea
Glossiphonia heteroclita	69	Hirudinea
Bithynia leachi	60	Gastropoda
Stagnicola palustris	59	Gastropoda
Glossiphonia complanata	59	Hirudinea
Gyraulis albus	59	Gastropoda
Theromyzon tessulatum	57	Hirudinea
Bathyomphalus contortus	49	Gastropoda
Planorbarius corneus	49	Gastropoda
Sphaerium corneum	47	Bivalvia
Stylaria lacustris	47	Oligochaeta
Planorbis carinatus	40	Gastropoda
Hippeutis complanatus	38	Gastropoda
Erpobdella testacea	37	Hirudinea
Piscicola geometra	32	Hirudinea
Gyraulus crista	29	Gastropoda
Lumbriculus variegatus	29	Oligochaeta
Valvata cristata	28	Gastropoda
Musculium lacustre	26	Bivalvia
Radix ovata	26	Gastropoda
Hemiclepsis marginata	23	Hirudinea
Limnodrilus hoffmeisteri	23	Oligochaeta
Ophidonais serpentina	20	Oligochaeta
Dugesia lugubris	19	Tricladida
Polycelis tenuis	19	Tricladida
Viviparus contectus	15	Gastropoda
Anisus vortex	14	Gastropoda
Dendrocoelum lacteum	13	Tricladida
Acroloxus lacustris	12	Gastropoda
Limnodrilus claperedeianus	12	Oligochaeta
Polycelis nigra	12	Tricladida

[a] The total number of samples investigated was 376 (=100%). Only arthropods are listed that could be found in at least 10% of the sampling localities. Data from Nijboer et al. (2003).

Agricultural Clay Ditches (total 254 taxa)

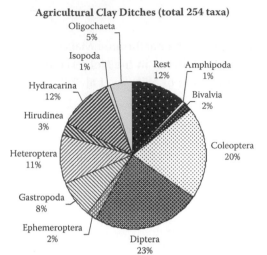

Agricultural Clay Ditches (total 12763 individuals)

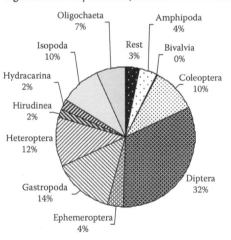

FIGURE 15.9 Relative distribution over main taxonomic groups of aquatic macroinvertebrate taxa (upper panel) and individuals (lower panel) in 19 samples from ditches with clay sediment and adjacent to agricultural fields. (From Arts et al. in preparation.)

proportion of the taxa (and individuals) could not yet properly be assigned to a certain voltinism category due to lack of ecological information. Another striking observation is that, when focusing on the percentage of individuals, the category of univoltine organisms is more abundant in the nature area ditches (34%) than in the agricultural area ditches (19%). In contrast, in the agricultural area ditches the bivoltine and multivoltine organisms have a larger share in the relative distribution of individuals.

Table 15.15 presents the relative distribution of voltinism categories over the individuals of the main taxonomic groups (Insecta, Macrocrustacea, Mollusca) sampled in

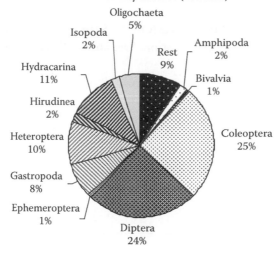

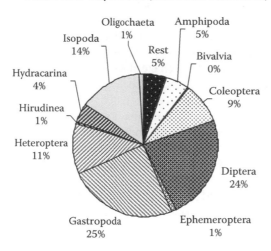

FIGURE 15.10 Relative distribution over main taxonomic groups of aquatic macroinvertebrate taxa (upper panel) and individuals (lower panel) in 18 samples from ditches with clay sediment and adjacent to forested or nature conservation areas.

the 2 types of ditches. It appears that in Dutch drainage ditches the multivoltine organisms mainly comprise insects and macrocrustaceans (predominantly *Asellus aquaticus*). In the clay ditches of both the agricultural area and the nature conservation area, the bivoltine organisms mainly comprise insects followed by molluscs. In the nature area ditches, molluscs have the highest relative contribution to the total number of univoltine individuals, while that is true of insects in the ditches of the agricultural area. In both types of ditches, all semivoltine macroinvertebrate individuals comprise insects.

TABLE 15.14

Relative Distribution of Macroinvertebrates Sampled in Clay Ditches of Agricultural and Nature Conservation Areas over Different Voltinism Categories[a]

	Taxa (%)		Individuals (%)	
	Agricultural Area	Nature Area	Agricultural Area	Nature Area
Univoltine	33	35	19	34
Bivoltine	20	18	25	16
Multivoltine	8	8	41	31
Semivoltine	2	2	<1	1
Unknown	37	37	15	18

[a] From Arts et al. in preparation.

An important question at stake is whether the different voltinism categories of freshwater species differ in sensitivity to pesticides. To investigate this, the available toxicity data for freshwater insects and a selected number of insecticides (data-rich compounds) were grouped into semi- or univoltine, bivoltine, and multivoltine taxa, and for each compound the HC50 (hazardous concentration to 50% of the taxa) was calculated for each voltinism group (see Figure 15.11). Insects were chosen to make this comparison since Maltby et al. (2005) demonstrated that in particular taxonomy drives the sensitivity of freshwater organisms to insecticides. In the case of insecticides, arthropods comprise the most sensitive taxonomic group. In addition, in freshwater ecosystems aquatic insects dominate the macroinvertebrate community (Figures 15.9 and 15.10) and have representatives in all voltinism categories (Table 15.15).

It appears from Figure 15.11 that the 95% confidence limits around the mean HC50 values are relatively large in case the number of taxa is relatively low (the number of toxicity data for different taxa can be found above the bars in Figure 15.11). Overall, a trend can be observed that multivoltine species are equally or more sensitive to insecticides than bivoltine and semi- or univoltine taxa, with the exception of the carbaryl data, in which multivoltine species tend to be less sensitive. These data suggest that the $NOEC_{community}$ (NOEC, no observed effect concentration) values for insecticides derived from micro/mesocosm test systems primarily populated with mutivoltine insects can be extrapolated to insect communities characterized by a larger variation in voltinism categories.

15.3.6 Mode of Action of Pesticides and Sensitive Aquatic Species

Agricultural pesticides are chemicals deliberately released into the environment to control pest species that harm agricultural crops. These agrochemicals can only be used if they do not harm the crop, crop rotation, and beneficial organisms in

TABLE 15.15

Relative Distribution of Voltinism Categories Over Individuals of Insecta, Marocrustacea, Mollusca, and Other Macroinvertebrates Sampled in Clay Ditches of Agricultural and Nature Conservation Areas Over Different Voltinism Categories[a]

	Agricultural Area (%)				Nature Conservation Area (%)			
	Multivoltine	Bivoltine	Univoltine	Semivoltine	Multivoltine	Bivoltine	Univoltine	Semivoltine
Insecta	68	59	64	100	29	65	32	100
Crustacea	32	0	0	0	69	0	0	0
Mollusca	<1	33	36	0	<1	31	67	0
Others	<1	8	<1	0	2	4	1	0
Total	100	100	100	100	100	100	100	100

[a] From Arts et al. in preparation.

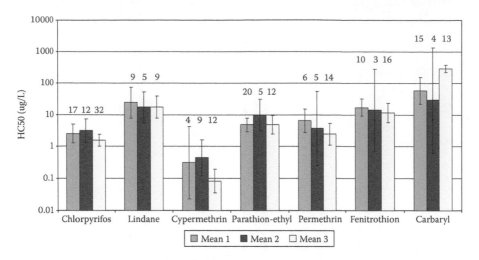

FIGURE 15.11 Hazardous concentration to 50% (mean HC50 and 95% confidence interval) of freshwater insects belonging to semi- or univoltine taxa (category 1), bivoltine taxa (category 2), and multivoltine taxa (category 3). The calculations are based on acute toxicity data (LC50 and EC50 values), and results are presented for the organophosphorus insecticides chlorpyrifos, parathion-ethyl, and fenitrothion; for the pyrethroids cypermethrin and permethrin; for the carbamate carbaryl; and for the organochlorine compound lindane. The numbers above the bars indicate the number of taxa within each voltinism category for which toxicity data were available.

agroecosystems (e.g., bees and earthworms). For this reason, the modern pesticides developed by the agrochemical industry usually are characterized by a specific toxic mode of action. Aquatic ecosystems, however, contain species related to the target organisms of pesticides. For example, the SSD curves presented in Figure 15.12 for herbicides reveal that algae or macrophytes comprise the most sensitive taxonomic groups. The SSD data for several insecticides indicate that arthropods in particular comprise the most sensitive taxonomic group (Figure 15.13). In contrast, SSD curves of fungicides (see Figure 15.14) may have vertebrates or nonarthropods as the most sensitive taxonomical group (see, e.g., the chlorothalonil data), while others can be characterized by biocidal properties in that representatives of several taxonomic groups are sensitive (see, e.g., the triphenyltinacetate data).

The data presented in Figure 15.13 suggest that in particular arthropods can be considered as the relevant sensitive group for insecticides. It should be noted, however, that different potentially sensitive taxonomic groups within arthropods may show a high in-between variability in sensitivity to insecticides. For example, Plecoptera and Ephemeroptera can be found over the whole range of the SSD curve for chlorpyrifos. This suggests that the WFD metric developed for macroinvertebrates cannot be used directly to evaluate toxic effects on invertebrate community composition. Most probably, the WFD metric is more related to overall vulnerability to multistressors than to sensitivity to pesticides per se.

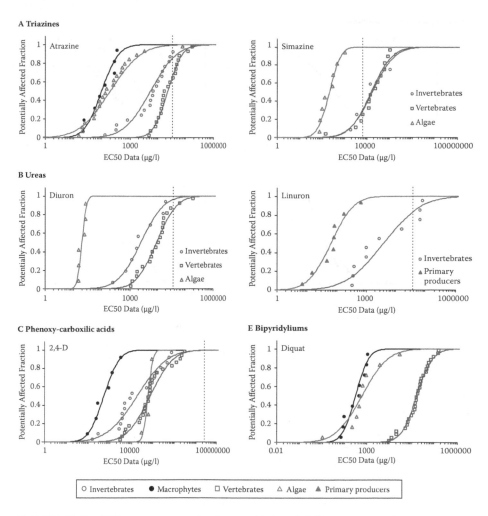

FIGURE 15.12 SSD curves comparing the sensitivity of different taxonomic groups to different classes of herbicides (---- indicates limit of solubility). (Data after Maltby et al. 2002 and van den Brink et al. 2006.)

From the data presented, it is clear that the specific toxic mode of action of insecticides, herbicides, and fungicides should not be ignored when assessing the risks of these chemicals for freshwater communities by means of the SSD approach and when selecting measurement endpoints in the conduct and evaluation of micro/mesocosm studies. Within this context, it should be noted that the toxicity metric for pesticides in ditches as described in STOWA (1993) is highly biased for relatively persistent insecticides and at most is indicative for pesticides used approximately 30 years ago. In recent years, many new and less-persistent substances were placed on the market, while the agricultural use of many old and persistent substances is no longer allowed.

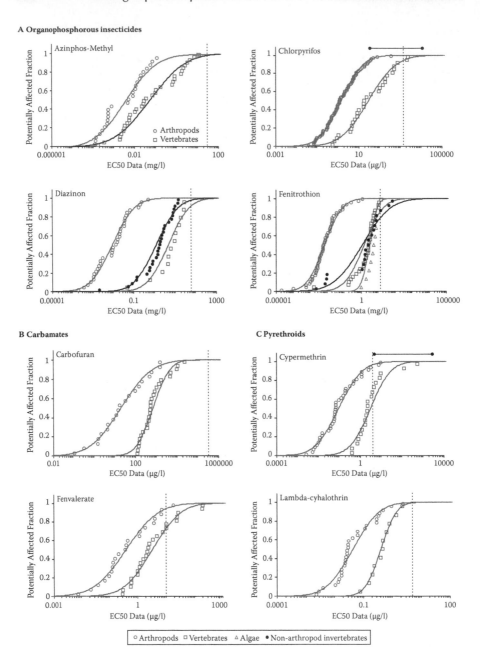

FIGURE 15.13   SSD curves comparing the sensitivity of different taxonomic groups to different classes of insecticide (----- indicates limit of solubility). (Data after Maltby et al. 2002, 2005.)

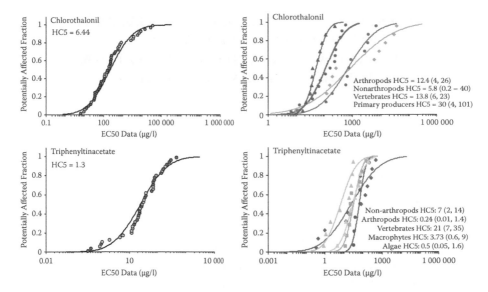

FIGURE 15.14 SSD curves comparing the sensitivity of different taxonomic groups to the fungicides chlorothalonil and triphenyltinacetate. The HC5 is the concentration below which 95% of the taxa tested will not be affected above their EC50. The 90% confidence limits of these HC5 values are presented between the brackets. The different grays represent different taxonomic groups of water organisms (see also Brock et al. 2006).

15.3.7 SUMMARY AND CONCLUSIONS

Macrophytes in eutrophic ditches are dominated by floating and submerged species. The diversity of the invertebrate community in macrophyte-dominated ditches is reported to be potentially high. The normal operating range of environmental and ecological conditions can be characterized as dynamic in space and time due to the mechanical cleaning and dredging regime. This management regime favors plant and animal species with pioneering properties. Invertebrates with 2 or many generations per year are widely distributed in Dutch drainage ditches of agricultural landscapes, but taxa with 1 generation per year are also common (approximately 33% of the taxa and 19% of the individuals). In drainage ditches many species sensitive to pesticides with different modes of action can be found: arthropods (sensitive to insecticides), macrophytes, and algae (sensitive to herbicides). For fungicides, several taxonomic groups (including worms, snails, and fish) might include sensitive organisms. The specific toxic mode of action and the sensitivity of different groups of taxa have to be taken into account when evaluating the effects of pesticides. For insects that differ in number of generations per year, no significant difference in overall sensitivity to insecticides could be demonstrated. This suggests that the $NOEC_{community}$ values for pesticides derived from micro/mesocosm test systems primarily populated by species with short life cycles can be extrapolated to communities characterized by a larger variation in length of life cycle. When considering recovery in the risk assessment, however, species-specific properties such as number of generations per year and dispersal abilities within the agricultural landscape cannot be ignored.

15.4 ECOLOGICAL CHARACTERIZATION OF PERMANENT AND EPHEMERAL STREAMS OF A TYPICAL MEDITERRANEAN AGRICULTURAL LANDSCAPE (EAST AND SOUTHEAST OF IBERIAN PENINSULA)

Elena Alonso Prados and Apolonia Novillo-Villajos

15.4.1 INTRODUCTION

This section gives a qualitative description of relatively biodiverse Mediterranean streams under high agriculture influence and provides some characteristics of biotic communities that can be found in these aquatic systems. In this case study, some typical water bodies of the agricultural landscape in the Iberian Peninsula have been selected. This area is heterogeneous; it includes a vast range of lithological and climatic contrasts, resulting in aquatic systems containing extraordinary richness and variety. The Iberian Peninsula is predominantly a semiarid region under the influence of Mediterranean climate. In the agricultural landscape, both permanent (influenced by melting snow) and ephemeral streams (e.g., ramblas) can be found.

15.4.2 CHARACTERISTICS OF AQUATIC SYSTEMS AND MACROINVERTEBRATE COMMUNITIES IN THE MEDITERRANEAN REGION

Aquatic systems are subject to the features of climate, relief, geology, and soil type. At the same time, the characteristics of the water bodies determine the composition and traits of the communities of organisms. Figure 15.15 summarizes some of the principal factors defining water body characteristics of the Mediterranean region.

The Mediterranean region is characterized not only by spatial variability but also by temporal variability related to seasonal (dry summers and cool wet winters with intense rainfalls) and climatic (long drought periods) phenomena. Thus, Mediterranean rivers are characterized by an annual and interannual discharge regime with annual floods and droughts. In that sense, Mediterranean rivers are subjected to 2 annual predictable perturbations (but with intensity and frequency unpredictable), implying the presence of permanent and temporary rivers (including intermittent and ephemeral ones). At the landscape level, the presence of permanent pools may be a key factor to maintain biodiversity in ephemeral streams. Recent research has shown that Mediterranean limnosystems do not comply with the current ecosystem paradigm based on research in the temperate region (reviewed by Álvarez-Cobelas et al. 2005). These authors outlined several differences between Mediterranean and temperate freshwater ecosystems. A summary of these differences is outlined next:

1) Many limnological processes depend in part on landscape features, and the catchment–lake ratio is 1 order of magnitude greater in Mediterranean areas than in colder temperate areas.
2) Many Mediterranean aquatic environments are shallow, increasing the importance of emergent and submerged plants and benthic–pelagic interactions in the overall ecological functioning.

FIGURE 15.15 Principal factors defining permanent and temporal (= ephemeral) water bodies in the Mediterranean region.

3) In the Mediterranean region, groundwater hydrology is as important as surface water hydrology for the functioning of aquatic systems. It is either lagged from rainfall seasonality or lacks any seasonality because it depends in turn on the hydrogeological features of the aquifer.
4) The number of sunny hours in the year affects the length of the growing season and hence the variability of annual phytoplankton biomass.
5) Chlorophyll responses to total phosphorus appear to be functionally distinct between temperate and Mediterranean areas.
6) The threshold of highest species richness at intermediate nutrient concentration seems to be shifted toward higher total phosphorous values in Mediterranean freshwaters.
7) Food webs are complex, and the role of cyprinids, the dominant fishes in Mediterranean environments, enhances interactions between sediment and the water columns by browsing activities.

In the context of placing plant protection products on the European market, 2 FOCUS exposure scenarios (R4 and D6) are defined for Mediterranean aquatic systems in agricultural landscapes. Taking into consideration some specific features (e.g., intermittent flowing, high salinity) of Mediterranean aquatic systems outlined, the necessity of developing new scenarios or improving the actual FOCUS scenarios for risk assessment in the Mediterranean region should be considered for the future. However, this question is outside of the scope of this report. Possible differences in vulnerability of aquatic species to pesticide stress between Mediterranean and temperate regions are an important question to be addressed. Studies focusing on ecological traits of aquatic macroinvertebrate communities of Mediterranean and temperate regions have been conducted by Bonada, Doledec, et al. (2007). These authors assembled the abundance of stream macroinvertebrate genera of 265 sites each from the Mediterranean Basin and from temperate Europe and linked these abundances to available information (published) on 61 categories of 11 biological traits reflecting the potential of resilience and resistance to disturbances. The more significant findings of this work were as follows:

1) Both local taxonomic and trait community composition differed between regions and were associated with a higher frequency of temporary streams in the Mediterranean region.
2) Concerning their taxonomic composition, permanent sites are dominated by Ephemeroptera, Plecoptera, and Trichoptera. If sites are temporary (ephemeral) and pools prevail during droughts, Odonata, Coleoptera, and Heteroptera dominate (Bonada, Doledec, et al. 2007; Bonada, Rieradevall, et al. 2007).
3) The trait composition of communities was less variable across geographical areas than their taxonomic composition. This is in agreement with works from Statzner et al. (2001, 2004) that reported relatively high invertebrate trait stability across European streams and large rivers. However, some differences between trait categories were observed between Mediterranean and temperate regions (see Table 15.16 after Bonada, Doledec, et al. 2007). The abundance of stream macroinvertebrates of 530 sites was assembled (265 M sites and 265 T sites), distributed from Scandinavia to Morocco and Turkey. The authors tested for regional differences in the trait categories using nonparametric Kruskal-Wallis tests. Several expected differences of trait categories between streams in Mediterranean and temperate regions were confirmed and summarized in this table. "M" indicates that the category proportion should be higher in the Mediterranean than in the temperate region, and "T" indicates the opposite.
4) Regional taxonomic richness, local trait richness, and diversity were significantly higher in the Mediterranean region.

The information presented suggests that M sites are characterized by macroinvertebrates with higher dispersion, colonization, and resilience capabilities when compared with similar systems in temperate regions of Europe. In the following sections, a compilation of ecological scenarios for pesticide risk assessment in the framework of 91/414/EEC is presented. These sections describe qualitative ecological characteristics of permanent streams (Júcar Basin) in east of Spain and ephemeral streams (ramblas) from the southeast of Spain. Data used for this report were compiled of public databases (www.chj.es) or scientific literature.

TABLE 15.16
Freshwater Macroinvertebrates Trait Categories from Mediterranean (M Sites) and Temperate Streams (T Sites)[a]

Biological Trait	Category	Differences between Streams in Mediterranean and Temperate Regions[b]
Maximum size (mm)	≤2.5	M
	>2.5 to 5	M
	>5 to 10	T
	>10 to 20	T
	>20 to 40	M
	>40 to 80	M
Life cycle duration	≤1	No significant differences between M and T
Aquatic stages	Larva	T
	Imago	M
Reproduction	Terrestrial clutches	M
Dispersal	Aquatic passive	T
	Aerial active	M
Resistance form	Diapause or dormancy	M

[a] Adapted from Bonada, Doledec, et al. 2007.
[b] The differences between M and T sites were at the macroclimate scale, ignoring mesoscale differences. The M and T sites were classified taking into account physical characteristics described in Köppen climatic maps (1931). In this article, the authors refer to the term "temperate" as sites covering several European biogeographical provinces from continental to oceanic microclimates.

15.4.3 STREAMS FROM EAST OF IBERIAN PENINSULA: JÚCAR BASIN

The area selected for this analysis is located in the Júcar River district (east of Spain); it belongs to the Iberic-Macaronesian Ecoregion. It should represent scenario R4 defined in the FOCUS SW. The Júcar River district extends to Comunidad Valenciana and part of Castilla-La Mancha, Aragón, and Cataluña. The average precipitation is 500 mm, varying from 250 mm in the south to 900 mm in the north. The annual average runoff is about 15% of the precipitation. Extreme hydrological events such as floods (during autumn) and droughts are quite common. The most important rivers in the basin are the Júcar, Turia, and Mijares, and numerous valuable wetlands are also frequent. According to CORINE 2000 land cover, agricultural use covers about 46% of the land use in the river district, predominant in the coastal and Castilla-La Mancha areas. Agricultural nonirrigated areas cover 36% of the territory, and agricultural irrigated areas cover 10%. In the low and middle parts of these watersheds, intensive agriculture is concentrated (Figure 15.16).

Some important abiotic and biotic properties of the different monitoring stations of the Mijares River (10 stations), the Jucar River (11 stations), and the Turia River (13 stations) are reported in Table 15.17. Some of the sampling points located in the agriculture-impacted area had relatively high conductivity levels (>1100 μS/cm), suggesting that besides pesticides other stressors may play a role. From Table 15.17, it is clear that with pH values

Source: Environmental atlas of the Mediterranean.

FIGURE 15.16   Common agriculture landscape in the east of Iberian Peninsula and low level of Mijares watercourse. The shaded gray in the box represents the citris crop throughout the map and the arrow indicates the dry bed at the zone of the mouth. The production of citrus in La Comunidad Valenciana represents 56% of the Spanish production. Scale 1:100.000.

typically ranging from 7.0 to 8.5, extremely acid or alkaline conditions are not common in these streams of the east part of the Iberian Peninsula. In addition, high changes in the annual stream flow were detected, as expected in a Mediterranean region.

The frequently observed aquatic macroinvertebrates (arthropods and nonarthropods) in the river stations selected (Júcar Basin) are reported in Table 15.18. A higher number of taxa than those summarized in Table 15.18 were found, but for further analysis we only used taxa that were recorded in at least 20% of all the samples from 2000 to 2005. Note that a frequently observed macroinvertebrate does not need to be dominant or to be present in high numbers. Taking this into consideration, the most frequently reported aquatic arthropods in Júcar Basin streams comprise the ephemeropteran *Caenis luctosa* and *Ephemerella ignita*, the trichopteran *Hydropsyche* sp., coleopterans from the Elmidae family, the Diptera taxa Chironomidae and Simulidae, and crustaceans belonging to the Gammaridae family. The most frequently recorded nonarthropods (Table 15.18) were the gastropods *Ancylus fluviatilis* and *Potamopyrgus antipodarum*, the leech *Dina lineata*, and Oligochaeta species.

We identified 5 stations with "good ecological status" to use them as reference stations. A comparative analysis between agriculture-impacted versus nonimpacted sites showed that

1) In agriculture-impacted areas (= 29 stations), a significantly lower number of macroinvertebrate taxa (mean = 36) per station was observed versus stations without a clear impact of agriculture (mean = 84).
2) Differences in the relative distribution of number of individuals over main taxonomic groups were observed. In the agriculture-impacted areas, the relative

TABLE 15.17

Summary of Abiotic Properties of Several Stations of the Júcar Basin[a]

| | Agriculture Impacted | | | | Nonagriculture Impacted | |
	Mijares Minimum to Maximum	Júcar Minimum to Maximum	Turia Minimum to Maximum	Mijares Average	Júcar Minimum to Maximum	Turia Average
Width (m)	1.04 to 16.38	7 to 16.66	4.34 to 14.75	2.94	3.11 to 12.55	5.09
Water layer (m)	0.18 to 0.80	0.34 to 15	0.10 to 1.31	0.15	0.31 to 0.66	0.25
Water flow (m³/s)	0.17 to 4.77	2.16 to 12.6	0.16 to 6.52	0.33	0.81 to 10.10	1.09
pH	7.89 to 8.36	7.66 to 8.24	7.61 to 8.56	7.75	7.92 to 8.04	8.00
Conductivity (µS/cm)	544 to 1114	626 to 1346	739 to 1625	1000	542 to 712	663
Dissolved oxygen (mg/L)	7.70 to 11	7.37 to 10.86	5.54 to 10.70	8.41	9.42 to 10.53	13.0

[a] Data compiled are minimum and maximum average values from 2000 to 2005.

TABLE 15.18
Frequently Observed Aquatic Arthropods and Nonarthropods in the Macroinvertebrate Community from Mediterranean Basin Streams (Júcar Basin) from East of Spain[a]

	Nonagriculture Impacted, Frequency (%)	Agriculture Impacted Sites, Frequency (%)
Arthropods		
Crustacea		
Atyaephyra desmarestii	[b]	31
Gammaridae	80	65
Procambarus clarkii	60	31
Ephemeroptera		
Baetis sp.	100	79
Caenis luctuosa	100	93
Ecdyonurus sp.	100	45
Ephemera danica	60	[b]
Ephemerella ignita	100	27
Ephoron virgo	[b]	24
Heptageniidae	40	[b]
Leptophlebiidae	60	[b]
Limnephilidae	100	[b]
Diptera		
Anthomyiidae	[b]	41
Atherix sp.	60	[b]
Ceratopogonidae	100	41
Chironomidae	100	100
Empididae	100	52
Psychodidae	[b]	27
Simuliidae	100	79
Stratiomyidae	80	[b]
Tabanidae	100	[b]
Trichoptera		
Cheumatopsyche lepida	40	[b]
Chimarra marginata	60	[b]
Hydropsyche sp.	100	76
Hydropsichidae	[b]	69
Hydroptila sp.	[b]	69
Philopotamidae	60	[b]
Psychomyiidae	60	[b]
Rhyacophila sp.	100	27
Sericostoma personatum	40	[b]
Odonata		
Calopterygidae	100	[b]
Onychogomphus uncatus	80	[b]

TABLE 15.18 (CONTINUED)
Frequently Observed Aquatic Arthropods and Nonarthropods in the Macroinvertebrate Community from Mediterranean Basin Streams (Júcar Basin) from East of Spain[a]

	Nonagriculture Impacted, Frequency (%)	Agriculture Impacted Sites, Frequency (%)
Coleoptera		
Dryopidae	[b]	24
Elmidae	100	62
Elmis sp.	100	41
Gyrinidae	60	35
Helodidae	60	[b]
Heteroptera		
Gerris sp.	[b]	55
Hydrometra stagnorum	60	38
Plecoptera		
Nemouridae	60	[b]
Perla marginata	40	[b]
Nonarthropods		
Gastropoda		
Ancylus fluviatilis	100	69
Hydrobiidae	100	45
Limoniidae	80	27
Lymnaea peregra	80	45
Physella acuta	80	79
Pisidium sp.	[b]	52
Planorbidae	[b]	31
Potamopyrgus antipodarum	100	79
Theodoxus fluviatilis	40	[b]
Hirudinea		
Dina lineata	60	38
Helobdella stagnalis	[b]	24
Turbellaria		
Dugesiidae	[b]	59
Oligochaeta		
Oligochaeta	100	100
Eiseniella tetraedra	60	62

[a] The total of locations sampled was 34 (5 nonagriculture impacted and 29 highly agriculture impacted) from 2000 to 2005. Only macroinvertebrates are listed that could be recorded on at least 20% of sampling localities from 2000 to 2005. Data from www.chj.es.
[b] Not recorded.

TABLE 15.19

Relative Distribution of Macroinvertebrates Sampled in Mediterranean Streams of the Júcar Basin[a]

	Taxa (%)		Individuals (%)	
	Agriculture Impacted Sites (29 Sites)	Nonagriculture Impacted Sites (5 Sites)	Agriculture Impacted Sites (29 Sites)	Nonagriculture Impacted Sites (5 Sites)
Crustacea	8	5	15	10
Ephemeropters	14	20	22	28
Diptera	16	15	18	17
Trichoptera	11	17	8	13
Odonata	0	5	0	1
Coleoptera	11	10	2	6
Heteroptera	5	2	0	0
Plecoptera	0	5	0	1
Gastropoda	22	17	31	24
Hirudinea	5	2	2	0
Turbellaria	3	0	0	0
Oligochaeta	5	2	2	1

[a] Data compiled from www.chj.es (from 2000 to 2005).

contribution of Crustaceans was 15%, while for Ephemeroptera it was 22%. In the areas not heavily impacted by agriculture, crustaceans accounted for 10% and Ephemeroptera for 29% (see Table 15.19). In both types of streams, the most abundant macroinvertebrate was the gastropod *Potamopyrgus antipodarum*.

   3) A higher relative contribution of species of Coleopteran (11%) and a lower relative contribution of Plecoptera (0.001%) and Ephemeroptera (14%) were observed in agriculture-impacted sites (see Table 15.19). In the nonimpacted sites, Plecoptera accounted for 5% and Ephemeroptera for 20% (see Table 15.19).

Macroinvertebrate communities from impacted agriculture sites (29 stations) were characterized using biological traits and ecological information available in the literature (see Usseglio-Polatera et al. 2000; Tachet et al. 2000). These traits describe life-cycle features (life-cycle duration, reproductive cycles per year, aquatic stages); resilience or resistance potentials (dispersal, locomotion, and resistance forms); physiology and morphology (respiration, maximum size); and reproduction and feeding behavior (reproduction, food and feeding habitats). Unfortunately, it was not possible to do a quantitative analysis of the different biological traits recorded in the area of study because of inconsistency of available data with respect to taxonomic resolution and ecological information available.

TABLE 15.20

Proposed Groups of Representative Species of Macroinvertebrates in Different Orders of Aquatic Macroinvertebrates from the Júcar Basin, East of Spain (Analysis of Data from www.chj.es at Confederación Hidrográfica Del Júcar, Years 2000 to 2005; Unpublished), for Risk Analysis

Order	Representative Species	Voltinism
	Nonarthropods	
Gastropoda (class)	*Potamopyrgus antipodarum*	Uni- or plurivoltine
	Ancylus fluviatilis	Univoltine
Oligochaeta	*Eiseniella tetraedra*	Univoltine
	Branchiura sowerbyi	Bi- or plurivoltine
Hirudinea	*Dina lineata, Glossiphonia complanata*	Univoltine
Turbellaria	*Dugessia polychroa*	Univoltine
	Arthropods	
Crustacea	*Austropotamobius pallipes*	Semivoltine
	Procambarus clarkii	Univoltine
	Asellus aquaticus	Bi- or plurivoltine
	Gammarus sp.	Plurivoltine
Coleoptera	*Elmis* sp.	Univoltine
	Graptodytes sp.	Uni- or plurivoltine
Diptera	Chironomidae ind.	Uni- or plurivoltine
	Tabanidae ind.	Univoltine
Odonata	*Calopteryx* sp.	Semivoltine
	Pyrrhosoma nymphula	Univoltine
Ephemeroptera	*Ephemera danica*	Semivoltine
	Baetis sp.	Uni- or plurivoltine
Trichoptera	*Beraea* sp.	Univoltine
	Cheumatopsyche lepida	Uni- or plurivoltine
Plecoptera	*Euleuctra geniculata*	Univoltine
	Isoperla sp.	Semivoltine or univoltine

The following macroinvertebrate taxa should be considered in risk assessment:

Nonarthropods: Gastropoda, Oligochaeta, Hirudinea, and Turbellaria.
Arthropods: Diptera, Odonata, Ephemeroptera, Trichoptera, Plecoptera, Malacostraca, and Coleoptera.

A list of representative species for each group is proposed in Table 15.20, indicating the more frequent class of voltinism for the group.

Fish are not generally included in mesocosm experiments but are of significant regulatory interest; therefore, we compiled information on species diversity (if available). A high diversity of fish species has been described in the rivers of the Júcar Basin (Hernando and Soriguer 1992). According to the studies conducted

TABLE 15.21

Distribution of Fish Species in the Different Rivers of the Júcar Basin (East of Iberian Peninsula)

Species	Júcar	Turia	Mijares	Notes
Family Ciprinidae				
Barbus bocagei bocagei	✓	✓	✓	Native
Chondrostoma toxoxtoma arrigonis	✓	✓	Nr	Endemic
Family Angullidae				
A. anguilla	✓	✓	✓	Native
Family Cyprinodontidae — Typical Litoral Freshwater Species				
Aphanius iberus	✓	✓	✓	Endangered and endemic species
Valencia hispanica	✓	✓	✓	Endangered and endemic species

by Confederación Hidrográfica del Júcar for the WFD, fish species of the family Ciprinidae are the most common and abundant. From this family, species belonging to the genera *Barbus, Chondrostoma,* and *Squalius* are predominant. Also, some endemic and endangered species are present in these aquatic systems. Some examples of fish species recorded in waters of Júcar Basin are summarized in Table 15.21. The potential exposure of species of the families Ciprinidae and Anguillidae should be considered in the risk assessment.

15.4.4   EPHEMERAL STREAMS: RAMBLAS IN SOUTHEAST IBERIAN PENINSULA

15.4.4.1   Ecological Characterization of Ramblas

As mentioned in Section 15.4.1, Mediterranean rivers are characterized by an annual and between year discharge regime with annual floods and droughts, implying the presence of permanent and temporary rivers (including intermittent and ephemeral ones). As an example of temporary rivers, we describe here a typical watershed from the Mediterranean arid area called "ramblas."

A "rambla" is defined as a ephemeral watercourse with specific geomorphological features that make it different from all other temporary streams. From a hydrological point of view, the term "ramblas" may also include ephemeral and intermittent rivers. Ramblas are characterized by incised channels with steep banks, a wide channel, and heterogeneous substrate (blocks, stones, gravel, sand, and silt). The low-depth water flow is limited to the middle of the channel. Ramblas are typical for Mediterranean regions (see review by Gomez et al. 2005). Ramblas are located principally in the southeast Iberian Peninsula.

Based on lithology, 3 types can be differentiated: ramblas of marl, limestone, and metamorphic basins (Table 15.22). These categories of ramblas differ in structural

TABLE 15.22

Principal Characteristics of Ramblas from the Southeast Iberian Peninsula[a]

Channel Features	Marl Basins	Limestone Basins	Metamorphic Basins
	Limonological and Hydrological Characteristics		
Channel width	10 to 80 m	2 to 10 m	4 to 10 m
Wetted width	0.5 to 2 m	0.5 to 2 m	0.5 to 2 m
Depth of the water	2 to 15 cm	2 to 50 cm	2 to 15 cm
Channel bank height	2 to 30 m	0.5 to 2m	0.5 to 2m
Channel bank slope	45° to 90°	45° to 90°	45 to 90°
Channel erosion	High	Low	Medium
Predominant sediments	Clay and silt	Stones, gravels	Gravels, sands
Subsurface–surface water interaction	Low	High	High
Superficial hydrology	Permanent or intermittent	Frequently intermittent, usually ephemeral	Frequently intermittent and flow discontinuous
	Physicochemical Characteristics		
Salinity (g/L)	3 to 30	0 to 2	2 to 4
Conductivity (μmhos/cm)	950 to 39309	Dry to 11000	Dry to 7000
Nitrates (mg/L)	0 to 159.1	0.1 to 4.0	0 to 89.3
Phosphates (μg/L)	0 to 1062	0 to 104.6	0 to 4.1
	Biota		
Channel and Riberian vegetation	*Tamarix* sp., *Juncus maritimus, Scirpus holoschoenus, Phragmites australis, Arthrochnemu macrostachium, Sarcornia fructicosa, Limonium* sp. Halophytic communities	*Tamarix* sp., *Pinus halepensis, Nerium oleander, Phragmites australis, Typha domingensis, Pistacia lentiscus, Rosmarinus officinalis*	*Nerium oleander, Arundo donax, Phragmites australis*
Location in Murcia (frequency from total, %)	Center, north, and southeast (65)	Northwest (27)	South and coastal area (8)

[a] Data from Gómez et al. (2005), Humedales y Ramblas de la Región de Murcia (2004), Moreno et al. (2001), and Vidal-Albarca et al. (2004).

parameters, hydrology, water chemistry, and biological communities. Marl ramblas are the more abundant and are located in agriculture landscapes.

A summary of information available for marl ramblas located in an agriculture landscape is compiled in Table 15.23. This information can be used to get a target image of these types of water bodies for risk assessment.

TABLE 15.23

Characteristics of Some Marl Ramblas Found in an Agriculture Landscape[a]

Name	Rambla Del Judio	Rambla Del Reventón	Rambla Salada	Rambla De Las Salinas
Water flow	30 L/s	6 L/s	0.02 to 0.2 L/s	Hipersaline
Hydrology	I and P	I and P	P	P
Macroinvertebrates	Dominant: Diptera, Heteroptera, Coleoptera Less abundant: Crustacea, Odonata, and Ephemeroptera	High species richness High number of endemic species	Dominant: Coleoptera, Diptera, Heroptera, and Odonata Rich in endemic species 25 taxones/16 families	Exclusive presence of coleopteran: *Octhebius glaber*
Macrophytes	*Cladophora* sp., *Enteromorpha* sp., *E. clathrata*, *Spyrogyra* sp.	*Chara* sp., *Ruppia maritime*, *Zanichellia pedunculata*, *Cladophora fracta*, *Enteromorpha flesuoxa*, *E. intestinalis*, *Rhizoclonium hieroglyphicum*, *Spyrogira* spp., *Zygnema* spp.	*Cladophora fracta*, *Enteromorpha intestinalis*, *Ruppia maritime*, *Vaucheria dichotoma*, *Calothrix braunii*, and *Lyngbya epiphytica*	Absence
Riberian vegetation: dominant species	*Phragmites australis*, *Tamarix canariensis*, *Juncus amritimus*, *Scirpus holoschoenus*	*Juncus maritimus*, *Cyperus distachyos*, *Phragmites australis*, *Tamarix canariensis*, *Salsola gensitoides*, and *Inula chritmoides*	*Phragmites australis*, *Tamarix canariensis*, *T. bovena*, *Juncus maritimus*, *Juncus subulatus*, *Sarcocornia fruticosa*, *Limonium* sp.	Absence
Land use	Irrigated land	Almond, olive, tomato, and cucurbit	Mediterranean scrub, citrus, and horticultural crops (intensive agriculture)	Irrigated land

[a] Data from Moreno et al. (2001), Humedales y Ramblas de la región de Murcia (2000) and Velasco et al. (2006).

Note: P = permanent. I = intermittent.

TABLE 15.24

Adaptations of Communities Inhabiting Ramblas[a]

Community	Taxon/Dominant Groups	Strategies	Community Characteristics
Algae and macrophytes	Zygnematales Oedogoniales Angiosperms Charales Vaucheriales	Macrophytes Monocarpic reproductive patterns Fast growth Early maturity Production of small propagules Algae Short life cycles Production of resistant cells	Low macrophyte species richness
Invertebrates	Diptera Coleoptera Heteroptera Odonata Ephemeroptera	High growth rate Short life cycles Small size larvae Active dispersion	Biomass value low After perturbation, the community is simplified

[a] Data from Moreno et al. (2001), Vidal-Abarca et al. (2004), and Velasco et al. (2006).

Intermittent water flow and salinity characterize many inland water bodies in semiarid regions. In aquatic systems such as ramblas, the transient nature of superficial hydrology and substrate typology are the principal factors that determine the biotic communities and their adaptations. In Table 15.24, the adaptations observed in the different communities found in ramblas are summarized.

In the Mediterranean area, the reduction of salinity in naturally saline systems is a problem that has increased in the last decades due to changes in agricultural practices, such as the expansion of irrigated agriculture in the watersheds. Velasco et al. (2006) showed that

1) Decreasing salinity results in an increased richness and diversity of macroinvertebrate species because of the colonization of freshwater species or low-salinity-tolerant species (such as *M. praemorsa* or some *Anax* sp., Simulidae, Ceratopogonidae, and Tanypodinae species).

2) In contrast, the abundance of individuals, evenness, and Simpson's index shows no significant response to changes in salinity. Three responses to salinity dilution were observed:

   a. An increase in the abundance of *Sigara selecta*, *Nebrioporus ceresyi*, *Nebrioporus baeticus*, and *Berosus hispanicus*

   b. A decrease in the abundance of *Ephydra flavipes* and *Stratiomys longicornis*

    c. No significant difference in the number of individuals of *Potamopyrgus antipodarum, Chironomus salinarius, Halocladius varians, Cloeon schoenemundi, Octhebius cuprescens,* and *Ochthebius delgadoi*

3) The biomass of *Cladophora glomerata* and *Ruppia maritima* increased as salinity levels dropped, while the biomass of epipelic algae decreased. Dilution changed the ecosystem's state from epipelic algae dominated to filamentous algae dominated.

15.4.4.2 Consequences for Risk Assessment

Ramblas, as examples of typical water bodies from the arid and semiarid regions of the southeast Iberian Peninsula, are a good illustration of temporary or ephemeral water bodies that are important in agricultural areas. Taking into account the scenarios defined in FOCUSsw, only water bodies with permanent hydrology are considered. Therefore, intermittent or ephemeral stream scenarios should be developed to protect water bodies typical of Mediterranean countries.

According to the information presented, the biological communities of ramblas are highly adapted to changing and extreme conditions (e.g., macroinvertebrates include very resilient species with short life cycles, high mobility, or desiccation-resistant resting stages), suggesting that these organisms are more resistant to a variety of disturbances (both physical and chemical).

Algae, macrophytes, and macroinvertebrates are highly adapted to these aquatic systems and are the dominant groups. In comparison, fish communities are less abundant and diverse (Humedales y Ramblas de la región de Murcia 2004). A high number of macroinvertebrates are endemic for these aquatic systems. Furthermore, in ramblas Coleoptera, Heteroptera, Odonata, and Ephemeroptera are common. Therefore, these groups of macroinvertebrates should be considered for risk assessment.

15.4.5 SUMMARY AND CONCLUSIONS

The Iberian Peninsula is predominantly a semiarid region under the influence of Mediterranean climate. In the agricultural landscape of this region, permanent streams, ephemeral streams (e.g., ramblas), wetlands, and ponds are the more representative water bodies. As a contribution to the ELINK aim to develop ecological scenarios for pesticide risk assessment, qualitative ecological characteristics of permanent streams (Júcar Basin) in the east of Spain and of ephemeral streams (ramblas) from the southeast of Spain are described.

In Júcar Basin streams, the diversity of macroinvertebrates and fish species is relatively high. Invertebrate species with 2 or more generations per year are important, but also macroinvertebrates with 1 generation per year are common. Fish species of the Ciprinidae family are the most common and abundant, but also some other endemic and endangered species are present. Unfortunately, we were not able to compile specific information about macrophyte communities in these streams.

Ramblas are a good illustration of temporary or ephemeral water bodies that are important in agricultural areas. Biological communities of ramblas are highly adapted to changing and extreme conditions (e.g., macroinvertebrates include very resilient species with short life cycles, high mobility, or desiccation-resistant resting stages), suggesting that these organisms are more resistant to a variety of disturbances (both physical and chemical). Algae, macrophytes, and macroinvertebrates are the dominant groups. A high number of endemic macroinvertebrate species can be found in these aquatic systems. Common macroinvertebrates inhabiting ramblas belong to Coleoptera, Heteroptera, Odonata, and Ephemeroptera. Therefore, these groups should be considered for risk assessment.

15.5 ECOLOGICAL CHARACTERIZATION OF WATER BODIES IN CLAY LANDSCAPES IN THE UNITED KINGDOM

Jeremy Biggs and Colin Brown

15.5.1 INTRODUCTION

A database has been generated containing biological, morphological, and physico-chemical properties of aquatic ecosystems in the United Kingdom.

Summary statistics for the morphological features of the defined water body types by landscape class include data on channel width, water body area, depth, flow characteristics, bank angle, substrate depth, and composition. Also included are bankside features such as distance from the bank top to an arable crop, abundance of aquatic vegetation, bank scrub and tree density, and adjacent land use. Up to 15 chemical parameters are represented in the database, including pH, conductivity, BOD, COD, suspended solids, and nutrients. Data sets were filtered to exclude monitoring sites potentially impacted from urban or industrial situations. Although the chemical determinants reported give an indication of the range of values for water quality indicators across the landscape classes (rivers, streams, and ponds), it should be recognized that the data set is limited. Caution should be used when attempting, for example, alignment of physicochemical properties with biological data. In the following, an emphasis is given to ponds in landscapes of landscape class 6 (LC6; clay landscapes) of the landscape classification that emerged from the national typology. As data also are available for other water bodies and other landscapes in the United Kingdom, some data are discussed and compared to the general database to provide elements for extrapolation purposes.

15.5.2 LANDSCAPE DESCRIPTION

A total of 13 landscapes were identified for England, Scotland, and Wales; these landscapes were based on general geological and soil descriptors. The landscapes are listed in Table 15.25 together with a general description and the area covered.

Landscapes from class 6 (i.e., clay landscapes) were selected for further illustration as they correspond to important crop areas where water bodies are rather vulnerable to pesticide entry. These landscapes correspond to level to gently sloping vales containing slowly permeable clays and heavy loams. The total area of this landscape in England and Wales is 19706 km^2. In this area, the average field size is 5.1 ha. A total of 7738 rivers, 348 streams, 11 ditches, and 55 ponds is recorded in this area.

On average, each square kilometer of this landscape contains

- 470 m of river (defined as >8.25 m wide)
- 830 m of stream (defined as <8.25 m wide)
- 4214 m of nonroad ditch
- 3.6 ponds

Average pond surface area is 1459.7 m^2.

TABLE 15.25

Summary of the Landscape Classes Derived for England, Scotland, and Wales

Number	Landscape	Description	Total Area (km²)	Crops (Minor Crops in Parentheses)
1	River floodplains and low terraces	Level to very gently sloping river floodplains and low terraces	7781	Permanent grass, some cereals, and oilseed rape, probably more intensive on terraces
2	Warplands, fenlands and associated low terraces	Level, broad "flats" with alluvial very fine sands, silts, clays, and peats	9017	Cereals (oilseed rape, beans), sugar beet, potatoes, peas, vegetables, top fruit
3	Sandlands	Level to moderately sloping, rolling hills and broad terraces; sands and light loams	10871	Cereals (oilseed rape, beans, and peas), sugar beet, potatoes (peas in East Anglia)
4	Till landscapes	Level to gently sloping glacial till plains; medium loams, clays, and chalky clays with high base status (eutrophic); some lighter textured soils on outwash	22151	Cereals, oilseed rape, and beans (peas in East Yorkshire), permanent and rotational grass (mainly in west)
5	Till landscapes	Level to gently sloping glacial till plains; medium loams and clays with low base status (oligotrophic); some lighter textured soils on outwash	15449	Permanent and rotational grass with some cereals and oilseed rape
6	Prequaternary clay landscapes	Level to gently sloping vales; slowly permeable, clays (often calcareous), and heavy loams; high base status (eutrophic)	19706	Permanent grass, cereals (>10 to 15%), leys, oilseed rape, maize (not in northeast or Weald), and beans
7	Chalk and limestone plateau and coombe valleys	Rolling "wolds" and plateaus with "dry" valleys; shallow to moderately deep loams over chalk and limestone	14197	Cereals (and oilseed rape, beans), sugar beet, potatoes, peas
8	Prequaternary loam landscapes	Gently to moderately sloping ridges, vales, and plateaus; deep, free-draining, and moderately permeable silts and loams	10072	Permanent and rotational grass, cereals and oilseed rape with some beans, grass, hops, and fruit

TABLE 15.25 (CONTINUED)
Summary of the Landscape Classes Derived for England, Scotland, and Wales

Number	Landscape	Description	Total Area (km²)	Crops (Minor Crops in Parentheses)
9	Mixed, hard, fissured rock and clay landscapes	Gently to moderately sloping hills, ridges, and vales; moderately deep free-draining loams mixed with heavy loams and clays in vales	12259	Permanent grass, rotational grass, and some cereals (<10 to 15%)
10	Hard rock landscapes	Gently to moderately sloping hills and valleys; moderately deep free-draining loams over hard rocks; some slowly permeable heavy loams on lower slopes and valleys	23342	Permanent grass, rotational grass, and some cereals (<10 to 15%)
11	Moundy morainic and fluvioglacial deposits	Gently and moderately sloping mounds, some terraces; free-draining morains, gravels, and sands on mounds, poorly draining gleys in hollows	2270	Permanent and rotational grass, some cereals
12	Foot slopes with loamy drift	Concave slopes or depressional sites, often with spring lines	1081	Permanent and rotational grass
13	Nonagricultural	All areas not cultivated with arable (including orchards, soft fruit, and horticultural) or maintained grassland	79690	No crops
Total	—	—	227886	—

15.5.3 Land Use in Clay Landscape (Landscape Class 6, LC6)

The photograph in Figure 15.17 shows the generalized nature of this landscape. The schematic plan (Figure 15.18) shows the land use, field size, and amount of water within an average 4-km² unit of the landscape.

This figure shows an important proportion of arable fields, together with grass fields in the surroundings of streams and ditches. Up to 9 ponds can be counted on this 4-km² area. Thus, this landscape illustrates quite well various water body types with relationships with cropped areas.

An overall picture of land use is given in Table 15.26, from data in England and Wales, more specifically in intensive arable lands. Land use is first separated as total arable, maintained grass, and nonagricultural lands. Arable lands are then detailed

FIGURE 15.17   Nature of clay landscapes.

at the level of major crop categories. It can be seen that the percentage of intensive arable land is highest in the eastern part of England, occupying 61.5% of the land area compared with a total of 38% arable land for the whole of England and Wales.

The main crops favored in intensive areas are wheat and oilseed rape. More specifically, for clay soils in these areas, maximum normal cultivation frequency for major crops is 2 years in 3 for winter wheat; and 1 year in 3 for winter barley, oilseed rape, and peas or beans. Potatoes and sugar beet are not major crops in this landscape.

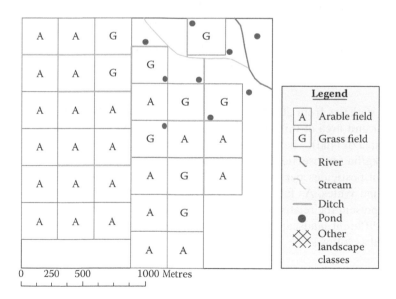

FIGURE 15.18   Land use, field size, and amount of water within an average 4-km² unit of the landscape.

TABLE 15.26
Land Use in England and Wales and in Intensive Arable Lands

	Land Use (%)	
	All of England and Wales	Intensive Arable (East Anglia)
Total arable	38.1	61.5
Total maintained grass	32.7	16.6
Total nonagricultural land[a]	29.2	21.9
Wheat	16.2	29.2
Winter barley	4.2	5.1
Spring barley	1.1	1.5
Other cereals	0.9	0.8
Oilseed rape	3.2	6.0
Linseed and flax	0.7	0.9
Potatoes	0.6	1.0
Peas and beans	2.0	4.0
Sugar beet	0.7	2.2
Setaside and follow	5.6	9.1
Fodder crops	1.8	0.4
Vegetables	0.5	0.9
Orchards and hops	0.4	0.2
Soft fruit	0.1	0.1
Bulbs and nursery	0.1	0.1

[a] Nonagricultural land includes built environments, forestry, rough grazing, seminatural vegetation, and large inland water bodies.

Usage statistics from 2000 have been used to calculate the average input of pesticides per unit area of this landscape (i.e., the total loading of pesticide divided by the total area of land). These are 0.76 kg/ha for herbicides, 0.31 kg/ha for fungicides, and 0.04 kg/ha for insecticides. On average, this landscape ranks 6th equal in terms of input of pesticides per unit area for the 10 agricultural landscapes defined for England and Wales. As far as entry routes are concerned, indices have been generated to represent the relative importance of different routes of entry for pesticides to water in different landscapes. For clay landscape, this vulnerability to entry routes is depicted as follows:

Spray drift	0.90 of 1.00	Most vulnerable landscape of 10
Drainage	1.00 of 1.00	Most vulnerable landscape of 10
Runoff	0.21 of 1.00	Arable fields likely to be drained

15.5.4 Hydrogeologic Conditions of Ponds

Hydrogeologic conditions within the clay landscapes of the United Kingdom correlate closely with FOCUS scenario D2 (Brimstone).

The topography of this landscape is generally flat to gently sloping. The substrate from which soils are formed is either slowly permeable or impermeable. This makes the soils vulnerable to formation of perched water tables, and many soils exhibit gleying (evidence of waterlogging) within a meter of the surface and even within the top 40 cm of the soil. Arable cultivation is dependent on intensive drainage systems to remove excess water from the profile and allow trafficking by agricultural machinery. The hydrological response of the soil is largely governed by saturated lateral and sublateral flow, with some potential for surface runoff. The heavy soils can be highly structured, making bypass flow an important feature for transfer of pesticides through the soil profile. Typical water flow in this landscape is detailed in Table 15.27. A detailed description of ponds in relation to landscape and hydrology and main is provided in Table 15.28. Each parameter is provided together with minimum and maximum values so that the uncertainty with regard to this description is provided quantitatively.

15.5.5 Abiotic Factors in Ponds

Water chemistry properties of ponds are described in Table 15.29. As with hydrology, each parameter is provided together with minimum and maximum values so that the uncertainty with regard to this description is provided quantitatively.

TABLE 15.27
Water Flow Classes in Clay Landscape

Inflow and Outflow Classes in Landscape 6	%
Inflow absent	52.7
Inflow present	47.3
Inflow volume 0 to 10 cm³/s	23.6
Inflow volume 11 to 100 cm³/s	9.1
Inflow volume 101 to 1000 cm³/s	10.9
Inflow volume 1000 to 10000 cm³/s	3.6
Inflow volume 10000+ cm³/s	0
Outflow absent	67.3
Outflow present	32.7
Outflow volume 0 to 10 cm³/s	16.4
Outflow volume 11 to 100 cm³/s	5.5
Outflow volume 101 to 1000 cm³/s	5.5
Outflow volume 1000 to 10000 cm³/s	5.5
Outflow volume 10000+ cm³/s	0
Number of samples	55

TABLE 15.28

Pond Hydrology and Relation to Landscape in Clay Landscapes ($n = 55$ Samples)

Property	Unit	Mean	Median	Standard Deviation	Minimum	Maximum
Pond Description						
Grazing intensity (subjective scale: 1 = low, 5 = very high)	0 to 5	0.5	0.0	1.0	0.0	4.0
Isolation (subjective scale: 1 = isolated from other water bodies, 5 = highly connected with other water bodies or wetlands)	0 to 5	2.0	2.0	1.0	0.0	3.5
Margin complexity rating (subjective scale: 1 = circular profile, 5 = highly complex or indented profile)	1 to 10	2.8	2.5	1.2	1.0	6.5
Margin grazed (average % of the pond margin grazed)	%	20.0	0.0	36.9	0.0	100.0
Margin overhung (average % vegetation overhang of pond margin)	%	40.8	38.0	31.4	0.0	100.0
Permanence rank (scale indicating ephemerality of pond: 1 = always contains water, 2 or 3 = sometimes dries up, 4 = dries up every summer)	1 to 4	1.8	1.5	1.0	1.0	4.0
Pond area	m²	1460	547	3116	24	20650
Pond grazed (average % of the pond area grazed)	%	9.5	0.0	23.0	0.0	100.0
Sediment depth	cm	40.3	31.0	34.2	0.0	150.0
Water depth (measured from center of water body)	cm	61.1	36.0	58.6	2.0	250.0
Water turbidity (subjective scale: 1 = clear, 4 = very turbid)	1 to 4	2.4	2.5	0.9	1.0	4.0
Average vegetation (%)	%	25.2	15.0	26.5	0.0	98.0
Catchment Geology						
Clay	%	55.9	50.0	36.4	0.0	100.0
Igneous and metamorphic	%	0.2	0.0	1.4	0.0	10.0
Limestone	%	12.4	0.0	27.1	0.0	100.0
Sandstone	%	31.5	20.0	34.2	0.0	100.0

TABLE 15.28 (CONTINUED)

Pond Hydrology and Relation to Landscape in Clay Landscapes ($n = 55$ Samples)

Property	Unit	Mean	Median	Standard Deviation	Minimum	Maximum
		Landscape within 100 m				
Arable	%	13.7	0.0	28.7	0.0	94.0
Bog	%	0.3	0.0	2.0	0.0	15.0
Buildings and concrete	%	2.9	0.0	8.3	0.0	41.0
Canals	%	0.0	0.0	0.0	0.0	0.0
Conifers	%	0.2	0.0	1.0	0.0	5.0
Deciduous trees	%	13.8	5.0	19.4	0.0	89.0
Dumps and waste	%	0.0	0.0	0.0	0.0	0.0
Heath and moor	%	2.3	0.0	11.9	0.0	85.0
Improved grassland	%	25.7	0.0	33.4	0.0	98.0
Marsh and fen	%	0.6	0.0	3.3	0.0	23.0
Orchards	%	0.7	0.0	5.4	0.0	40.0
Parks and gardens	%	5.4	0.0	13.6	0.0	72.0
Paths	%	0.4	0.0	0.8	0.0	3.0
Ponds and lakes	%	1.3	0.0	5.1	0.0	30.0
Railways	%	0.0	0.0	0.1	0.0	1.0
Rank vegetation	%	3.8	0.0	8.3	0.0	45.0
Roads	%	2.0	0.0	5.5	0.0	39.0
Rock and stone	%	0.3	0.0	1.0	0.0	5.0
Salt marshes	%	0.1	0.0	0.7	0.0	5.0
Sand dunes	%	0.0	0.0	0.0	0.0	0.0
Scrub and hedge	%	7.8	5.0	8.5	0.0	38.0
Semiimproved grassland	%	4.9	0.0	17.4	0.0	85.0
Streams and springs	%	0.4	0.0	1.4	0.0	8.0
Unimproved grassland	%	13.5	0.0	28.3	0.0	90.0
		Landscape within 5 m				
Arable	%	2.3	0.0	9.1	0.0	60.0
Bare earth	%	1.2	0.0	8.8	0.0	65.0
Bog	%	0.4	0.0	2.7	0.0	20.0
Buildings and concrete	%	0.3	0.0	1.1	0.0	5.0
Conifers	%	1.5	0.0	7.8	0.0	50.0
Deciduous trees	%	15.7	0.0	25.0	0.0	100.0
Dumps and waste	%	0.0	0.0	0.0	0.0	0.0
Heath and moor	%	2.9	0.0	15.0	0.0	95.0
Improved grassland	%	8.0	0.0	20.5	0.0	100.0
Marsh and fen	%	4.2	0.0	18.9	0.0	100.0
Orchards	%	0.3	0.0	2.0	0.0	15.0
Parks and gardens	%	5.6	0.0	15.9	0.0	70.0
Paths	%	0.6	0.0	2.8	0.0	20.0

(*continued*)

TABLE 15.28 (CONTINUED)
Pond Hydrology and Relation to Landscape in Clay Landscapes ($n = 55$ Samples)

Property	Unit	Mean	Median	Standard Deviation	Minimum	Maximum
Landscape within 5 m						
Ponds and lakes	%	0.0	0.0	0.0	0.0	0.0
Rank vegetation	%	19.3	2.0	27.9	0.0	100.0
Roads	%	1.5	0.0	5.9	0.0	30.0
Rock and stone	%	0.6	0.0	2.5	0.0	15.0
Sand dunes	%	0.0	0.0	0.0	0.0	0.0
Scrub and hedge	%	20.7	15.0	22.8	0.0	78.0
Semiimproved grassland	%	1.6	0.0	7.9	0.0	50.0
Streams and springs	%	0.4	0.0	2.7	0.0	20.0
Unimproved grassland	%	13.1	0.0	24.8	0.0	80.0
Pollution Source						
Agricultural land	0 to 10	1.5	1.5	1.5	0.0	4.5
Overall pollution rating (subjective pollution score [all by the same person] incorporating all pollutant sources)	0 to 10	3.7	3.5	2.3	0.5	9.0
Road runoff	0 to 10	0.4	0.0	0.9	0.0	4.0
Stream plus other inflow (subjective pollution score for inflowing sources)	0 to 10	0.4	0.0	1.0	0.0	4.5
Urban (a subjective pollution score for urban sources)	0 to 10	0.6	0.0	1.1	0.0	4.5
Water Source						
Flood	%	0.1	0.0	6.7	0.0	5.0
Flush and spring	%	11.2	0.0	25.9	0.0	95.0
Ground water	%	28.4	0.0	36.4	0.0	95.0
Land drains	%	0.9	0.0	2.0	0.0	50.0
Rainfall	%	4.5	5.0	2.1	0.0	10.0
Stream or ditch	%	12.4	0.0	25.8	0.0	98.0
Surface water	%	42.2	40.0	33.7	1.0	95.0
Upper pond	%	0.3	0.0	38.6	0.0	15.0

15.5.6 BIOTA OF PONDS

15.5.6.1 Macroinvertebrates

For macroinvertebrates, data are available in clay landscapes for streams, ditches, and ponds; no data are available for rivers or lakes.

TABLE 15.29

Water Chemistry Properties in Ponds in Clay Landscapes

Property	Unit	Mean	Median	Standard Deviation	Minimum	Maximum	Number of Samples
Al	mg/L	0.01384	0.00008	0.04640	0.00000	0.28690	43
Ca	mg/L	83.71	69.14	70.37	3.15	440.90	53
Electrolytic conductivity	mS/cm	487.65	421.00	319.36	7.00	1748.00	52
Cu	µg/L	0.00362	0.00003	0.00834	0.00001	0.03856	43
Fe	mg/L	0.45932	0.00045	2.29554	0.00000	14.90000	43
K	mg/L	11.035	4.700	18.895	0.500	100.000	53
Mg	mg/L	0.01101	0.00846	0.01059	0.00066	0.05108	43
Na	mg/L	29.03	24.17	21.37	2.66	96.20	53
Ni	mg/L	0.0036	0.0010	0.0060	0.0010	0.0270	43
Pb	mg/L	0.0045	0.0000	0.0094	0.0000	0.0405	43
pH	mg/L	7.26	7.40	0.76	4.20	9.00	50
Soluble reactive phosphorous	mg/L	0.242	0.084	0.553	0.000	3.339	49
Suspended solids	mg/L	32.16	10.70	70.05	2.50	421.50	42
Total organic N	mg/L	0.322	0.041	0.534	0	1.895	33
Zn	mg/L	0.0192	0.0002	0.0461	0.0000	0.2014	43

TABLE 15.30

Species Richness from Samplings in the Water Bodies in LC6 Compared to Richness in All Classes

	Mean Number of Species	
	Clay Landscape	All Land Classes
Streams	14.2	14.6
Rivers	12.1	—
Ditches	No data	17.5
Ponds	32.0	30.2

Table 15.30 provides a picture of species richness from samplings in the water bodies in clay landscape compared to richness in all classes. Pond species richness in clay landscapes is similar to the average estimated for all land classes. Stream species richness is slightly below average in clay landscapes, and ditch richness is considerably lower than the average.

In ponds, fauna is dominated by water beetles, bugs, and molluscs (Figure 15.19). In comparison, in streams the fauna is dominated, in terms of species richness, by caddis flies, water beetles, molluscs, and crustaceans. Ditches are intermediate between these 2 habitat types, having a relatively high proportion of beetles, caddis flies, and molluscs. Pesticide-sensitive biota (e.g., mayflies, caddis flies, stoneflies, amphipod crustaceans) make up 25% to 30% of species in streams and a smaller proportion in ditches and ponds. Data on invertebrate abundance are available for ponds and ditches (Figure 15.20). In ditches, the fauna is dominated by crustaceans, water snails, and beetles. In ponds, the fauna is dominated by snails and crustaceans.

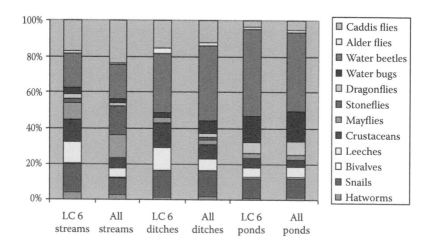

FIGURE 15.19 Invertebrate species composition of clay landscape (LC6) streams, ditches, and ponds compared to streams and ditches in all landscape types.

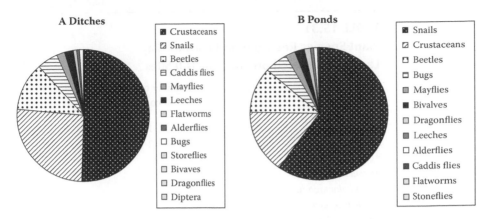

FIGURE 15.20 Invertebrate abundance in ditches and ponds in clay landscapes.

15.5.6.2 Plants

The composition of the flora varies strongly in different habitat types in clay landscape. Streams and rivers support the highest proportion of submerged and floating-leaved plants, with ditches having few submerged and floating-leaved species (Figure 15.21).

Wetland plant data are available in clay landscape for 5 habitat types: streams, rivers, ponds, lakes, and ditches. In streams, rivers, and lakes, a clay landscape has relatively species-rich plant assemblages. Pond and ditch plant assemblages are comparatively species poor in this landscape. A detailed description of plant species being recorded in ponds is available in Table 15.31.

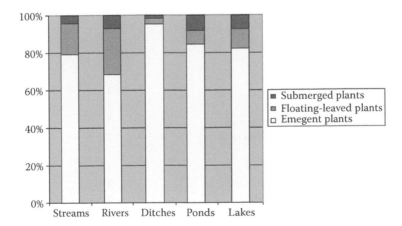

FIGURE 15.21 Proportions of emergent, submerged, and floating-leaved plants in 5 water body types in clay landscapes.

TABLE 15.31
Plant Species Recorded in Ponds of Clay Landscape, Based on Data for 160 Sites

Species Name	Number of Times Recorded
Acornia spp.	1
Acorus calamus	2
Agrostis stolonifera	82
Alisma lanceolatum	4
Alisma plantago-aquatica	36
Alopecurus geniculatus	16
Angelica sylvestris	3
Apium nodiflorum	15
Azolla filiculoides	1
Berula erecta	1
Bidens cernua	2
Bidens tripartita	6
Callitriche hamulata	4
Callitriche obtusangula	3
Callitriche platycarpa	1
Callitriche spp.	17
Callitriche stagnalis	6
Callitriche stagnalis/platycarpa agg.	3
Caltha palustris	5
Cardamine amara	2
Cardamine pratensis	9
Carex acutiformis	1
Carex curta	1
Carex otrubae	13
Carex paniculata	1
Carex pendula	25
Carex pseudocyperus	6
Carex riparia	6
Carex spp.	1
Carex vesicaria	1
Carex vulpina	1
Ceratophyllum demersum	7
Ceratophyllum submersum	1
Cirsium palustre	1
Crassula helmsii	1
Deschampsia cespitosa	14
Eleocharis palustris	7
Elodea nuttallii	6
Epilobium ciliatum/adenocaulon	1

TABLE 15.31 (CONTINUED)
Plant Species Recorded in Ponds of Clay Landscape, Based on Data for 160 Sites

Species Name	Number of Times Recorded
Epilobium hirsutum	9
Epilobium obscurum	6
Epilobium palustre	1
Epilobium parviflorum	8
Epilobium spp.	1
Epilobium tetragonum	4
Equisetum fluviatile	5
Equisetum palustre	2
Eupatorium cannabinum	2
Filipendula ulmaria	7
Fontinalis antipyretica	3
Galium palustre	36
Glyceria fluitans	40
Glyceria maxima	11
Glyceria maxima (varigated)	3
Glyceria notata	4
Glyceria spp.	1
Hippuris vulgaris	2
Hottonia palustris	2
Hydrocotyle vulgaris	1
Hypericum tetrapterum	6
Impatiens noli-tangere	1
Iris spp. (exotic)	2
Iris pseudacorus	25
Juncus acutiflorus	4
Juncus articulatus	24
Juncus bufonius agg.	4
Juncus bulbosus	1
Juncus compressus	1
Juncus conglomeratus	13
Juncus effuses	64
Juncus inflexus	40
Lagarosiphon major	4
Lemna gibba	1
Lemna minor	59
Lemna minuscule	4
Lemna trisulca	7
Lotus pedunculatus	10
Lychnis flos-cuculi	2

(continued)

TABLE 15.31 (CONTINUED)
Plant Species Recorded in Ponds of Clay
Landscape, Based on Data for 160 Sites

Species Name	Number of Times Recorded
Lycopus europaeus	32
Lysimachia nummularia	3
Lysimachia vulgaris	3
Lythrum portula	1
Lythrum salicaria	8
Mentha aquatica	26
Menyanthes trifoliate	4
Molinia caerulea	1
Myosotis laxa	13
Myosotis scorpioides/laxa	1
Myosotis scorpioides	8
Myriophyllum aquaticum	1
Myriophyllum spicatum	1
Nuphar lutea	3
Nymphaea alba	7
Nymphaea spp. (exotic)	3
Oenanthe aquatica	7
Oenanthe crocata	18
Persicaria amphibia	2
Persicaria hydropiper	10
Persicaria lapathifolia	4
Petasites hybridus	3
Phalaris arundinacea	17
Phragmites australis	2
Potamogeton spp. (fine leaved)	2
Potamogeton berchtoldii	1
Potamogeton crispus	6
Potamogeton natans	16
Potamogeton pectinatus	3
Potamogeton pusillus	5
Pulcaria dysenterica	5
Ranunculus flammula	4
Ranunculus lingua	2
Ranunculus omiophyllus	1
Ranunculus peltatus	3
Ranunculus sceleratus	25
Ranunculus spp.	8
Ranunculus trichophyllus	1
Rorippa amphibia	1

TABLE 15.31 (CONTINUED)
Plant Species Recorded in Ponds of Clay
Landscape, Based on Data for 160 Sites

Species Name	Number of Times Recorded
Rorippa microphylla	1
Rorippa nasturtium-aquaticum	4
Rorippa nasturtium-aquaticum	3
Rorippa palustris	6
Rumex hydrolapathum	1
Rumex maritimus	1
Sagittaria sagittifolia	1
Schoenoplectus lacustris	3
Schoenoplectus tabernaemontani	1
Scirpus maritimus	1
Scrophularia auriculata	7
Scutellaria galericulata	1
Solanum dulcamara	8
Sparganium emersum	1
Sparganium erectum	33
Spirodela polyrhiza	2
Stellaria uliginosa	1
Stratiotes aloides	2
Symphytum officinale	1
Typha angustifolia	2
Typha latifolia	42
Veronica anagallis-aquatica	3
Veronica beccabunga	20
Veronica catenata	4
Veronica scutellata	2
Zannichellia palustis	2

15.5.7 CONSEQUENCES FOR RISK ASSESSMENT

This ecological data set provides a realistic picture of the ponds that may be encountered in a typical landscape of England.

In the first instance, the description of landscape gives information about its representativeness with regard to density of water body in the landscape, size, and proximity to cropped area. Land use figures also inform an assessment of representativeness with regard to agronomic pressure and pesticide pressure. Vulnerability to drift, drainage, and runoff as entry routes was also provided to ascertain the representativeness of data collected with regard to agronomic situations.

Ponds were described in most detail for the purpose of this case study, although similar data are available for the other water body types. A typical pond could be described with regard to its permanency, size, water column and sediment depth, margins' nature and complexity, catchment geology, and relation to landscape, the last described at 5- and 100-meter distances from water bodies. Water and pollution sources are described. Some of the parameters are described as subjective indices (e.g., 0-to-5 scale) with efforts to involve a limited number of evaluators. All this information is essential to discuss the realism of the risk assessment performed particularly about exposure quantification and related estimates. Abiotic parameters also are provided together with minimum and maximum values so that the uncertainty with regard to this description is provided quantitatively.

Both macroinvertebrate and macrophyte data are available in some detail. Macroinvertebrate data include number of species; relative abundance among invertebrate groups in LC6 ditches, streams, and ponds; and comparisons with pooled data in all landscape classes. Macrophyte data are provided at species richness level and for ponds at species level. These data allow comparisons between water body types within the landscape class and can be used in risk assessment to extrapolate from local to wider geographic scales.

Species distribution data can help to focus on widely occurring representative (and sensitive) species that may require further investigation in higher-tier ecotoxicological or ecological studies.

16 References for Work Group Reports (Part II)

Aangepast maaibeheer en de visstand. Ervaringen met minder maaien in de Tieler- en Culemborgerwaarden. 1998. OVB-Bericht 20:92–96.

Adriaanse PI. 1996. Fate of pesticides in field ditches: the TOXSWA simulation model. Wageningen (NL): DLO Winand Staring Centre. Report 90.

Allesina S, Bondavallia C, Scharler UM. 2005. The consequences of the aggregation of detritus pools in ecological networks. Ecol Modell 189:221–232.

Álvarez Cobelas M, Rojo C. Angeler DG. 2005. Mediterranean limnology: current status, gaps and future. J Limnol 64:13–29.

Ankley GT, Erickson RJ, Phipps GL, Mattson VR, Kosian PA, Sheedy BR, Cox JS. 1995. Effects of light intensity on the phototoxicity of fluoranthene to a benthic macroinvertebrate. Environ Sci Technol 29:2828–2833.

Arts GHP, Brock TCM, Bloemendaal FHJL, Roelofs JGM. 1988. Beheer. In: Bloemendaal FHJL, Roelofs JGM, editors. Waterplanten en waterkwaliteit. Utrecht (NL): Stichting Uitgeverij Koninklijke Nederlandse Natuurhistorische Vereniging. p 177–188.

Ashauer R, Boxall ABA, Brown CD. 2006a. Predicting effects on aquatic organisms from fluctuating or pulsed exposure to pesticides. Environ Toxicol Chem 25:1899–1912.

Ashauer R, Boxall ABA, Brown CD. 2006b. Uptake and elimination of chlorpyrifos and pentachlorophenol into the freshwater amphipod *Gammarus pulex*. Arch Environ Contam Toxicol 51:542–548.

Ashauer R, Boxall ABA, Brown CD. 2007a. Modelling combined effects of pulsed exposure to carbaryl and chlorpyrifos on *Gammarus pulex*. Environ Sci Technol 41:5535–5541.

Ashauer R, Boxall ABA, Brown CD. 2007b. New ecotoxicological model to simulate survival of aquatic invertebrates after exposure to fluctuating and sequential pulses of pesticides. Environ Sci Technol 41:1480–1486.

Ashauer R, Boxall ABA, Brown CD. 2007c. Simulating toxicity of carbaryl to *Gammarus pulex* after sequential pulsed exposure. Environ Sci Technol 41:5528–5534.

Balzter H, Braun PW, Kohler W. 1998. Cellular automata models for vegetation dynamics. Ecol Modell 107:113–125.

Barnthouse LW. 2004. Quantifying population recovery rates for ecological risk assessment. Environ Toxicol Chem 23:500–508.

Bartell SM, Campbell KR, Lovelock CM, Nair SK, Shaw JL. 2000 Characterizing aquatic ecological risks from pesticides using a diquat dibromide case study III. Ecological process models. Environ Toxicol Chem 19:1441–1453.

Bartell SM, Gardner RH, O'Neill RV. 1992. Ecological risk estimation. Chelsea (MI): Lewis Publishers. 233 p.

Bartell SM, Lefebvre G, Kaminski G, Carreau M, Campbell KR. 1999. An ecosystem model for assessing ecological risks in Québec rivers, lakes, and reservoirs. Ecol Modell 124:43–67.

Beckett DC, Aartila TP, Miller AC. 1992. Contrast in density of benthic invertebrates between macrophyte beds and open littoral patches in Eau Galle Lake, Wisconsin. Am Midl Nat 127:77–90.

Billoir E, Péry ARR, Charles S. 2007. Integrating the lethal and sublethal effects of toxic compounds into the population dynamics of *Daphnia magna*: a combination of the DEBtox and matrix population models. Ecol Modell 203:204–214.

Boesten JJTI, Köpp H, Adriaanse PI, Brock TCM, Forbes VE. 2007. Conceptual model for improving the link between exposure and effects in the aquatic risk assessment of pesticides. Ecotoxicol Environ Saf 66:291–308.

Bonada N, Doledec S, Statzner B. 2007. Taxonomic and biological trait differences of stream macroinvertebrate communities between Mediterranean and temperate regions: implications for future climatic scenarios. Biol Conserv 13:1658–1671.

Bonada N, Rieradevall M, Prat N. 2007. Macroinvertebrate community structure and biological traits related to flow permanence in a Mediterranean network. Hydrobiologia 589:91–106.

Briers RA, Warren PH. 2000. Population turnover and habitat dynamics in Notonecta (Hemiptera: Notonectidae) metapopulations. Oecologia 123:216–222.

Brock T, Van Campen L, Edlinger B. 1980. Een oriëntatie op de functie, biologie en beheer van sloten in Nederland. Literatuurscriptie no 14. Nijmegen (NL): Laboratorium voor Aquatische Oecologie.

Brock TCM. 1988. De invloed van waterplanten op hun omgeving. In: Bloemendaal FHJL. Roelofs JGM, editors. Utrecht (NL): Waterplanten en waterkwaliteit. Stichting Uitgeverij Koninklijke Nederlandse Natuurhistorische Vereniging. p 27–41.

Brock TCM, Arts GHP, Maltby L, Van den Brink PJ. 2006. Aquatic risks of pesticides, ecological protection goals, and common aims in European Union legislation. Integr Environ Assess Manag 2:e20–e46.

Brock TCM, Van der Velde G. 1996. Aquatic macroinvertebrate community structure of a Nymphoides peltata-dominated and macrophyte-free site in an oxbow lake. Neth J Aquat Ecol 30:151–163.

Brock TCM, Van Wijngaarden RPA, Van Geest GJ. 2000. Ecological risks of pesticides in freshwater ecosystems. Part 2: insecticides. Wageningen (NL): Alterra Green World Research. Alterra-Rapport 089.

Campbell PJ, Arnold DJS, Brock TCM, Grandy NJ, Heger W, Heimbach F, Maund SJ, Streloke M. 1998. Guidance document on higher tier risk assessment for pesticides (HARAP). Proceedings from the HARAP workshop. Brussels: SETAC Publications.

Caswell H. 2001. Matrix population models. 2nd ed. Sunderland (MA): Sinauer Associates Inc.

Davies B, Biggs J, Williams P, Whitfield M, Nicolet P, Sear D, Bray S, Maund S. Forthcoming 2008. Comparative biodiversity of aquatic habitats in the European agricultural landscape. Agric Ecosyst Environ 125:1–8.

DeAngelis DL, Gross LJ, editors. 1992. Individual-based models and approaches in ecology: populations, communities and ecosystems. New York: Chapman and Hall.

De Kroon H, Plaisier A, Van Groenendael J, Caswell H. 1986. Elasticity: the relative contribution of demographic parameters to population growth rate. Ecology 67:1427–1431.

De Kroon H, Van Groenendael J, Ehrlen J. 2000. Elasticities: a review of methods and model limitations. Ecology 81:607–618.

Driever SM, van Nes EH, Roijackers RMM. 2005. Growth limitation of *Lemna minor* due to high plant density. Aquat Bot 81:245–251.

Ducrot V, Péry ARR, Mons R, Queau H, Charles S, Garric J. 2007. Dynamic energy budgets as a basis to model population-level effects of zinc-spiked sediments in the gastropod *Valvata piscinalis*. Environ Toxicol Chem 26:1774–1783.

[EC] European Commission. 1997. Commission proposal for a council objective establishing annex VI to directive 91/414/EEC concerning the placing of plant protection products on the market. Off J Eur Comm C 240:1–23.

[EC] European Commission. 2000. Directive 2000/60/EC of the European Parliament and of council of 23 October 2000 establishing a framework for community action in the field of water policy ("Water Framework Directive"). Off J Env Comm L 327/1.

[EFSA] European Food Safety Authority. 2005. Opinion of the Scientific Panel on Plant Health, Plant Protection Products and Their Residues on a request from the EFSA related to the evaluation of dimoxystrobin. EFSA J 178:1–45.

[EFSA] European Food Safety Authority. 2006. Opinion of the Scientific Panel on Plant Health, Plant Protection Products and Their Residues on a request from the EFSA on the final report of the FOCUS Working Group on Landscape and Mitigation Factors in Ecological Risk Assessment. EFSA J 437:1–30.

European and Mediterranean Plant Protection Organization. 2003a. Environmental risk assessment scheme for plant protection products. Chapter 6: surface water and sediment. EPPO Bull 33:169–181.

European and Mediterranean Plant Protection Organization. 2003b. Environmental risk assessment scheme for plant protection products. Chapter 7: aquatic organisms. EPPO Bull 33:183–194.

Forbes VE, Calow P. 1999. Is the per capita rate of increase a good measure of population-level effects in ecotoxicology? Environ Toxicol Chem 18:1544–1556.

Forbes VE, Hommen U, Thorbek P, Heimbach F, Van den Brink PJ, Wogram J, Thulke HH, Grimm V. 2009. Ecological models in support of regulatory risk assessments of pesticides: developing a strategy for the future. Integr Environ Assess Manag 5:167–172.

[FOCUS] Forum for the Co-ordination of Pesticide Fate Models and Their Use. 2001. FOCUS surface water scenarios in the EU evaluation process under 91/414/EEC. Report of the FOCUS Working Group on Surface Water Scenarios. EC Document Reference SANCO/4802/2001-rev. 2.

[FOCUS] Forum for the Co-ordination of Pesticide Fate Models and Their Use. 2002. Generic guidance for FOCUS groundwater scenarios. Version 1.1, April 2002 Report of the FOCUS Groundwater Scenarios Workgroup.

[FOCUS] Forum for the Co-ordination of Pesticide Fate Models and Their Use. 2006. Guidance document on estimating persistence and degradation kinetics from environmental fate studies on pesticides in EU registration. Report of the FOCUS Work Group on Degradation Kinetics, EC Document reference Sanco/10058/2005 version 2. Brussels.

[FOCUS] Forum for the Co-ordination of Pesticide Fate Models and Their Use. 2007a. Landscape and mitigation factors in aquatic ecological risk assessment. Volume 1. Extended summary and recommendations, the final report of the FOCUS Working Group on Landscape and Mitigation Factors in Ecological Risk Assessment. EC Document Reference Sanco/10422/2005, version 2.0, September. 169 p.

[FOCUS] Forum for the Co-ordination of Pesticide Fate Models and Their Use. 2007b. Landscape and mitigation factors in aquatic risk assessment. Volume 2. Detailed technical reviews. Report of the FOCUS Working Group on Landscape and Mitigation Factors in Ecological Risk Assessment. EC Document Reference SANCO/10422/2005, version 2.0, September.

Giddings JM, Brock TCM, Heger W, Heimbach F, Maund SJ, Norman SM, Ratte HT, Schafers C, Steloke M. 2001. Community-level aquatic system studies — interpretation criteria. Proceedings from the CLASSIC workshop. Pensacola (FL): SETAC.

Gleason T, Nacci DE. 2001. Risks of endocrine-disrupting chemicals to wildlife: extrapolating from effects on individuals to population responses. Hum Ecol Risk Assess 7:1027–1042.

Gómez R, Hurtado I, Suárez ML, Vidal-Abarca MR. 2005. Ramblas in south-east Spain: threatened and valuable ecosystems. Aquat Conserv Marine Freshw Ecosyst 15:387–402.

Grant A. 1998. Population consequences of chronic toxicity: incorporating density dependence into the analysis of life table response experiments. Ecol Modell 105:325–335.

Greenwood R, Salt DW, Ford MG. 2007. Pharmacokinetics: computational versus experimental approaches to optimize insecticidal chemistry. In: Ishaaya I, Nauen R, Horowitz AR, editors. Insecticide design using advanced technologies. New York: Springer-Verlag.

Grimm V. 1999. Ten years of individual-based modelling in ecology: what have we learned, and what could we learn in the future? Ecol Modell 115:129–148.

Guidance document on aquatic ecotoxicology in the context of the Directive 91/414/EEC. 2002. Sanco/3268/2001 rev. 4 (final). Brussels.

Hallam TG, Clark CE, Lassiter RR. 1983. Effects of toxicants on populations: a qualitative approach I. Equilibrium environmental exposure. Ecol Model 18:291–304.

Hanski I. 1999. Metapopulation ecology. New York (NY): Oxford University Press.

Hernando JA, Soriguer MC. 1992. Biogeography of the freshwater fish of the Iberian Peninsula. Limnetica 8:243–253.

Hesen PLGM, Buijs JNJ, Blok J. 1994. Kroos onder controle? H2O 27:12–15, 24.

Hommen U, Poethke HJ, Dülmer U, Ratte HT. 1993. Simulation models to predict ecological risks of toxins in freshwater systems. ICES J Mar Sci 50:337–347.

Hulot FD, Lacroix G, Lescher-Moutoué F, Loreau M. 2000. Functional diversity governs ecosystem response to nutrient enrichment. Nature 405:340–343.

Humedales y Ramblas de la región de Murcia. 2004. Varios autores. Consejería de Agricultura, Agua y Medio Ambiente-Fundación Universidad Empresa. Murcia (ES): Universidad de Murcia.

Jager T, Kooijman SALM. 2005. Modelling receptor kinetics in the analysis of survival data for organophosphorus pesticides. Environ Sci Technol 39:8307–8314.

Jagers op Akkerhuis GAJM, Kjær C, Damgaard C, Elmegaard N. 1999. Temperature-dependent, time-dose-effect model for pesticide effects on growing, herbivorous arthropods: bioassays with dimethoathe and cypermethrin. Environ Toxicol Chem 18:2370–2378.

Janse JH. 1997. A model of nutrient dynamics in shallow lakes in relation to multiple stable states. Hydrobiologia 342:1–8.

Jaworska JS, Rose KA, Brenkert AL. 1997. Individual-based modelling of PCBs effects on young-of-the-year largemouth bass in southeastern USA reservoirs. Ecol Modell 99:113–135.

Jeffries M. 1994. Invertebrate communities and turnover in wetland ponds affected by drought. Freshw Biol 32:603–612.

Justus J. 2006. Loop analysis and qualitative modelling: limitations and merits. Biol Philos 21:647–666.

Kaiser H. 1976. Quantitative description and simulation of stochastic behaviour in dragonflies (*Aeschna cyanea*, Odonata). Acta Biotheor 25:163–210.

Kaiser H. 1979. The dynamics of populations as results of the properties of individual animals. Fortschr Zool 25:109–136.

Klok C, De Roos AM. 1996. Population level consequences of toxicological influences on individual growth and reproduction in *Lumbricus rubellus* (Lumbricidae, Oligochaeta). Ecotoxicol Environ Saf 33:118–127.

Koelmans AA, Van Der Heijde A, Knijfe L, Aalderink RH. 2001. Integrated modelling of eutrophication and organic contaminant fate and effects in aquatic ecosystems. A review. Water Res 15:3517–3536.

Koh HL, Hallam TG, Lee HL. 1997. Combined effects of environmental and chemical stressors on a model *Daphnia* population. Ecol Modell 103:19–32.

Kooijman SALM, Bedaux JJM. 1996. The analysis of aquatic toxicity data. Amsterdam: VU University Press.

Kooijman SALM, Jager T, Kooi BW. 2004. The relationship between elimination rates and partition coefficients of chemical compounds. Chemosphere 57:745–753.

Lagadic L, Leicht W, Ford MG, Salt DW, Greenwood R. 1993. Pharmacokinetics of cyfluthrin in *Spodoptera littoralis* (Boisd.). I. In vivo distribution and elimination of [14C]cyfluthrin in susceptible and pyrethroid-resistant larvae. Pestic Biochem Physiol 45:105–115.

Lagadic L, Weile M, Leicht W, Salt DW, Greenwood R, Ford MG. 1994. Pharmacokinetics of cyfluthrin in *Spodoptera littoralis* (Boisd.). II. Effect of lindane pretreatment on the toxicity and in vivo metabolism of [14C]cyfluthrin in susceptible larvae. Pestic Biochem Physiol 48:173–184.

Lande R. 1993. Risks of population extinction from demographic and environmental stochasticity and random catastrophes. Am Nat 142:911–927.

Lee JH, Landrum PF, Koh CH. 2002. Prediction of time-dependent PAH toxicity to *Hyalella azteca* using a damage assessment model. Environ Sci Technol 36:3131–3138.

Lek S, Dimopoulos I, Derraz M, El Ghachtoul Y. 1996. Modélisation de la relation pluie-débit à l'aide des réseaux de neurones artificiels. Rev Sci l'Eau 9:319–331.

Lennartz B, Louchart X, Voltz M, Andrieux P. 1997. Diuron and simazine losses to runoff water in Mediterranean vineyards. J Environ Qual 26:1493–1502.

Leslie PH. 1945. On the use of matrices in certain population mathematics. Biometrika 33:183–212.

Leslie PH. 1948. Some further notes on the use of matrices in population dynamics. Biometrika 35:213–245.

Levins R. 1969. Some demographic and genetic consequences of environmental heterogeneity for biological control. Bull Entomol Soc Am 15:237–240.

Levins R. 1970. Extinction. In: Gesternhaber M., editor. Some mathematical problems in biology. Providence (RI): American Mathematical Society. p 77–107.

Liess M, Brown C, Dohmen P, Duquesne S, Hart A, Heimbach F, Krueger J, Lagadic L, Maund S, Reinert W, Streloke M, Tarazona JV. 2005. Effects of pesticides in the field. Brussels: SETAC Press.

Liess M, Schulz R, Berenzen N, Drees J, Wogram J. 2001. Pflanzenschutzmittel-Belastung und Lebensgemeinschaften in Fließgewässern mit landwirtschaftlich genutztem Umland. Berlin (Germany): Umweltbundesamt. UBA-Texte 65/01.

Liess M, Von der Ohe P. 2005. Analyzing effects of pesticides on invertebrate communities in streams. Environ Toxicol Chem 24:954–965.

Louchart X, Voltz M, Andrieux P, Moussa R. 2001. Herbicide transport to surface waters at field and watershed scales in a Mediterranean vineyard area. J Environ Qual 30:982–991.

Macan TT. 1958. Methods of sampling the bottom fauna in stony streams. Verh Int Ver Theor Angew Limnol 8:1–21.

Maltby L, Blake N, Brock TCM, Van den Brink PJ. 2002. Addressing interspecific variation in sensitivity and the potential to reduce this source of uncertainty in ecotoxicological assessments. UK Department for Environment, Food and Rural Affairs. DEFRA project code PN0932.

Maltby L, Blake N, Brock TCM, Van den Brink PJ. 2005. Insecticide species sensitivity distributions: the importance of test species selection and relevance to aquatic ecosystems. Environ Toxicol Chem 24:379–388.

Meier C, Böhmer J, Biss R, Feld C, Haase P, Lorenz A, Rawer-Jost C, Rolauffs P, Schindehütte K, Schöll F, Sundermann A, Zenker A, Hering D. 2006. Weiterentwicklung und Anpassung des nationalen Bewertungssystems für Makrozoobenthos an neue internationale Vorgaben (Extension and adaptation of the national assessment system for benthic invertebrates to international requirements). Web site of the University of Duisburg-Essen. Available from: http://www.fliessgewaesserbewertung.de/en/download/berechnung/ [Accessed 16 April 2008].

Moreno JL, Aboal M, Vidal-Abarca MR, Suárez ML. 2001. Macroalgae and submerged macrophytes from fresh and saline waterbodies of ephemeral streams (Ramblas) in semiarid south-eastern Spain. Mar Freshw Res 52:891–905.

Munns WR Jr, Gervais J, Hoffman AA, Hommen U, Nacci DE, Nakamaru M, Sibly R, Topping CJ. 2007. Modeling approaches to population-level Ecological risk assessment. In: Barnthouse LW, Munns WR Jr, Sorensen MT, editors. Population-level ecological risk assessment. Boca Raton (FL): CRC Press. p 179–210.

Naito W, Mayamamoto K-I, Nakanishi J, Masunaga S, Bartell SM. 2002. Application of an ecosystem model for aquatic ecological risk assessment of chemicals for a Japanese lake. Water Res 36:1–14.

Naito W, Mayamamoto K-I, Nakanishi J, Masunaga S, Bartell SM. 2003. Evaluation of an ecosystem model in ecological risk assessment of chemicals. Chemosphere 53:363–375.

Nijboer R, Verdonschot P, Van den Hoorn M. 2003. Macrofauna en vegetatie van de Nederlandse sloten. Alterra-rapport 688.

Ortega F, Parra G, Guerrero F. 2006. Usos del suelo en las cuencas hidrográficas de los humedales del Alto Guadalquivir: Importancia de una adecuada gestión. Limnetica 25:723–732.

Pantel S. 2003. Vorschläge zur multivariaten Strukturanalyse varianzreicher Tiergemeinschaften am Beispiel von Makroinvertebraten-Zönosen landwirtschaftlicher Fließgewässer in Südostniedersachsen [dissertation]. [Aachen (DE)]: Technical University of Aachen.

Park RA, Clough JS, Wellman MC. 2004. AQUATOX (Release 2). Modeling Environmental Fate and Ecological Effects in Aquatic Ecosystems, Volume 1 [user's manual]. Report EPA-823-R-04-001. Washington (DC): Environmental Protection Agency.

Park RA, Clough JS, Wellman MC. 2004. AQUATOX for Windows: a modular fate and effects model for aquatic ecosystems-volume 1: release 2 user's manual. EPA-823-R-04-001.

Park RA, O'Neil RV, Bloomfield JA, Shugart HH, Booth RS, Goldstein RA, Mankin JB, Koonce JF, Scavia D, Adams MS, Clesceri LS, Colon EM, Dettmann EH, Hoopes J, Huff DD, Katz S, Kitchell JF, Kohberger RC, LaRow EJ, McNaught DC, Peterson JL, Scavia D, Titus JE, Weiler PR, Wilkinson JW, Zahorcak CS. 1974. A generalized model for simulating lake ecosystems. Simulation 23:33–50.

Pastorok RA, Bartell SM, Ferson S, Ginzburg LR, editors. 2002. Ecological modeling in risk assessment: chemical effects on populations, ecosystems, and landscapes. Boca Raton (FL): Lewis Publishers.

Peeters ETHM, Scheffer M. 2004. Schonen, baggeren en biodiversiteit in sloten. Wageningen (NL): Wageningen Universiteit. Rapportnummer M331, LSG Aquatische Ecologie en Waterkwaliteitsbeheer.

Péry ARR, Bedaux JJM, Zonneveld C, Kooijman SALM. 2001. Analysis of bioassays with time varying concentrations. Water Res 35:3825–3832.

Pieters BJ, Jager T, Kraak MHS, Admiraal W. 2006. Modeling response of *Daphnia magna* to pesticide pulse exposure under varying food conditions: intrinsic versus apparent sensitivity. Ecotoxicology 15:601–608.

Poethke HJ, Oertel D, Seitz A. 1994. Modelling effects of toxicants on pelagic food-webs: many problems — some solutions. Ecol Modell 75-76:511–522.

Preuss TG, Hammers-Wirtz M, Hommen U, Rubach MN, Ratte HT. 2009. Development and validation of an individual based *Daphnia magna* population model: the influence of crowding on population dynamics. Ecol Modell 220:310–329.

Rashleigh B. 2007. Assessment of lake ecosystem response to toxic events with the AQUATOX model. In: Gonenc IE, Koutitonsky V, Rashleigh B, Ambrose RA, Wolfin JP, editors. Assessment of the fate and effects of toxic agents on water resources. Dordrecht (NL): Springer. p 293–299.

Rautmann D, Streloke M, Winkler R. 2001. New basic drift values in the authorisation procedure for plant protection products. In: Forster R, Streloke M, editors. Workshop on risk assessment and risk mitigation measures in the context of the authorization of plant protection products (WORMM). BBA 27 to 29 September 1999. Biologischen Bundesanstalt für Land- und Forstwirtschaf Berlin-Braunschweig, Heft 383, Parey, Berlin, 2001. p 133–141.

Sagova M, Adams MS, Butler MG. 1993. Relationship between plant roots and benthic animals in three sediment types of a dimictic mesotrophic lake. Arch Hydrobiol 128:423–436.

Scavia D. 1980. An ecological model of Lake Ontario. Ecol Modell 8:49–78.

Schäfer RB, Caquet T, Siimes K, Mueller R, Lagadic L, Liess M. 2007. Effects of pesticides on community structure and ecosystem functions in agricultural streams of three bio-geographical regions in Europe. Sci Total Environ 382:272–285.

Scheffer M, Achterberg AA, Beltman B. 1984. Distribution of macroinvertebrates in a ditch in relation to the vegetation. Freshw Biol 14:367–370.

Schulz R, Elsaesser D, Ohlige R, Stehle S, Zenker K. 2007. Umsetzung der georeferenzierten prob-abilistischen Risikobewertung in den Vollzug des PflSchG — Pilotphase — Dauerkulturen. Endbericht zum F & E Vorhaben 206 63 402 des Umweltbundesamtes. Landau (DE). Institut für Umweltwissenschaften, Universität Koblenz-Landau.

Sherratt TN, Jepson PC. 1993. A metapopulation approach to modelling the long-term impact of pesticides on invertebrates. J Appl Ecol 30:696–705.

Sourisseau S, Bassères A, Perié F, Caquet Th. 2008. Calibration, validation and sensitivity analysis of an ecosystem model applied to artificial streams. Water Res doi:10.1016/j.watres.2007.08.039.

Spencer M, Ferson S. 1998. RAMAS ecotoxicology, Version 1.0a. User's manual. Volume 2: ecological risk assessment for structured populations. Setauket (NY): Applied Biomathematics. p. 29–35.

Spray Draft Task Force. 1999. AgDrift, spray drift task force spray model, version 1.11.

Spromberg JA, John BM, Landis WG. 1998. Metapopulation dynamics: indirect effects and multiple distinct outcomes in ecological risk assessment. Environ Toxicol Chem 17:1640–1649.

Statzner B, Bis B, Doledec S, Usseglio-Polatera P. 2001. Perspectives for biomonitoring at large spatial scales: a unified measure for the functional composition of invertebrate communities in European running waters. Basic Appl Biol 2:73–85.

Statzner B, Dolédec S, Hugueny B. 2004. Biological traits composition of European stream invertebrate communities: assessing the effects of various trait filter types. Ecography 27:470–488.

STOWA. 1993. Ecologische beoordeling en beheer van oppervlaktewateren. Wetenschappelijke verantwoording van het beoordelingssysteem voor sloten. Utrecht (NL): Stichting Toegepast Onderzoek Waterbeheer, Nr 93–15.

Sturm A, Wogram J, Hansen PD, Liess M. 1999. Potential use of cholin-esterase in monitor-ing low levels of organophosphates in small streams: natural variability in three-spined stickleback (*Gasterosteus aculeatus* L.) and relation to pollution. Environ Toxicol Chem 18:194–200.

Tachet H, Richoux P, Bournaud M, Usseglio-Polatera P. 2000b. Invertebrates d'eau douce; systematique, biologie, ecologie. Paris: CNRS.

Thomann RV, Di Toro DM, Winfield RP, O'Connor DJ. 1975. Mathematical modeling of phytoplankton in Lake Ontario, part 1. Model development and verification. Bronx (NY): Manhattan College, for US Environmental Protection Agency EPA-600/3-75-005.

Traas TP, Janse JH, Aldenberg T, Brock TCM. 1998. A food web model for fate, direct, and indirect effects of Dursban 4E (active ingredient chlorpyrifos) in freshwater micro-cosms. Aquat Ecol 32:179–190.

Traas TP, Janse JH, Van den Brink PJ, Brock TCM, Aldenberg T. 2004. A freshwater food web model for the combined effects of nutrients and insecticide stress and subsequent recovery. Environ Toxicol Chem 23:521–529.

Traas TP, Stäb JA, Kramer PRG, Cofino WP, Aldenberg T. 1996. Modeling and risk assessment of tributyltin accumulation in the food web of a shallow freshwater lake. Environ Sci Technol 30:1227–1237.

Traas TP, Van den Brink PJ, Janse JH, Aldenberg T. 1998. Modeling of ecosystem function in aquatic microcosms as influenced by chronic levels of a fungicide and an herbicide. Bilthoven (NL): National Institute of Public Health and the Environment. RIVM report 607601002.

Twisk W, Noordervliet MAW, Ter Keurs WJ. 2000. Effects of ditch management on caddisfly, dragonfly and amphibian larvae in intensively farmed peat areas. Aquat Ecol 34:397–411.

Twisk W, Noordervliet MAW, Ter Keurs WJ. 2003. The nature value of the ditch vegetation in peat areas in relation to farm management. Aquat Ecol 37:191–209.

Usseglio-Polatera P, Bournaud M, Richoux P, Tachet H. 2000. Biological and ecological traits of benthic freshwater macroinvertebrates: relationships and definition of groups with similar traits. Freshw Biol 43:175–205.

Van den Brink PJ, Baveco J, Verboom J, Heimbach F. 2007. An individual-based approach to model spatial population dynamics of invertebrates in aquatic ecosystems after pesticide contamination. Environ Toxicol Chem 26:2226–2236.

Van den Brink PJ, Blake N, Brock TCM, Maltby L. 2006. Predictive value of species sensitivity distributions for effects of herbicides in freshwater ecosystems. Hum Ecol Risk Assess 12:645–674.

Van den Brink PJ, Brown CD, Dubus IG. 2006. Using the expert model PERPEST to translate measured and predicted pesticide exposure data into ecological risks. Ecol Model 191:1, 106–117.

Van den Brink PJ, Rolesma J, Van Nes EH, Scheffer M, Brock TCM. 2002. PERPEST, a case-based reasoning model to predict ecological risks of pesticides. Environ Toxicol Chem 21:2500–2506.

Van den Brink PJ, Van Wijngaarden RPA, Lucassen WGH, Brock TCM, Leeuwangh P. 1996. Effects of the insecticide Dursban 4E (active ingredient chlorpyrifos) in outdoor experimental ditches: II Invertebrate community responses and recovery. Environ Toxicol Chem 15:1143–1153.

Van der Gaast JWJ, Van Bakel, PJT. 1997. Differentiatie van waterlopen ten behoeve van het bestrijdingsmiddelenbeleid in Nederland. Wageningen (NL). DLO Winand Staring Centre, Rapport 526.

Van Geest GJ, Zwaardemaker NG, Van Wijngaarden RPA, Cuppen JGM. 1999. Effects of a pulsed treatment with the herbicide afalon (active ingredient linuron) on macrophyte-dominated mesocosms II. Structural responses. Environ Toxicol Chem 18:2866–2874.

Van Vierssen W. 1982. The ecology of communities dominated by Zannichellia taxa in western Europe [thesis]. Nijmegen (NL): University of Nijmegen.

Van Wijngaarden RPA, Brock TCM, Van den Brink TCM, Gylstra R, Maund SJ. 2006. Ecological effects of spring and late summer applications of lambda-cyhalothrin on freshwater microcosms. Arch Environ Contam Toxicol 50:220–239.

Velasco J, Millán A, Hernández J, Gutiérrez C, Abellán Sánchez D, Ruiz M. 2006. Response of biotic communities to salinity changes in a Mediterranean hypersaline stream. Saline Syst 2:12.

Verhulst PF. 1845. Recherches mathématiques sur la loi d'accroissement de la population. Brussels. Mémoires de l'Académie Royale des Sciences et Belles Lettres de Bruxelles, Band 18. p 1–41.

Vicari A, Rossi Pisa P, Catione P. 1999. Tillage effects on runoff losses of atrazine, metolachlor, prosulfuron and trisulfuron. 11th EWRS Symposium, Basel, Switzerland, 28 June to 1 July 1999.

Vidal-Abarca M, Gómez R, Suárez ML. 2004. Los ríos de las regiones semiáridas. Alicante (ES): Ecosistemas 1(Enero/Abril).

Weber D, Preuß TG, Ratte HT, Bruns E, Görlitz G, Schäfer D, Reinken G, Ottermanns R, Claßen S, Agatz A. 2007. Modellierung der Effekte von Pflanzenschutzmitteln auf ausgewählte Phytoplankter — -Modellentwicklung. Paper presented at SETAC Europe GLB annual meeting. 12 to 14 September. Leipzig, (DE).

Widianarko B, Van Straalen N. 1996. Toxicokinetics-based survival analysis in bioassays using non-persistent chemicals. Environ Toxicol Chem 15:402–406.

Wogram J. 1996. Zur Ökologie des Dreistachligen Stichlings (*Gasterosteus aculeatus* L.) in Agrarfließgewässern [thesis]. [Braunschweig (DE)]: Technical University of Braunschweig.

Wogram J. 2001. Auswirkungen der Pflanzenschutzmittel-Belastung auf Lebensgemeinschaften in Fließgewässern des landwirtschaftlich geprägten Raumes [dissertation]. [Braunschweig (DE)]: Technical University of Braunschweig.

Wogram J, Liess M. 2001. Rank ordering of the sensitivity of macroinvertebrate species to toxic compounds, by comparison with that of *Daphnia magna*. Bull Environ Contam Toxicol 67:360–367.

Wogram J, Sturm A, Segner H, Liess M. 2000. Effects of parathion on acetylcholinesterase, butyrylcholinesterase and carboxylesterase in three-spined stickleback (*Gasterosteus aculeatus* L.) following short-term exposure. Environ Toxicol Chem 20:1528–1531.

Part III

Appendices and Glossary

Appendix 1

This is a list of workshop participants and their participation in ELINK work groups (WGs; for WG themes, see Chapter 2, Section 2.1), expertise, and affiliation (A = academia; R = regulator; B = business).

Name	WG Theme	Expertise	Affiliation	Institute
Aagaard, Alf	3	Fate	R	Danish EPA, Copenhagen, Denmark; EFSA, Parma, Italy
Adriaanse, Paulien	4	Fate	A	Alterra, Wageningen, the Netherlands
Alix, Anne	5	Effects	R	AFSSA-DIVE, Maisons Alfort, France
Alonso Prados, Elena	5	Fate	R	INIA, Madrid, Spain
Arnold, David	3	Fate	B	Consultant, Cambridge, United Kingdom
Ashauer, Roman	4	Fate and effects	A	Eawag, Dübendorf, Switzerland
Boesten, Jos	3	Fate	A	Alterra, Wageningen, the Netherlands
Brock, Theo	5	Effects	A	Alterra, Wageningen, the Netherlands
Brown, Colin	2	Fate	A	University of York, Heslington, York, United Kingdom
Bruns, Eric	4	Effects	B	Bayer CropScience, Monheim, Germany
Campbell, Peter	3	Effects	B	Syngenta, Jealotts Hill, Bracknell, Berks, United Kingdom
Capri, Ettore	1	Fate	A	Univiversità Cattolica del Sacro Cuore, Piacenza, Italy
Dohmen, G. Peter	3	Effects	B	BASF-AG, Limburgerhof, Germany
Douglas, Mark	3	Effects	B	Dow AgroSciences, Abingdon, United Kingdom
Dubus, Igor G.	1	Fate	A	BRGM, Orleans, France
Ericson, Gunilla	3	Effects	R	Kemi, Sundbyberg, Sweden
Finizio, Antonio	3	Fate	A	University of Milan, Italy
Görlitz, Gerhard	2	Fate	B	Bayer CropScience, Monheim, Germany

(*continued*)

Name	WG Theme	Expertise	Affiliation	Institute
Gottesbueren, Bernhard	4	Fate	B	BASF-AG, Limburgerhof, Germany
Hamer, Mick	3	Effects	B	Syngenta, Jealotts Hill, Bracknell, Berks, United Kingdom
Heimbach, Fred	3	Effects	B	RIFCON GmbH, Leichlingen, Germany
Hendley, Paul	2	Fate	B	Syngenta, Greensboro, North Carolina, United States
Hingston, James	1	Fate	R	PSD, York, United Kingdom
Hommen, Udo	5	Effects	B	Fraunhofer IME, Schmallenberg, Germany
Howard, Karen	4	Fate	B	Exponent International, Harrogate, United Kingdom
Huber, Andreas	5	Fate	B	DuPont de Nemours, Bad Homburg, Germany
Knauer, Katja	4	Effects	A	University of Basel, Switzerland
Kondzielski, Igor	5	Fate	R	Institute of Environmental Protection, Warsaw, Poland
Kreuger, Jenny	2	Fate	A	SLU, Uppsala, Sweden
Kubiak, Roland	1	Fate	A	Inst. für Agrarökologie, Neustadt, Germany
Lagadic, Laurent	4	Effects	A	INRA, Rennes, France
Liess, Matthias	4	Effects	A	UFZ, Leipzig, Germany
Lythgo, Chris	4	Fate	R	EFSA, Parma, Italy
Mackay, Neil	1	Fate	B	DuPont Crop Protection, Stevenage, United Kingdom
Maltby, Lorraine	2	Effects	A	University of Sheffield, United Kingdom
Maund, Steve	5	Effects	B	Syngenta, Basel, Switzerland
Mitchel, Gary C.	1	Effects	B	FMC Agricultural Products, Princeton, New Jersey, United States
Moody, Aidan	4	Fate	R	Pesticide Control Service, Celbridge, Ireland
Neumann, Michael	2	Fate	R	UBA, Dessau, Germany
Norman, Steve	1	Effects	B	Makhteshim Agan, Brussels, Belgium
Novillo Villajos, Apolonia	5	Effects	R	INIA, Madrid, Spain
O'Leary Quinn, Jo	4	Effects	R	PSD, York, United Kingdom

Name	WG Theme	Expertise	Affiliation	Institute
Pol, Werner	4	Fate	R	Ctgb, Wageningen, the Netherlands
Ratte, Toni	4	Effects	A	RWTH Aachen University, Germany
Redolfi, Elena	3	Effects	R	ICPS, Milano, Italy
Schäfer, Helmut	1, 2	Fate	B	Bayer CropScience, Monheim, Germany
Schulz, Ralf	1	Effects	A	University Koblenz-Landau, Germany
Streloke, Martin	3	Effects	R	BVL, Braunschweig, Germany
Sweeney, Paul	1, 4	Fate	B	Syngenta, Jealotts Hill, Bracknell, Berks, United Kingdom
Szentes, Csaba	5	Fate	R	Agricultural Office Tolna, Hungary; EFSA, Parma, Italy
Van den Brink, Paul	2	Effects	A	Alterra, Wageningen, the Netherlands
Van Vliet, Peter	4	Effects	R	Ctgb, Wageningen, the Netherlands
Wogram, Jörn	5	Effects	R	UBA, Dessau, Germany

Appendix 2

The following sponsors generously supported the ELINK workshop and the printing of the ELINK proceedings:

The Chemical Company

 Bayer CropScience

 Dow AgroSciences

 agriculture, nature and food quality

Netherlands Ministry of Agriculture, Nature and Food Quality

MAKHTESHIM
A G A N

The publication of the ELINK guidance document was kindly supported by Università Cattolica del Sacro Cuore, Piacenza (Italy)

Appendix 3

There were 3 case studies used during the 2 workshop meetings to aid problem formulation and discussions, those for ELINKstrobin, ELINKmethrin, and ELINKsulfuron.

1 CASE STUDY FOR ELINKstrobin

OBJECTIVE OF THE CASE STUDY

The case study ELINKstrobin explores how to link the fate and effects of a medium degrading compound that may be applied on several occasions and has several routes of exposure. Key questions are as follows:

- How should the exposure patterns with repeated sharp or tailing peaks with overlap be included in the assessment, particularly for chronic toxicity?
- How can ecological recovery potential be included in light of modeled repeated events?
- How can variable exposure patterns with a baseline concentration and repeated tailing peaks with overlap to chronic effect studies of various designs be compared?

PROBLEM FORMULATION

ELINKstrobin — like several other strobilurine fungicides — shows broad aquatic toxicity, mainly via acute effects, and it has a generally very narrow acute-to-chronic (A/C) ratio. Thus, in most studies the dominant endpoint is mortality, and there are no other sublethal effects at concentrations significantly (i.e., < factor 5) below those causing mortality. This was shown for ELINKstrobin in several chronic fish studies (28-day juvenile growth test, a standard [35-day flow-through] and higher-tier [static] early life-stage toxicity [ELS] study) showing this kind of impact and indicating that the very young fish — at the time shortly after hatching and time of swim up — are the most sensitive life stages.

However, from 1 chronic fish study, a 97-day flow-through ELS study with trout, a significantly lower endpoint was derived that was based on a very small, but in this case statistically significant, reduction in fish biomass at the end of the study. It may thus not be excluded that very long exposure durations may exert an impact (although being very small and likely of little ecological relevance) beyond the general very narrow A/C ratio. This background (i.e., being aware that the main impact is due to acute toxicity but also trying to cover a potential additional negative impact from long-term chronic exposure) provides the basis for the risk assessment approach shown here to try to cover peak exposure periods as well as potential chronic exposure.

TABLE A3.1
Use Pattern of ELINKstrobin

Crop	Application Pattern	Growth Stage(s)
Winter cereals	2 × 100 g/ha	BBCH 25-61 (minimum interval 14 days)

GENERAL INFORMATION

ELINKstrobin is a fungicide of the chemical class of the strobilurines with a wide use pattern.

The mode of action of strobilurines — the first fungicides of this group were identified in the fungus *Strobilurus tenacellus* — is inhibition of mitochondrial respiration by binding to the ubihydroquinone oxidation center of the mitochondrial bc1 complex (complex III), thereby blocking electron transfer. The block of electron transfer at the site of quinol oxidation (the Qo site) in the cytochrome bc_1 complex prevents adenosine triphosphate (ATP) formation (therefore they are also referred to as "Q_oI fungicides." The relevant pattern in cereals is listed in Table A3.1.

ENVIRONMENTAL FATE AND EXPOSURE

E-Fate Profile

ELINKstrobin degrades[1] moderately fast in soil and in water. The main physicochemical properties are given in Table A3.2.

Further laboratory and outdoor studies under natural light conditions and different depths of water layers are available (see following). The dissipation behavior of ELINLKstrobin in the aquatic environment has been investigated with different study types. ELINKstrobin is hydrolytically stable. However, under influence of sunlight and in the presence of photosensitizers in natural waters,

TABLE A3.2
Main Physicochemical Properties for Substance: ELINKstrobin

Molar mass (g·mol⁻¹)	300
Saturated vapor pressure (Pa) measured at 20 °C	1.00E-09
Water solubility (mg·L⁻¹):	2.00E+00
Half-life (DegT50) in water (days) measured at 20 °C[a]	60
Half-life (DegT50) in sediment (days) at 20 °C (default worst case)	1000
Half-life (DegT50) in soil (days) at 20 °C and pF2	80
*K*om (coefficient for sorption on organic matter) (L·kg⁻¹)	174.01
Freundlich exponent (−)	0.9

[a] Stems from dark water sediment study.

ELINKstrobin is degraded in water bodies. The results of the different studies are as follows[2]:

Aqueous photolysis
  Nonsensitized DegT50:   65 days
  Natural water photolysis DegT50:   15 days

Water/sediment study (dark) (6 cm water layer, 2 cm sediment layer)
  DegT50 at 20 °C in water: 60 days
  DegT50 at 20 °C in sediment: 1000 days (default worst case)

Outdoor water sediment study (20 cm water layer, 2 cm sediment layer)
  DisT50 water at 20 °C
  (including sorption to sediment):   15.5 days
  DegT50 water at 20 °C: 25 days
  DegT50 sediment at 20 °C: 30 days

Outdoor pond study with fish golden orfe (50 cm water layer, 2 cm sediment layer)
  DisT50 at ambient temperature: 19.7 days
  FOMC DisT50 at ambient temperature: 15.7 days (FOMC [first order multi-compartment dissipation] DisT90 100 days)

Outdoor mesocosm (100 cm water depth, 5 cm natural sandy sediment)
  DisT50 water at ambient temperatures: 47 days

EXPOSURE PROFILE

The exposure of ELINKstrobin in FOCUS (Forum for the Co-ordination of Pesticide Fate Models and Their Use) surface water bodies can generally be characterized by repeated entry into surface water by spray drift and runoff or drainage events. The modeled concentration pattern shows baseline concentrations from drainage and tailing peaks with overlaps or sharp peaks depending on the scenario and the associated water body (see Attachment A3.1).

The predicted global maximum concentrations in surface water for the different FOCUSsw scenarios are listed in Table A3.3. Details on individual actual and time-weighted average (TWA) concentrations are to be found in Attachment A3.1. As ELINKstrobin poses no risk for sediment-dwelling organisms, the PECsed (predicted environmental concentration in sediment) values are not given to simplify the case study.

ECOTOXICITY PROFILE

Overview

ELINKstrobin is mainly characterized by acute toxicity to aquatic organisms; however, a single chronic standard (flow-through) fish study provided a particularly low endpoint.

TABLE A3.3

Summary of Global Maximum and Time-Weighted Average PECsw Values for a 7-Day Period for the Different FOCUSsw Scenarios

Crop	Scenario	Global Maximum PECsw (µg L⁻¹)	PECsw, TWA 7 days (µg/L)
Winter cereals	D1 ditch	2.035	1.978
	D1 stream	1.269	1.231
	D2 ditch	2.714	1.677
	D2 stream	1.692	0.992
	D3 ditch	0.554	0.0973
	D4 pond	0.310	0.308
	D4 stream	0.473	0.297
	D5 pond	0.188	0.186
	D5 stream	0.515	0.117
	D6 ditch	0.565	0.420
	R1 pond	0.292	0.271
	R1 stream	2.181	0.250
	R3 stream	3.699	0.482
	R4 stream	4.049	0.734

In addition, a number of metabolites (degradation products) have been tested with fish, daphnia, and algae; all endpoints were more than 100 mg/L, showing the ecotoxicological nonrelevance of the degradation products. A summary of the main ecotoxicological endpoints is provided in Table A3.4.

EFFECTS DATA

Acute Fish Studies

Fish studies provide the lowest endpoints (both acute and chronic). Additional fish species have been tested to address the species sensitivity distribution; the results for all acute fish tests (6 species) are presented in Table A3.5.

The acute studies allow the following conclusions:

• The dose–response curve is very steep (median lethal concentration/no observed effect concentration [LC50/NOEC] ~ 1.5 [±0.3]).
• Sensitivity differences are small, mean LC50 is about 36 µg a.s./L (±16.8 µg a.s./L), the mean NOEC is 25 µg a.s./L (±11 µg a.s./L) (each species considered only once).
• The impact of the formulation on the result is low (ratios of a.s. versus formulation = 0.65 − 1.79).
• Golden orfe and fathead minnow (additional species in chronic tests) are not less sensitive than trout (in the acute tests).

TABLE A3.4
Aquatic Ecotoxicology Endpoints of ELINKstrobin (Results for the Higher-Tier Static Studies Are Based on Initial Test Concentrations)

Test Species	Test Substance	Study Type	Result (µg a.s./L)	
			LC50	NOEC
Oncorhynchus mykiss	a.s.	Static — 96 hours	43.4	31.6
O. mykiss	a.s.	Flow through — 96 hours	44.4	21.8
O. mykiss	Formulation A	Static — 96 hours	30	25
O. mykiss	Formulation B	Static — 96 hours	24	20
Lepomis macrochirus	a.s.	Static — 96 hours	51.2	46.4
L. macrochirus	a.s.	Flow through — 96 hours	52	36
L. macrochirus	Formulation B	Static — 96 hours	81	36
Cyprinus carpio	Formulation B	Static — 96 hours	54	—
Danio rerio	Formulation B	Static — 96 hours	28	—
Pimephales promelas	Formulation B	Static — 96 hours	28	—
Leuciscus idus mel.	Formulation A	Static — 96 hours	25	—
O. mykiss	a.s.	Juvenile fish 28 days, flow-through	n.d.[a]	10
O. mykiss	a.s.	ELS — 97 days flow through	n.d.	1
P. promelas	a.s.	ELS — 36 days flow through	n.d.	16
L. idus	Formulation A	ELS — 66 days, static under outdoor conditions	n.d.	15
D. magna	a.s.	Static — 48 hours	100	40
D. magna	a.s.	Semistatic — 21 days	n.d.	15
P. subcapitata	a.s.	Static — 96 hours	153[b]	13
Chironomus riparius	a.s.	Spiked water, 28 days	>316	30
Mesocosm	Formulation A	Static — 4 months	—	15[c]

[a] n.d., not determined.
[b] Growth rate.
[c] NOEAEC (based on Effect classes 1 and 2 for all endpoints).

TABLE A3.5
Sensitivities of Different Fish Species in Acute Toxicity Tests

Species	LC50 (µg a.s./L)	NOEC (µg a.s./L)
Oncorhynchus mykiss	44	27
Lepomis macrochirus	51	41
Leuciscus idus	25	17
Cyprinus carpio	44	24
Pimephales promelas	20	16
Danio rerio	27	16

- The number of fish species tested allows a calculation of a sensitivity distribution. The respective HC5 (hazardous concentration to 5% of the tested taxa) based on LC50 data is 17.1 μg a.s./L (lower limit of 8.1 μg a.s./L, upper limit of 23.8 μg a.s./L), and based on the acute NOEC data, the HC5 is 11.4 μg a.s./L (lower limit 5.5 μg/L).

ELINKstrobin has a log Pow greater than 3; however, the respective bioconcentration factor (BCF) study shows low potential for bioaccumulation, with a low BCF and very rapid metabolization and dissipation:

$$BCF < 100 \ (48)$$
$$CT50 = 0.5$$
$$k_1 \ 130 \ d^{-1}, k_2 \ 1.4 \ d^{-1}$$

Chronic Fish Studies

A range of different chronic fish studies have been performed; the results in more detail are as follows:

1) The 28-day juvenile fish test (flow through) results are shown in Table A3.6. Mortality occurred only at the highest concentration and only during the first week of the study; thereafter, no further mortality or any visual symptoms were observed.
2) The 97-day ELS study with trout (flow through) results are shown in Table A3.7. Mortality occurred only at the highest concentration during the hatching period; some effects on growth were observed at the next 2 highest concentrations, behavioral (reduced feed consumption effects were seen at these 2 concentrations, developing over time [after about 60 days]).
3) The 36-day ELS study with fathead minnow (flow through) results are shown in Table A3.8. The main endpoint was mortality, which occurred during the hatching period and shortly afterward in the highest concentration; no significant effects on growth, behavioral, or morphological observation occurred (however, there was a slight tendency of impaired growth at 16 and 32 μg a.s./L in this chronic flow-through study).

TABLE A3.6

28-Day Juvenile Growth Test with Rainbow Trout under Flow-Through Conditions

End Point	Control	0.316	1	3.16	10	31.6
Mortality (%)	0	0	0	0	0	55
Symptoms	None	None	None	None	None	A, P, Z
Mean weight (g)	5.26	5.67	5.53	5.3	5.33	5.08
Mean length (cm)	7.67	7.85	7.77	7.7	7.67	7.44

TABLE A3.7
ELS Study with Trout

End Point	Control	0.316	1	3.16	10	31.6
Start of hatch (days)	32 to 33	32 to 33	32 to 33	32 to 33	32 to 33	36
End of hatch (days)	39	39	39	39	39	—
Time to swim-up (days)	55	55	55	55	55	—
Young fish survival rate (%)	58	55	61	52	55	0[a]
Mean wet weight (% of control)	100	96	92	63[a]	70[a]	—
Total wet weight (g)	12.99	11.82	12.1	7.38[a]	8.66[a]	—
Mean body length (% of control)	100	97	95[a]	86[a]	87[a]	—
Total length (cm)	63.4	58.6	63.6	48.8 4[a]	52.4[a]	—

[a] Statistically significant difference.

4) The 66-day ELS study with golden orfe (static outdoor) results are shown in Table A3.9. The main endpoint was mortality, which occurred during the hatching period; no other effects were observed. Test concentrations decreased during the course of the static study; concentrations at time of hatch were approximately 10 μg a.s./L, which may thus be used as the NOEC for this study.

TABLE A3.8
ELS Study with Fathead Minnow

End Point	Control	Solvent Control	2.2	4.2	8	16	32
Start of hatch (day)	4	4	4	4	4	4	4
Fish survival rate 32 days posthatch (%)	90	88	98	78	93	78	38[a]
Sublethal effects 32 days posthatch (%)	None	None	None	None	None	None	D
Mean wet weight (mg)	89.3	92.3	82.3	97.6	81.7	76.5	71.4
Mean body length (mm)	24.6	24.1	23.6	24.8	23.9	23.5	22.4

[a] Statistically significant difference.

TABLE A3.9
Outdoor ELS Study with Golden Orfe

End Point	Control	1.7 µg/L	5 µg/L	15 µg/L	66.7 µg/L
Start of hatch (days)	7	7	7	7	7
Survival end of hatch (%)	87	89	87	87	0%
Survival before reduction[a] (%)	82	84	87	87	—
Survival end of test (66 days) (%)	72	84	87	62	—
Mean weight (g)	0.282	0.245	0.297	0.283	—
Mean length (cm)	3.19	3.06	3.26	3.20	—

[a] The number of fish was impartially reduced to 20 per replicate 21 DAT.

The results of the 4 different chronic fish studies with 3 different species are very much in line (Table A3.10). Mortality is observed at concentrations of 30 µg a.s./L and above, and no mortality occurs at 10 µg a.s./L or less. In general, the same is true for sublethal effects; in 3 of these studies, no sublethal impact was observed at concentrations without mortality (i.e., 10 µg a.s./L), which is in agreement with the steep dose–response relationship in acute (and chronic) studies or from information for other substances in this fungicidal class and this mode of action.

In the 97-day flow-through study with trout, the same is true with respect to mortality. However, here a 10- to 30-fold lower endpoint was derived based on a sublethal parameter (i.e., a very minor, but statistically significant reduction in fish growth at the end of the study).

Under the assumption that the small effect observed in the 97-day study is a true effect and considering all information from all acute and chronic studies, it may not be excluded that a long-term continuous exposure at a concentration of 1 µg a.s./L and higher may result in a negative impact on sensitive fish species.

TABLE A3.10
Summary of Chronic Fish Studies with ELINKstrobin

Test System	Species	LOEC/NOEC (mortality) (µg a.s./L)	LOEC/NOEC (sublethal) (µg a.s./L)
28-day juvenile growth test, flow through	Oncorhynchus mykiss	31.6/10.0	31.6/10.0
36-day ELS, flow through	Pimephales promelas	32/16	32/16
97-day ELS, flow through	Oncorhynchus mykiss	31.6/10.0	3.16/1.0[a]
			1.0/0.316
66-day ELS, static outdoor	Leuciscus idus	40/10[b]	40/10[b]

[a] NOEAEC.
[b] Based on the concentration level at the time of hatching.

ATTACHMENT A3.1 FOCUSsw Scenarios: Exposure
for ELINKstrobin in Winter Cereals

The *y*-axes of the figures have different scales to illustrate the patterns rather than the levels of exposure.

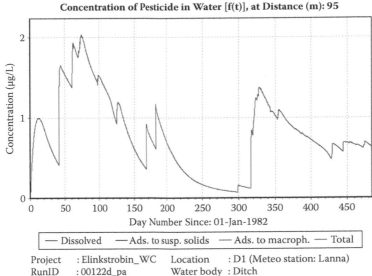

Project	: Elinkstrobin_WC	Location	: D1 (Meteo station: Lanna)
RunID	: 00122d_pa	Water body	: Ditch
Substance	: ELINK "Strobilurin"	Crop	: Cereals, winter

FIGURE A3.1   ELINKstrobin–winter cereals: D1 Ditch.

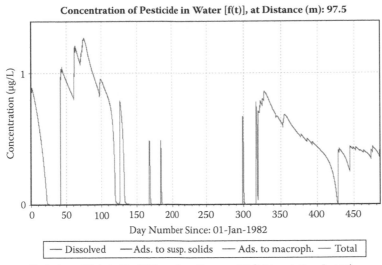

Project	: Elinkstrobin_WC	Location	: D1 (Meteo station: Lanna)
RunID	: 00122s_pa	Water body	: Stream
Substance	: ELINK "Strobilurin"	Crop	: Cereals, winter

FIGURE A3.2   ELINKstrobin–winter cereals: D1 Stream.

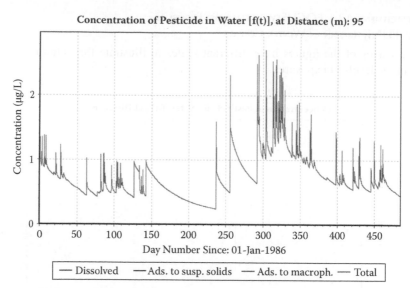

FIGURE A3.3   ELINKstrobin–winter cereals: D2 Ditch.

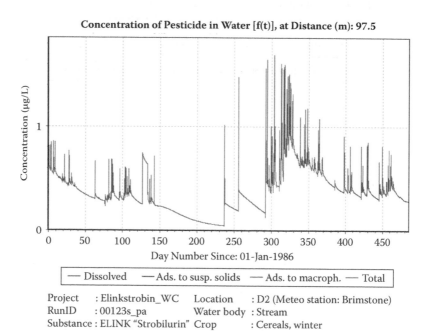

FIGURE A3.4   ELINKstrobin–winter cereals: D2 Stream.

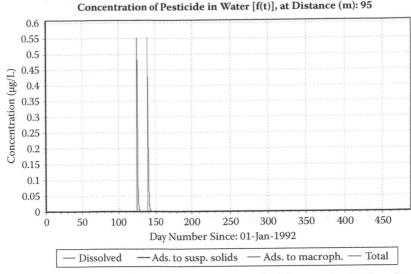

Project   : Elinkstrobin_WC   Location   : D2 (Meteo station: Vredepeel)
RunID   : 00124d_pa   Water body : Ditch
Substance : ELINK "Strobilurin"   Crop   : Cereals, winter

FIGURE A3.5   ELINKstrobin–winter cereals: D3 Ditch.

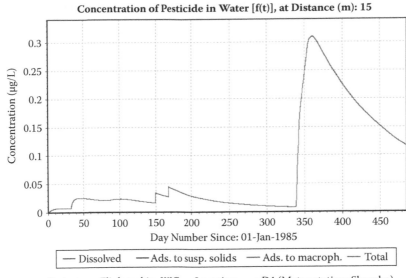

Project   : Elinkstrobin_WC   Location   : D4 (Meteo station: Skousbo)
RunID   : 00125p_pa   Water body : Pond
Substance : ELINK "Strobilurin"   Crop   : Cereals, winter

FIGURE A3.6   ELINKstrobin–winter cereals: D4 Pond.

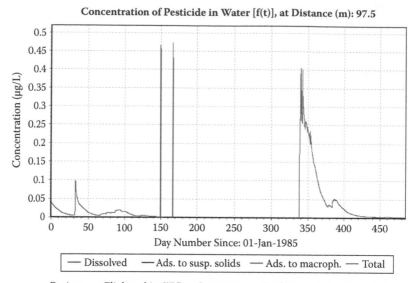

FIGURE A3.7   ELINKstrobin–winter cereals: D4 Stream.

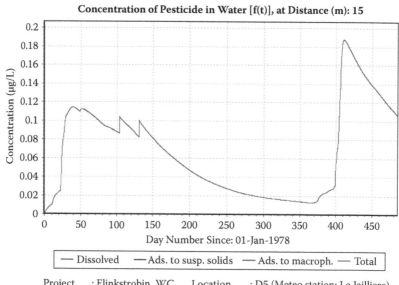

FIGURE A3.8   ELINKstrobin–winter cereals: D5 Pond.

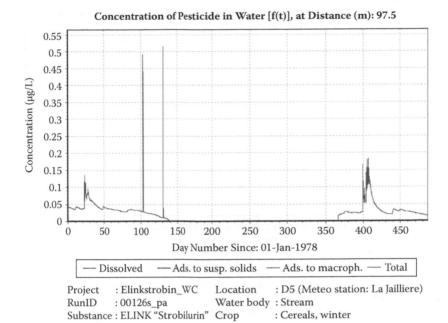

Project   : Elinkstrobin_WC   Location   : D5 (Meteo station: La Jailliere)
RunID   : 00126s_pa   Water body : Stream
Substance : ELINK "Strobilurin" Crop   : Cereals, winter

FIGURE A3.9   ELINKstrobin–winter cereals: D5 Stream.

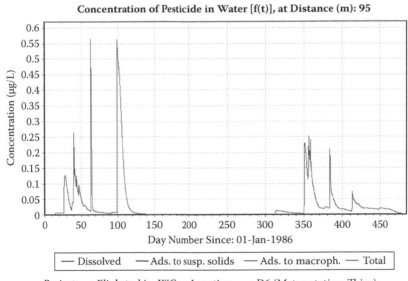

Project   : Elinkstrobin_WC   Location   : D6 (Meteo station: Thiva)
RunID   : 00127d_pa   Water body : Ditch
Substance : ELINK "Strobilurin" Crop   : Cereals, winter

FIGURE A3.10   ELINKstrobin–winter cereals: D6 Ditch.

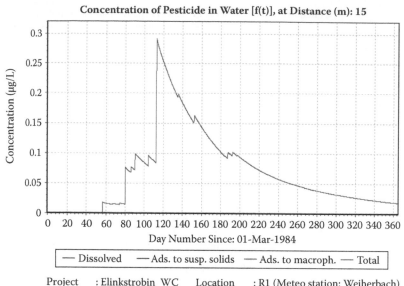

FIGURE A3.11   ELINKstrobin–winter cereals: R1 Pond.

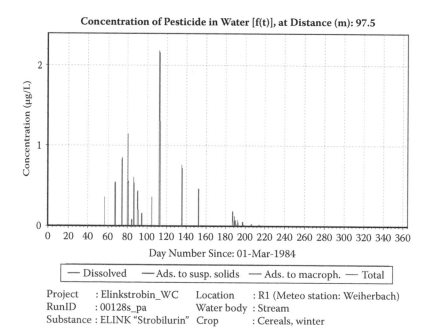

FIGURE A3.12   ELINKstrobin–winter cereals: R1 Stream.

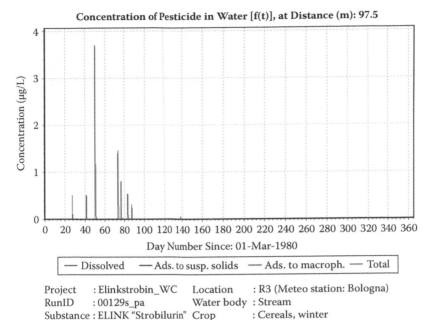

FIGURE A3.13   ELINKstrobin–winter cereals: R3 Stream.

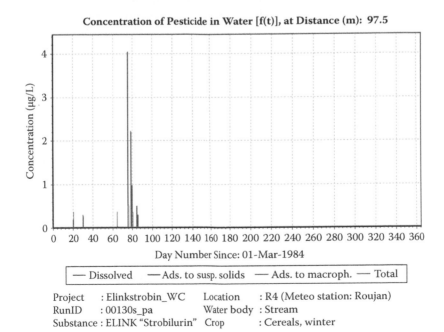

FIGURE A3.14   ELINKstrobin–winter cereals: R4 Stream.

2   CASE STUDY OF ELINKmethrin

OBJECTIVE OF THE CASE STUDY

The case study of ELINKmethrin explores how to link the fate and effects of a rapidly dissipating compound that may be applied on several occasions. The case study focuses on potential for effects on aquatic invertebrates. Key questions are as follows:

- How should rapid dissipation be included in the assessment, particularly for chronic toxicity?
- What modeling or experimental approaches can be used to evaluate effects of short-term, repeated exposures?
- How can ecological recovery potential be included in light of modeled repeated events?

GENERAL INFORMATION

Chemical class: Pyrethroid (type II alpha-cyano)

Mode of action: Primary mode of action is through interference with ion channels in the nerve axon, resulting in hyperactivity of the nervous system with subsequent lack of normal function. The type II pyrethroids appear to have a primary mechanism of action that involves an action at the presynaptic membrane that involves increased release of synaptic vesicles through an effect on voltage-dependent ion channels. Symptoms of poisoning appear rapidly in all pyrethroids, and this is characteristic of this class of compounds. The rapid onset of poisoning and the lack of persistence of the pyrethroids in the environment emphasizes the importance of acute toxicity data for assessing environmental exposures.

   In addition to their action in the nervous system, pyrethroids have been reported to interfere with certain adenosine triphosphatase (ATPase) enzymes associated with maintaining ionic concentration gradients across membranes. It has been speculated that this may increase the sensitivity of saltwater aquatic organisms to these insecticides through the addition of osmotic stress. This has also been suggested as the reason why pyrethroids generally appear to be more toxic at median (isotonic) salinities in euryhaline species than at either high or low salinities. In general, susceptibility to the pyrethroids is dependent on sensitivity at the site of action and toxicokinetics. Included in the latter are bioavailability and rates of biological transformation. It has been suggested that biotransformation may play a role in differential toxicity of the pyrethroids to fish, which are generally more sensitive than mammals to pyrethroids. This same mechanism may explain the general lower sensitivity of fish compared to arthropods.

USE PATTERN

ELINKmethrin is used to control a wide range of insects, especially Lepidoptera, but also Coleoptera, Diptera, Hemiptera, and other classes.

Good agricultural practice (GAP):	Field crops (e.g., spring cereals, carrots, potatoes, etc.)
Application rate:	10 g ai/ha
Number of applications:	5 applications
Application interval:	7 days

ENVIRONMENTAL FATE PROFILE

Log Pow:	7.2
Koc (coefficient for sorption on organic matter):	300 000
Mean aerobic soil DT50 (period required for 50% disappearance):	20 days
Water solubility:	5 μg/L
Water–sediment half-life (whole system):	50 days
Hydrolytic half-life:	pH 7.0 stable, pH 9.0, DT50 0.5 day

ECOTOXICITY PROFILE

In standard laboratory tests, ELINKmethrin is highly acutely and chronically toxic to aquatic Crustacea and Insecta, highly acutely and chronically toxic to fish, but not toxic to aquatic plants. Toxicity data are included in Table A3.11.

Although potential risks to fish may also be indicated from these data, the case study focuses on potential risks to water column invertebrates. Potential risks to sediment-dwelling invertebrates are also not considered in this case study.

TABLE A3.11
Aquatic Toxicity Data of ELINKmethrin

Species	Test Duration and Type	Toxicity[a] (μg ai L^{-1})
Daphnia magna (water flea)	48-hour flow through	EC50 = 0.12
Oncorhynchus mykiss (rainbow trout)	96-hour flow through	LC50 = 1.3
Lepomis macrochirus (bluegill sunfish)	96-hour flow through	LC50 = 1.4
Selenastrum capricornutum (green alga)	96-hour static	ErC50 > 5000
Daphnia magna (water flea)	21-day static replacement	NOEC = 0.001 (based on growth)
Pimephales promelas (fathead minnow)	30-day flow through	NOEC = 0.40

[a] EC50, median effective concentration; LC50, median lethal concentration; NOEC, no observed effect concentration.

TABLE A3.12
Maximum and 7-Day TWA PEC Surface Water Values for FOCUS Step 3

Scenario	Water Body	Max PECsw (µg/L)	7-Day TWAEC (µg/L)
D1	Ditch	0.041	0.015
D1	Stream	0.029	0.003
D3	Ditch	0.032	0.004
D4	Pond	0.001	0.001
D4	Stream	0.028	0.001
D5	Pond	0.001	0.001
D5	Stream	0.03	0.001
R4	Stream	0.022	0.001

EXPOSURE PROFILE

FOCUS step 3 modeling has been performed for uses in spring cereals (10 g ai/ha, maximum 5 applications, 7-day interval). The peak exposure concentrations for the scenarios are presented in Table A3.12.

Outputs from the modeling (Attachment A3.2) demonstrate that exposure profiles are dominated by spray drift events, followed by rapid dissipation. Runoff and drainage inputs are negligible. Exposure can therefore be characterized as a series of pulses that occur around the times of application.

DATA FOR A HIGHER-TIER RISK ASSESSMENT
Time-to-Effect Data

Observations from the *Daphnia* flow-through chronic study indicate that there is a time dependency of effect on mortality in the study (Table A3.13). The NOEC from the study (1.0 ng/L) is derived from effects on growth (measured at the end of the study).

In a study with *Gammarus pulex*, the effects of short-term exposures of 1, 3, 6, 12 (followed by return to clean water), and 96 hours were compared for their effects on the organisms after 96 hours (Table A3.14). Short-term exposures had significantly less effect than long-term exposures, with 1- and 3-hour exposures more than an order of magnitude less toxic than 96-hour exposures. This confirms

TABLE A3.13
Toxicity of ELINKmethrin to *Daphnia magna* after Various Exposure Periods in the 21-Day Chronic Study

Day	3	5	7	10	12	14	17	19	21
LC_{50} (ng/L)	54	29	25	12	9.6	7.2	6.6	3.6	3.0

TABLE A3.14
Toxicity of ELINKmethrin to *Gammarus pulex* at 96 Hours after Exposure of Different Durations

Hours	1	3	6	12	96
EC_{50} (ng/L)	230	180	53	31	13

---

that the effects of short-term exposures to ELINKmethrin are very much less than longer-term exposures. Under conditions of rapid dissipation, it seems that effects will be significantly mitigated by depuration.

SPECIES SENSITIVITY DISTRIBUTIONS

A range of arthropod species has been tested with ELINKmethrin in acute toxicity tests (static without renewal of test medium). Effect concentrations were based on mean measured concentrations in the studies. Data are included in Table A3.15 along with a species sensitivity distribution in Figure A3.15.

OUTDOOR MESOCOSM EXPERIMENT

The fate and effects of ELINKmethrin were compared in mesotrophic (macrophyte-dominated) and eutrophic (phytoplankton-dominated) ditch microcosms (0.5 m³). The experimental design was in accordance with guidance documents. The study endpoints were the population and community dynamics of macroinvertebrate and zooplankton taxa. Macroinvertebrates were sampled using substrate samplers, and zooplankton were collected using a water column sampler.

ELINKmethrin was applied 3 times at 1-week intervals to achieve concentrations of 10, 25, 50, 100, and 250 ng/L in the water column. The first application was made on May 16. The rate of dissipation of ELINKmethrin in the water

---

TABLE A3.15
Species Sensitivity Data for ELINKmethrin

Species	48-Hour EC50 (ng/L)
Hyalella azteca	2.9
Chaoborus sp.	3.1
Gammarus pulex	17
Asellus aquaticus	28
Corixa sp.	31
Cloeon dipterum	37
Hydracarina	49
Ischnura elegans	132
Cyclops sp.	310
Chironomus riparius	2300
Ostracoda	3500

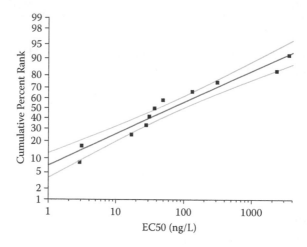

FIGURE A3.15   Species sensitivity distribution of aquatic arthropods with ELINKmethrin.

column of the 2 types of test systems was similar (see Figure A3.16 for 250-ng/L treatments; other treatments showed similar dissipation profiles). After 1 day, only 30% of the amount applied remained in the water phase. Initial direct effects were observed primarily on arthropod taxa. By 7 days after treatment, negligible residues of ELINKmethrin remained.

The most sensitive species was the phantom midge (*Chaoborus obscuripes*). Threshold levels for slight and transient direct toxic effects were similar (10 ng/L) between types of test systems. At treatment levels of 25 ng/L and higher, apparent population and community responses occurred. At treatments of 100 and 250 ng/L, the rate of recovery of the macroinvertebrate community was lower in the macrophyte-dominated systems, primarily because of a prolonged decline of the amphipod *Gammarus pulex*. This species occurred at high densities only in the

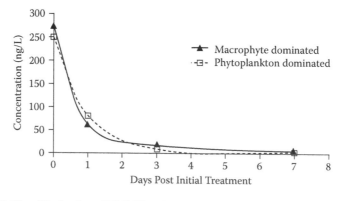

FIGURE A3.16   Dissipation of ELINKmethrin from the water phase.

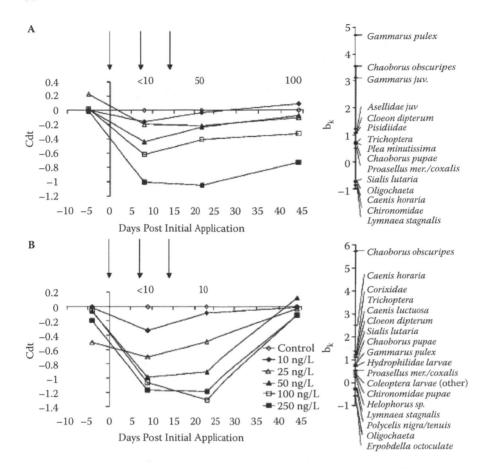

FIGURE A3.17 Principal response curves for the macrophyte-dominated (A) and phyto-plankton-dominated (B) systems.

macrophyte-dominated enclosures. Indirect effects (e.g., increase of rotifers and microcrustaceans) were more pronounced in the plankton-dominated test systems, particularly at treatment levels of 25 ng/L and higher. Principle response curves (PRCs) for the macrophyte-dominated and phytoplankton-dominated systems are shown in Figure A3.17. Calculated NOEC values at the community level are plotted above the figures. Arrows depict application dates. The vertical axis represents the difference in community structure between treatments and the control expressed as regression coefficients (Cdt) of the PRC model. The species weight (b_k) can be interpreted as the affinity of the taxon to the PRC.

Responses to ELINKmethrin were driven by effects on *Gammarus pulex* and *Chaoborus obscuripes* in macrophyte-dominated systems and *Chaoborus obscuripes* and *Cloeon dipeterum* in phytoplankton-dominated systems. The dynamics of these organisms are shown in Figure A3.18.

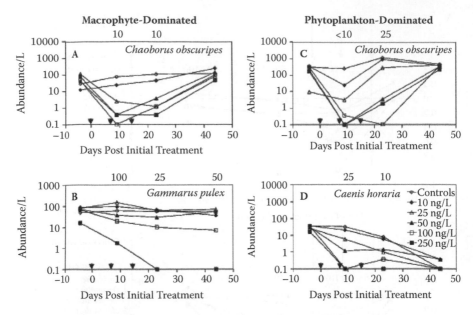

FIGURE A3.18 Dynamics of macroinvertebrate species in the macrophyte-dominated (A and B) and phytoplankton-dominated (C and D) microcosms including responses of the phantom midge (*Chaoborus obscuripes*; A and C), the amphipod *Gammarus pulex* (B) and the ephemeropteran *Caenis horaria* (D). Calculated no-observed-effect concentrations are plotted above the figures. Arrowheads depict application dates.

Table A3.16 details the effect classes observed in the experiment for the different taxonomic groups. The definitions of the effects classes are contained in Table A3.17.

TABLE A3.16
Summary of Effects Observed in the Macrophyte-Dominated and Phytoplankton-Dominated Enclosures

	10 ng/L	25 ng/L	50 ng/L	100 ng/L	250 ng/L
Macrophyte Dominated					
Macrocrustaceans	1	2↓	2↓	4↓	4↓
Insects	2↓	3↓	3↓	3↓	3↓
Other macroinvertebrates	1	1	1	1	2↑↓
PRC macroinvertebrates	2	2	3	3	4
Microcrustaceaus	1	2↓	2↓	2↓	4↓
Rotifers	2↑↓	2↑↓	2↑↓	2↑↓	2↓; 3↓
PRC zooplankton	1	1	2	2	2
Phytoplankton chlorophyll[a]	1	1	1	1	1
Macroptryte biomass	1	1	1	1	1
Community metabolism	1	1	1	1	1
Phytoplankton Dominated					
Macrocrustaceans	—[b]	—[b]	—[b]	—[b]	—[b]
Insects	2↓	3↓	3↓	3↓	3↓
Other macroinvertebrates	1	1	1	2↑	2↑
PRC macroinvertebrates	2	3	3	3	3
Microcrustaceans	2–3↑	4↑	4↑	4↑	4↑↓
Rotifers	2↑	3↑	3↑	3↑	3↑
PRC zooplankton	2	2	2	2	2
Phytoplankton chlorophyll[a]	1	1	1	1	2↑[c]
Macrophyte biomass	—	—	—	—	—

[a] The numbers in die table refer to K > preselected effete elates (Brock et al.). 1 = no effect; 2 = flight effect; 3 = apparent short-term effects, full recovery observed (4 to 8 weeks); 4 = apparent effects, no full recovery observed at the end of the experiment; ↑ = increase; ↓ = decrease: ↑↓ = increase and decrease on species or sampling date; PCR = principal response curve.

[b] Low abundance of free-living population.

[c] Trend of an increase.

TABLE A3.17
Effect Classes Adapted after Brock et al. (2000, 2005) and SANCO (2002) to Evaluate Aquatic Micro/Mesocosm Experiments

Effect Class	Description	Criteria
1	Effects could not be demonstrated ($NOEC_{micro/mesocosm}$)	• No (statistically significant) effects observed as a result of the treatment • Observed differences between treatment and controls show no clear causal relationship
2	Slight effects	• Effects reported as "slight" or "transient" or other similar descriptions • Short-term or quantitatively restricted response of one or a few sensitive endpoints and only observed at individual samplings
3 (A or B)	Pronounced short-term effects	• Clear response of sensitive endpoints but full recovery of affected endpoints within 8 weeks after the first (Effect class 3A) or last (Effect class 3B) application (due to recovery in treated test systems) • Effects reported as "temporary effects on several sensitive species," "temporary effects on less sensitive species/endpoints," or other similar descriptions • Effects observed at some subsequent sampling instances
4	Pronounced effects in short-term study	• Clear effects (such as large reductions in densities of sensitive species) observed, but the study is too short to demonstrate complete recovery within 8 weeks after the (last) application
5 (A or B)	Pronounced long-term effects in long-term study but full recovery not observed within 8 weeks after the last application	• Clear response of sensitive endpoints, and recovery time is longer than 8 weeks after the last application, but full recovery is demonstrated to occur in the year of application (Effect class 5A) or full recovery cannot be demonstrated before termination of the experiment or before the start of the winter period (Effect class 5B)

ATTACHMENT A3.2 OUTPUT FROM FOCUS STEP 3 AND DETAILS OF CONCENTRATION PROFILES FROM SWASH (SURFACE WATER SCENARIOS HELP SOFTWARE) OUTPUT

Application timings derived for the various scenarios (5 applications at 10 g ai/ha in spring cereals)

EXAMPLE FOCUSsw SCENARIOS: EXPOSURE FOR ELINKMETHRIN IN SPRING CEREALS

The *y*-axes of Figures A3.19–A3.27 have different scales to illustrate the patterns rather than the levels of exposure.

		Applications Timing Derived for the Various Scenarios (5 Applications at 10g as/ha in Spring Cereals)				
Scenario	Water Body	Application 1	Application 2	Application 3	Application 4	Application 5
D1	Ditch	25 Apr 1982	14 May 1982	17 Jun 1982	24 Jun 1982	02 Jul 1982
D1	Stream	25 Apr 1982	14 May 1982	17 Jun 1982	24 Jun 1982	02 Jul 1982
D3	Ditch	17 Mar 1992	04 Apr 1992	20 Apr 1992	04 May 1992	14 May 1992
D4	Pond	18 Apr 1985	30 May 1985	04 Jul 1985	11 Jul 1985	18 Jul 1985
D4	Stream	18 Apr 1985	30 May 1985	04 Jul 1985	11 Jul 1985	18 Jul 1985
D5	Pond	07 Mar 1978	08 Apr 1978	22 Apr 1978	11 May 1978	27 May 1978
D5	Stream	07 Mar 1978	08 Apr 1978	22 Apr 1978	11 May 1978	27 May 1978
R4	Stream	05 Mar 1984	12 Mar 1984	21 Mar 1984	04 May 1984	11 May 1984

Summary of PECsw (predicted environmental concentration in surface water) for the FOCUS Step 3 scenarios

Global

Scenario	Water Body	Max (μg/L) Water	Date	1-Day TWA (μg/L)	2-Day TWA (μg/L)	4-Day TWA (μg/L)	7-Day TWA (μg/L)
D1	Ditch	0.041	02 July 1982	0.025	0.021	0.018	0.015
D1	Stream	0.029	02 July 1982	0.014	0.009	0.006	0.003
D3	Ditch	0.032	14 May 1992	0.015	0.011	0.006	0.004
D4	Pond	0.001	18 July 1985	0.001	0.001	0.001	0.001
D4	Stream	0.028	18 July 1985	0.007	0.004	0.002	0.001
D5	Pond	0.001	27 May 1978	0.001	0.001	0.001	0.001
D5	Stream	0.03	27 May 1978	0.006	0.003	0.002	0.001
R4	Stream	0.022	11 May 1984	0.004	0.003	0.002	0.001

Scenario	Water Body	14-Day TWA (μg/L)	21-Day TWA (μg/L)	28-Day TWA (μg/L)	42-Day TWA (μg/L)	50-Day TWA (μg/L)	100-Day TWA (μg/L)
D1	Ditch	0.014	0.013	0.012	0.01	0.009	0.006
D1	Stream	0.003	0.003	0.003	0.002	0.001	0.001
D3	Ditch	0.004	0.003	0.003	0.003	0.002	0.002
D4	Pond	0.001	0.001	0.001	0.001	0.001	< 0.001
D4	Stream	0.001	0.001	0.001	0.001	0.001	< 0.001
D5	Pond	0.001	0.001	0.001	0.001	0.001	< 0.001
D5	Stream	0.001	< 0.001	< 0.001	< 0.001	< 0.001	< 0.001
R4	Stream	0.001	0.001	0.001	0.001	< 0.001	< 0.001

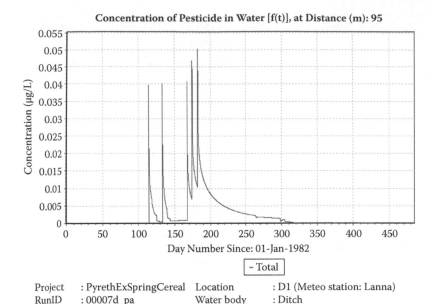

Project   : PyrethExSpringCereal   Location   : D1 (Meteo station: Lanna)
RunID   : 00007d_pa   Water body   : Ditch
Substance   : PyrethroidExample   Crop   : Cereals, spring

FIGURE A3.19 ELINKmethrin–spring cereals: D1 Ditch.

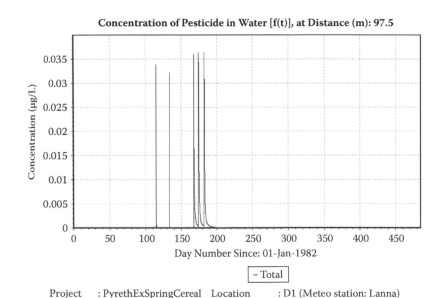

Project   : PyrethExSpringCereal   Location   : D1 (Meteo station: Lanna)
RunID   : 00007s_pa   Water body   : Stream
Substance   : PyrethroidExample   Crop   : Cereals, spring

FIGURE A3.20 ELINKmethrin–spring cereals: D1 Stream.

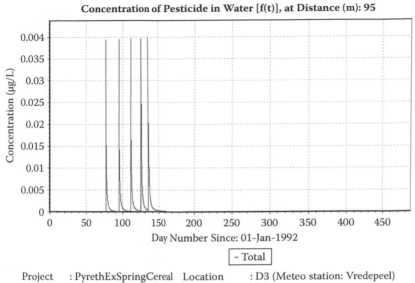

Project   : PyrethExSpringCereal   Location   : D3 (Meteo station: Vredepeel)
RunID   : 00008d_pa   Water body   : Ditch
Substance : PyrethroidExample   Crop   : Cereals, spring

FIGURE A3.21   ELINK methrin–spring cereals: D3 Ditch.

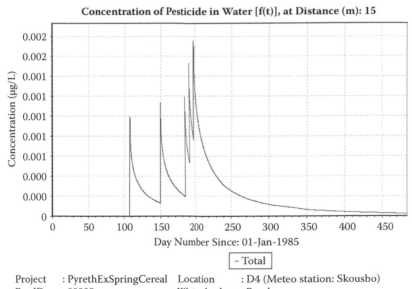

Project   : PyrethExSpringCereal   Location   : D4 (Meteo station: Skousbo)
RunID   : 00009p_pa   Water body   : Pond
Substance : PyrethroidExample   Crop   : Cereals, spring

FIGURE A3.22   ELINK methrin–spring cereals: D4 Pond.

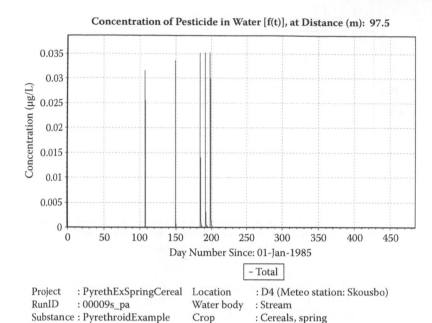

Concentration of Pesticide in Water [f(t)], at Distance (m): 97.5

Project	: PyrethExSpringCereal	Location	: D4 (Meteo station: Skousbo)
RunID	: 00009s_pa	Water body	: Stream
Substance	: PyrethroidExample	Crop	: Cereals, spring

FIGURE A3.23 ELINKmethrin–spring cereals: D4 Stream.

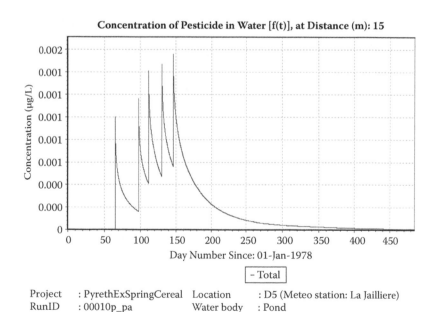

Concentration of Pesticide in Water [f(t)], at Distance (m): 15

Project	: PyrethExSpringCereal	Location	: D5 (Meteo station: La Jailliere)
RunID	: 00010p_pa	Water body	: Pond
Substance	: PyrethroidExample	Crop	: Cereals, spring

FIGURE A3.24 ELINKmethrin–spring cereals: D5 Pond.

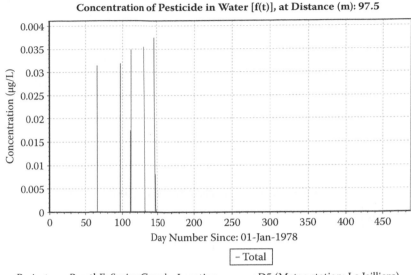

FIGURE A3.25   ELINKmethrin–spring cereals: D5 Stream.

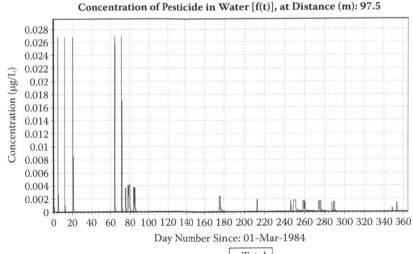

FIGURE A3.26   ELINKmethrin–spring cereals: R4 Stream.

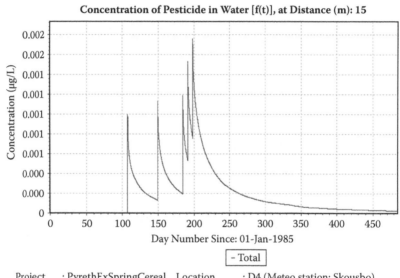

(a)

FIGURE A3.27 Example of estimated dissipation for spring cereals D4 Pond.

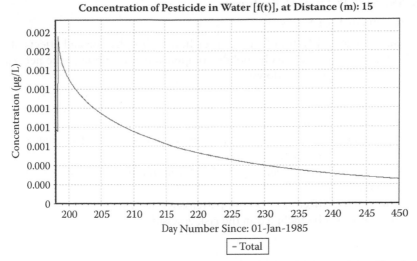

Concentration of Pesticide in Water [f(t)], at Distance (m): 15

- Total

Project   : PyrethExSpringCereal   Location   : D4 (Meteo station: Skousbo)
RunID   : 00009p_pa   Water body   : Pond
Substance : PyrethroidExample   Crop   : Cereals, spring

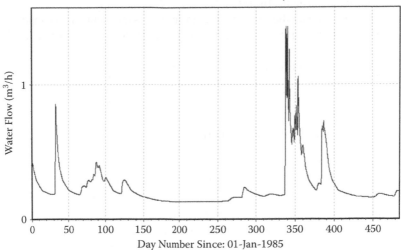

Water Flow Out of Water Body

Project   : PyrethExSpringCereal   Location   : D4 (Meteo station: Skousbo)
RunID   : 00009p_pa   Water body   : Pond
Substance : PyrethroidExample   Crop   : Cereals, spring

(b)

FIGURE A3.27   (Continued)

3   CASE STUDY OF ELINKsulfuron

OBJECTIVE OF THE CASE STUDY

The case study for ELINKsulfron explores how to link the fate and effects of a rapidly degrading and relatively mobile compound that is applied once per season. Key questions are as follows:

- How should rapid dissipation be included in the assessment, particularly for chronic toxicity?
- What modeling or experimental approaches can be used to evaluate effects caused by different exposure patterns?
- What are the appropriate approaches for including ecological recovery potential?
- How can the different exposure patterns resulting from different exposure pathways be included in the risk assessment?

SOME CHALLENGES OF THE CASE STUDY

- How to relate different exposure conditions predicted from FOCUS models to the ecotoxicological properties, namely, mode of action (growth inhibition) and demonstrated recovery potential
- How to relate the exposure predicted by FOCUS models to the exposure in the effects studies
- How to evaluate the probabilistic drain flow assessment and its applicability to general European conditions (can the results be extrapolated to other sites?)
- How to account for the recovery potential in the risk assessment

GENERAL INFORMATION

Chemical Class

ELINKsulfuron is a sulfonylurea herbicide.

Mode of Action

The compound acts by inhibition of the enzyme acetolactate synthase (ALS) and thus blocks cell division. The rapid effect on cell division is not accompanied by measurable effects on energy metabolism and biosynthetic processes. As a result, the plants show arrested growth as the first visible symptom, while the basic functions of cell metabolism continue. The plants usually remain between 1 and several weeks in this phase before tissue necrosis, indicating cell death, occurs.

Use Pattern

- 15 g a.s./ha in cereals
- 1 application per year, in spring or autumn
- Early postemergent (BBCH 12-25)

Environmental Fate Profile

- Water solubility: 480 mg/L
- Vapor pressure: 5×10^{-11} Pa
- *Koc* value: 43 L/kg; $1/n = 0.90$
- Soil DT50:   24 days
- Aquatic DT50:   40 days (degradation, total system); 36 days (dissipation, water)

Ecotoxicological Profile

Summary

In standard laboratory studies, ELINKsulfuron is highly toxic to algae and aquatic plants. It has a (very) low toxicity to fish, aquatic invertebrates, and sediment-dwelling organisms. The endpoint for the most sensitive standard test species, *Lemna gibba*, in a 7-day growth inhibition test is 1.5 µg a.s./L (E_bC50, median effective concentration on biomass).

Lemna gibba is the most sensitive of 10 tested aquatic plant species (E_bC50 values 1.5 to more than 5.3 µg/L, HC5 1.43 µg/L). *Lemna gibba* shows pronounced recovery from short-term exposure to ELINKsulfuron (7-day recovery after 7-day exposure: $E_bC50 > 9.4$ µg/L). The experimental data indicate a high recovery potential also for plants other than *Lemna*.

Standard Laboratory Studies

An overview of the standard aquatic ecotoxicology data for ELINKsulfuron is given in Tables A3.18 to A3.21.

Higher-Tier Studies

Recovery Studies with *Lemna gibba*

Two additional studies with *Lemna gibba* were conducted to prove the potential for recovery. Duckweed was exposed to ELINKsulfuron over a period of 4 days (first study) or 7 days (second study). Thereafter, subsamples of the treated plants (4 plants with 3 fronds each) were transferred to untreated nutrient medium, where growth was monitored for another 7 days (Table A3.22 and Figure A3.28).

TABLE A3.18
ELINKsulfuron Acute Tests with Fish

Test Organism	Test Item	Study Type	Test Duration (hours)	LC/EC50 (mg a.s./L)	NOEC (mg a.s. /L)
Oncorhynchus mykiss (rainbow trout)	a.s.	Static	96	>100	100
Lepomis macrochirus (bluegill sunfish)	a.s.	Static	96	>100	100
Cyprinodon variegatus (sheepshead minnow)	a.s.	Static	96	>100	100

TABLE A3.19
ELINKsulfuron Acute Tests with Aquatic Invertebrates

Test Organism	Test Item	Study Type	Test Duration (hours)	LC/EC50 (mg a.s./L)	NOEC (mg a.s./L)
Daphnia magna (water flea)	a.s.	Static	48	>1.0	100
Mysidopsis bahia (shrimp)	a.s.	Static	96	>100	100
Crassostrea virginica (eastern oyster)	a.s.	Flow through	96	>100	100

TABLE A3.20
ELINKsulfuron Growth Inhibition Tests with Algae

Test Organism	Test Item	Test Duration (hours)	LC/EC50 (mg a.s./L)	NOEC (mg a.s./L)
Pseudokirchneriella subcapitata (green alga)	a.s.	72	E_bC50: 0.18 E_rC50: >0.29	0.018
Pseudokirchneriella subcapitata (green alga)	Formulation	72	E_bC50: 0.065 E_rC50: 0.089	0.042
Anabaena flos-aquae (blue-green alga)	a.s.	72	E_bC50: 2.8 E_rC50: 5.6	1.0
Navicula pelliculosa (diatom)	a.s.	72	E_bC50: >75 E_rC50: >75	75
Skeletonema costatum (marine diatom)	a.s.	72	E_bC50: 82 E_rC50: >100	36

TABLE A3.21
ELINKsulfuron Standard 7-Day Growth Inhibition Tests with *Lemna gibba*

Test Item	Test Duration (days)	Exposure	LC/EC50 (µg a.s./L)	NOEC (µg a.s./L)	Remark on Analytical Results
a.s.	7	Static renewal (renewal on days 3 and 5)	E_bC50: 1.5 E_rC50: 1.8	0.42	Nominal concentrations (measured concentration 93 to 124%)
Formulation	7	Static renewal (renewal on days 3 and 5)	E_bC50: 2.1 E_rC50: 1.5	0.40	Nominal concentrations (measured concentration > 80%)

TABLE A3.22
Recovery of *Lemna gibba* after Transfer from a Standard Growth Inhibition Test into Freshwater[a]

Duration of Exposure and Recovery Phase	EC50 (µg a.s./L)		NOEC (µg a.s./L)	
4-day exposure + 7-day recovery	4-day exposure E_rC50: 1.5 E_bC50: > 3.8	7-day recovery E_rC50: > 3.8 E_bC50: > 3.8	4-day exposure NOEC: 0.25	7-day recovery NOEC: 0.44
7-day exposure + 7-day recovery	7-day exposure ErC50: 1.8 E_bC50: 1.5	7-day recovery E_rC50: > 9.4 E_bC50: > 9.4 Recovery, days 10 to 14 E_rC50: > 9.4 E_bC50: > 9.4	7-day exposure NOEC: 0.77	7-day recovery NOEC: 0.77 Recovery, days 10 to 14 NOEC: 1.4

[a] All endpoints are based on mean measured concentrations.

After a 4-day exposure to the highest treatment level (3.8 µg/L, mean measured), plants were inhibited by 9% according to frond number (Figure A3.28, left) and by 20% according to biomass, compared to the control, during the 7-day recovery phase. Plants previously exposed to a mean measured concentration of 0.44 µg/L were not significantly inhibited, compared to the control, during the recovery phase. After 7 days of exposure to the second-lowest treatment level (nominal 1.8 µg/L)

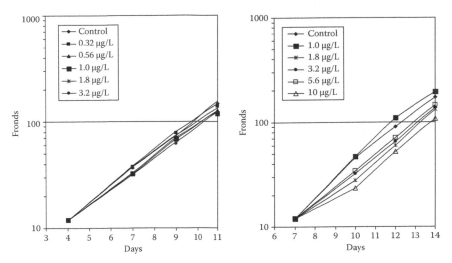

FIGURE A3.28 Growth of *Lemna gibba* (frond numbers) during a 7-day recovery period, after exposure to ELINKsulfuron over 4 days (left graph) or 7 days (right graph). The (nominal) concentrations given in the legend refer to the treatment period.

and above, duckweed showed retarded growth during the first 3 days of the recovery period (Figure A3.28, right). Nevertheless, growth curves ran parallel to the control between day 3 and day 7 of the recovery period, indicating recovery of the growth rate after an initial lag phase. Growth rate regarding frond number was not significantly inhibited at the highest concentration tested (9.4 µg/L, mean measured). Both recovery studies led to the conclusion that duckweed is able to recover from exposure to concentrations of ELINKsulfuron that are far above the standard 7-day EC50. Thus, short-term exposure should have no long-term impact on duckweed populations.

Toxicity Data for Higher Aquatic Plants Other than Lemna

Nine higher aquatic plant species (other than *Lemna gibba*) were tested in a greenhouse study using artificial sediment (quartz sand with fertilizer) and well water. The species included submersed and emergent species, as well as monocotyledon and dicotyledon species. The plants were exposed to ELINKsulfuron for 7 days, followed by a 14-day recovery phase in untreated water.

The endpoints are EbC50 values after 7 days of exposure or after 7 days of exposure plus 14 days of recovery (Table A3.23, Figure A3.29). Since chemical

TABLE A3.23
ELINKsulfuron Growth Inhibition and Recovery Tests with 9 Higher Aquatic Plant Species

Test Organism	Classification	7-Day Treatment Phase E_bC50 (µg a.s./L)[a]	14-Day Recovery Phase E_bC50 (µg a.s./L)
Lemna gibba (for comparison)		1.5	>9.4 (7-day recovery)
Lagarosiphon major (curly water thyme)	Submerged monocot	1.7	>8.0
Myriophyllum heterophyllum (red foxtail)	Submerged dicot	2.0	>11.0
Ceratophyllum demersum (coontail)	Submerged dicot	3.0	>5.7
Potamogeton pectinatus (sago pondweed)	Submerged monocot	3.2	5.5
Mentha aquatica (water mint)	Emergent dicot	3.4	>10.0
Vallisneria americana (water celery)	Submerged monocot	>3.8	>3.8
Elodea canadensis (Elodea)	Submerged monocot	>5.0	>5.0
Ranunculus lingua (grand spearwort)	Emergent dicot	>5.1	>5.1
Glyceria maxima (reed sweet-grass)	Emergent monocot	>5.3	>5.3

[a] Measured concentrations at day 0 and day 7 were within the range of 80% and 120% of the mean, respectively.

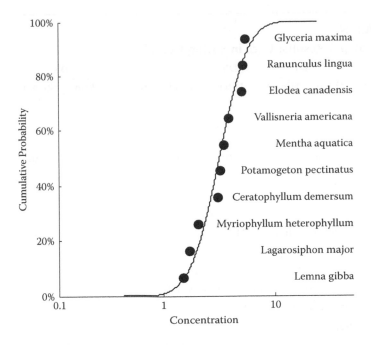

FIGURE A3.29   Species sensitivity distribution for ELINKsulfuron and 10 higher aquatic plants (including Lemna), based on 7-day E_bC50 values.

analysis showed no significant decline of ELINKsulfuron concentrations during the treatment phase (measured concentrations at day 0 and day 7 within a range of 80% and 120% of mean), the results are expressed as mean measured concentrations.

The HC5 (based on all 7-day E_bC_{50} values) is 1.43 µg/L, with a lower limit 0.82 µg/L. Because of the pronounced recovery from short-term effects of most species, the E_bC50 values are generally above the highest treatment level (3.8 to 11 µg/L).

TABLE A3.24
FOCUS Step 3 Results: Uses in Spring Cereals[a]

Scenario	D1		D3	D4		D5		R4
Water Body	Ditch	Stream	Ditch	Pond	Stream	Pond	Stream	Stream
Maximum PEC$_{sw}$ (µg/L)	0.243	0.153	0.095	0.005	0.075	0.005	0.074	0.063
7-day TWA$_{sw}$ (µg/L)	0.240	0.149	0.013	0.005	0.004	0.005	0.001	0.002
TER for maximum PEC$_{sw}$	6.2	9.8	15.8	300	20.0	300	20.3	23.8
TER for 7-day TWA$_{sw}$	6.3	10.1	115	300	375	300	1500	750

[a] Only this subset of the FOCUS scenarios is relevant for cultivation of spring cereals.

EXPOSURE PROFILE

Typical aquatic exposure profiles of ELINKsulfuron at the edge of the field, for the 3 FOCUS water bodies, are given in Attachment A3.4.

An overview of the results of the FOCUS step 3 calculations and comparisons of the maximum PECsw values and of the 7-day TWA concentrations with the E_bC50 (Lemna) of 1.5 µg/L is given in Tables A3.24 and A3.25.

HC5 = 1.43 µg/L

Lower-limit HC5 = 0.82 µg/L

TABLE A3.25

FOCUS Step 3 Results: Uses in Winter Cereals

	D1		D2		D3	D4	
Water Body	**Ditch**	**Stream**	**Ditch**	**Stream**	**Ditch**	**Pond**	**Stream**
Maximum PEC_{sw} (µg/L)	0.679	0.537	1.835	1.148	0.095	0.078	0.100
7-day TWA_{sw} (µg/L)	0.633	0.405	0.889	0.455	0.010	0.078	0.078
TER for maximum PEC_{sw}	2.2	2.8	0.8	1.3	15.7	19.2	15.0
TER for 7-day TWA_{sw}	2.4	3.7	1.7	3.3	144.2	19.2	19.2

	D5		D6	R1		R3	R4
Water Body	**Pond**	**Stream**	**Ditch**	**Pond**	**Stream**	**Stream**	**Stream**
Maximum PEC_{sw} (µg/L)	0.097	0.089	0.655	0.010	0.626	0.929	0.063
7-day TWA_{sw} (µg/L)	0.095	0.051	0.326	0.009	0.037	0.110	0.003
TER for maximum PEC_{sw}	15.5	16.9	2.3	150	2.4	1.6	23.8
TER for 7-day TWA_{sw}	15.8	29.4	4.6	167	40.5	13.6	500

Attachment A3.3 FOCUSsw Scenarios: Exposure for ELINKsulfuron in Spring Cereals

The *y*-axes of the figures have different scales to illustrate the patterns rather than the levels of exposure.

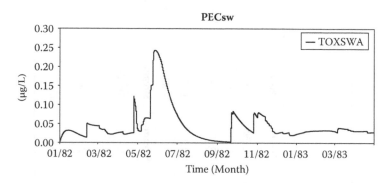

FIGURE A3.30   ELINKsulfuron–spring cereals: D1 Ditch.

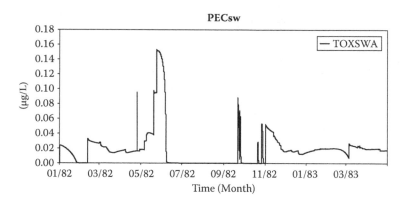

FIGURE A3.31   ELINKsulfuron–spring cereals: D1 Stream.

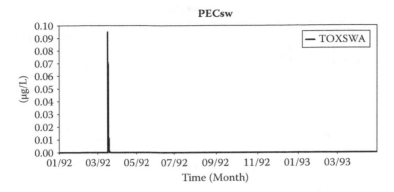

FIGURE A3.32   ELINKsulfuron–spring cereals: D3 Ditch.

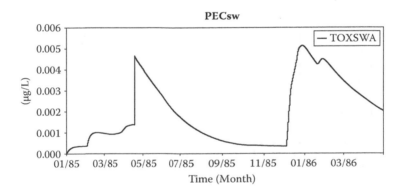

FIGURE A3.33   ELINKsulfuron–spring cereals: D4 Pond.

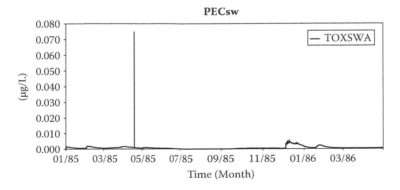

FIGURE A3.34   ELINKsulfuron–spring cereals: D4 Stream.

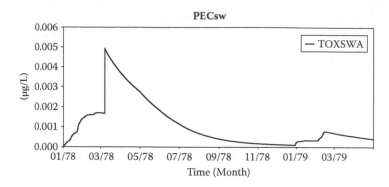

FIGURE A3.35   ELINKsulfuron–spring cereals: D5 Pond.

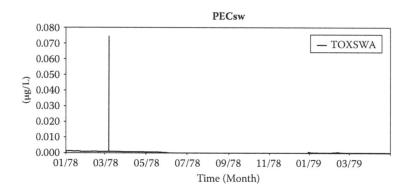

FIGURE A3.36   ELINKsulfuron–spring cereals: D5 Stream.

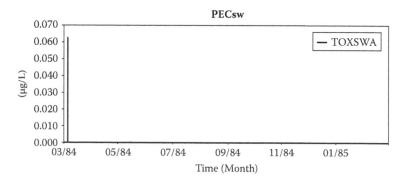

FIGURE A3.37   ELINKsulfuron–spring cereals: R4 Stream.

ATTACHMENT A3.4 FOCUSsw SCENARIOS: EXPOSURE
FOR ELINKSULFURON IN WINTER CEREALS

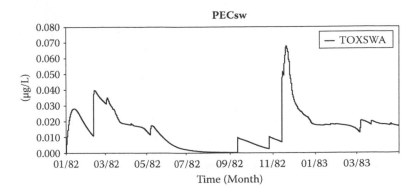

FIGURE A3.38   ELINKsulfuron–winter cereals: D1 Ditch.

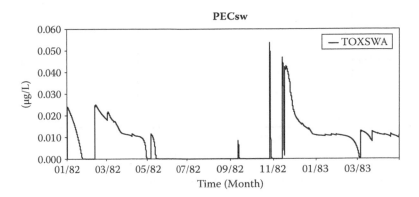

FIGURE A3.39   ELINKsulfuron–winter cereals: D1 Stream.

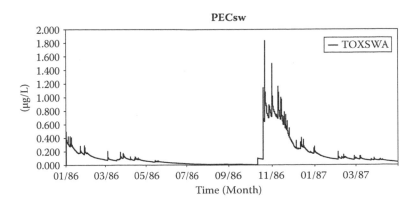

FIGURE A3.40   ELINKsulfuron–winter cereals: D2 Ditch.

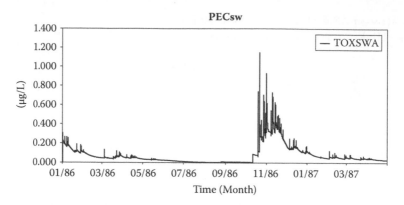

FIGURE A3.41   ELINKsulfuron–winter cereals: D2 Stream.

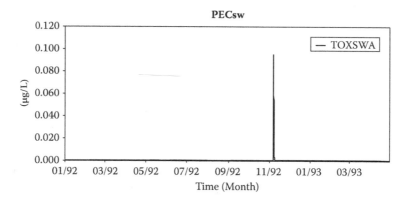

FIGURE A3.42   ELINKsulfuron–winter cereals: D3 Ditch.

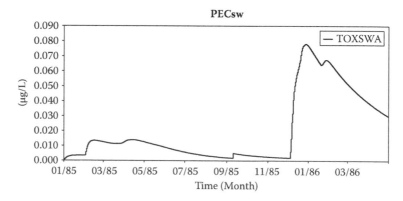

FIGURE A3.43   ELINKsulfuron–winter cereals: D4 Pond.

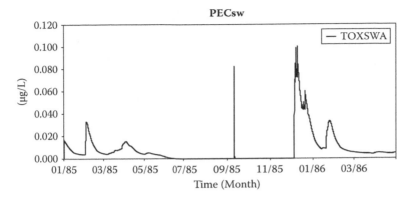

FIGURE A3.44   ELINKsulfuron–winter cereals: D4 Stream.

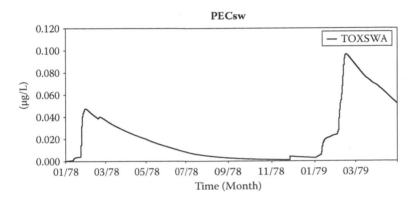

FIGURE A3.45   ELINKsulfuron–winter cereals: D5 Pond.

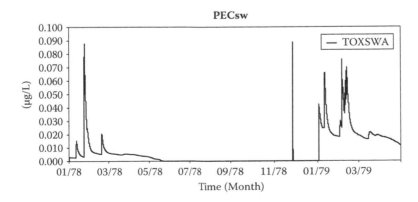

FIGURE A3.46   ELINKsulfuron–winter cereals: D5 Stream.

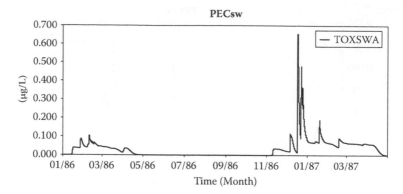

FIGURE A3.47   ELINKsulfuron–winter cereals: D6 Ditch.

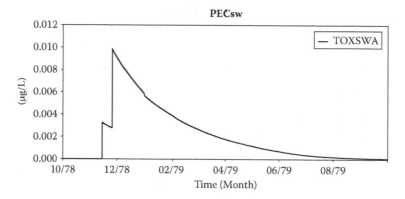

FIGURE A3.48   ELINKsulfuron–winter cereals: R1 Pond.

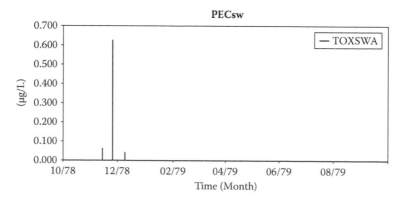

FIGURE A3.49   ELINKsulfuron–winter cereals: R1 Stream.

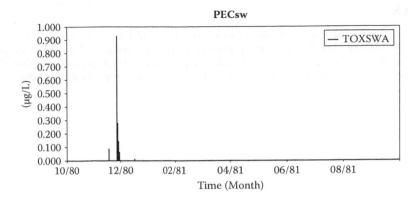

FIGURE A3.50   ELINKsulfuron–winter cereals: R3 Stream.

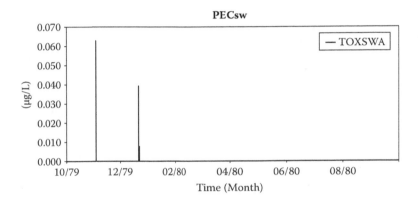

FIGURE A3.51   ELINKsulfuron–winter cereals: R4 Stream.

NOTES

1. DT50 values that describe degradation (DegT50) and those that describe dissipation (DisT50) are to be distinguished. DT50 values are in general given for simple first-order (SFO) kinetics. Exceptions are specifically marked like DT50 according to first-order multiple compartment model (FOMC). The terminology of the FOCUS kinetics work group (2005) is used.
2. DegT50 is the DT50 for degradation, DisT50 is the DT50 for dissipation. See Note 1.

Appendix 4: Foldout Page of the ELINK Decision Scheme: Explanatory Notes for Figure 2.1

BOX 1

- The PEC_{max} can be the highest actual concentration from a range of exposure profiles from different scenarios or from a single exposure profile (e.g., $PEC_{global\ max}$ from FOCUS [Forum for the Co-ordination of Pesticide Fate Models and Their Use] step 3) (see Chapters 3 and 4).
- A summary of the current first-tier effects endpoints for acute and chronic risk assessment is presented in Tables 3.1 and 3.2, respectively (Chapter 3).

BOX 2

- Situations in which a time-weighted average (TWA) approach is not appropriate in the chronic risk assessment are described in Section 3.4 of Chapter 3.

BOX 3

- The default value of 7 days for the TWA predicted environmental concentration (PEC) is a general recommendation if no specific information about the relation between exposure pattern and time to onset of effect (TOE) for the (relevant life stages of the) organisms that trigger the chronic risk are available. This default value helps to avoid lengthy discussion about the time window for TWA calculations. On the one hand, it is very unlikely that the risks from compounds with a short TOE are overlooked. On the other hand, simply using the duration of the effects test that triggered the chronic risk might be too liberal if the TOE is short. Information on criteria to determine the length of the time window of the TWA PEC and the corresponding TWA regulatory acceptable concentration (RAC) is given in Section 3.4 of Chapter 3.

- Usually, the highest TWA PEC for the relevant time period, either from each individual scenario or water body or the highest from the range of available pertinent exposure profiles should be used for risk assessment (Chapters 3 and 4).

BOX 4

- Information on the species sensitivity distribution (SSD) approach in the aquatic risk assessment can be found in Chapter 5. To date, SSDs for plant protection products have been generated mainly with toxicity values from acute tests. Often, not enough chronic no observed effect concentration (NOEC) or chronic EC_{10} (effective concentration that affects 10% of the test organisms) values are available to construct a chronic SSD, partly due to the fact that test methods for nonstandard test species are lacking. The HC_5 or HC_1 (hazardous concentration to 5% or 1%, respectively, of the tested taxa) from the distribution curve may be used for setting the RAC (Section 5.4 of Chapter 5).
- For further information on refined exposure tests with the relevant (standard) test organisms or populations (or sensitive life stages), refer to Chapter 6. The exposure regime tested should be guided by exposure predictions (e.g., maximum duration of pulse exposure) or the relevant generalized exposure profile (Chapter 4).
- If the adopted approach comprises a combination of experimental laboratory studies and computer simulation models to extrapolate time-variable exposure, further information on toxicokinetic and toxicodynamic (TK/TD) and population models can be found in Chapters 7 and 9, respectively.

BOX 5

- Information on the model ecosystem approach can be found in Chapter 8 and on linking the threshold concentration for effects to exposure predictions in Section 8.4 of that chapter.
- In the microcosm and mesocosm experiment, the tested exposure regime should have been guided by exposure predictions or the relevant generalized exposure profile (Chapter 4 and Section 8.3 of Chapter 8). Note that especially due to the different intended uses it might only be possible to test a representative exposure pattern. Furthermore, importance of exposure routes and different mitigation options depends on the regional context.

- Consider if micro/mesocosms contain the relevant populations and community potentially at risk (compare with ecological scenario derived from ecological field data; Chapter 10).
- Ecological models (e.g., population and food web models) may be used for spatiotemporal extrapolation of the threshold levels for toxic effects derived from micro/mesocosm tests (Chapter 9).

BOX 6

- Information on factors that affect recovery of sensitive populations in micro/mesocosm experiments and how to link exposure and effects when recovery is taken into account is presented in Sections 8.5 to 8.7 of Chapter 8.
- In the micro/mesocosm experiment, the tested exposure regime should have been guided by exposure predictions or the relevant generalized exposure profile (Chapter 4 and Section 8.3 of Chapter 8).
- Consider if micro/mesocosms contain the relevant populations and community potentially at risk (compare with ecological scenario derived from ecological field data; Chapter 10).
- Consider if external recovery is important for potentially sensitive populations; if yes, apply modeling (e.g., metapopulation models) or experimentation with a spatial component to refine RAC (see Section 8.6 and Chapters 9 and 10).
- Compare relevant RAC from Boxes 1 to 5 and the RAC based on the NOEAEC (no observed ecologically adverse effect concentration) with exposure profile and evaluate differences in terms of field relevance (see Section 8.7 and Chapter 10).

BOX 7

- Even if the triggers are not met in the procedures described in Boxes 1 to 6 (indicating unacceptable risks), there might be further tools available to refine the risk assessments, for example, the combination of probabilistic exposure and effects assessment approaches. Since these methods, and the implications for linking exposure and effects, were not discussed during the ELINK workshops, no specific guidance on these methods is given here.

Problem Formulation: Check Validity of Exposure; Define Critical Eco(Toxico)Logical Issues

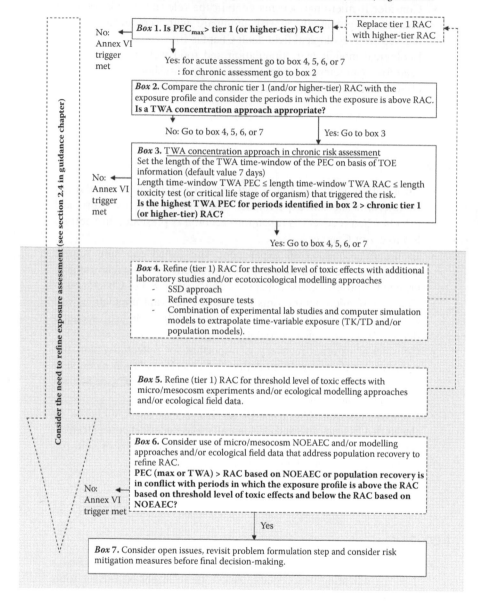

FIGURE 2.1 ELINK decision scheme for acute and chronic aquatic risk assessment for plant protection products (for explanatory notes, see text).

Glossary

AA-EQS: Annual average environmental quality standard

Abiotic factor: Physical, chemical, and other nonliving environmental factors

Abundance: The total number of individuals or taxa in an area, sample, or volume

A/C ratio: Ratio of acute (usually EC50 or LC50) to chronic (usually NOEC) effects

AChE: Acetylcholinesterase

Acute: Responses occurring within a short period in relation to the life span of the organism. It can be used to define either the exposure (acute test) or the response to an exposure (acute effect).

AF: Assessment factor (= uncertainty factor)

Applicant: Person desirous of obtaining registration of or use permit for formulated pesticides and active ingredients

AQUATOX: Simulation model for aquatic systems for performing ecological risk assessment

Area under the curve (AUC): The integral under the concentration time function for a given period

Assemblage: A group of organisms occurring together in the same habitat

Bayesian statistics: Statistics based on Bayesian inference. Bayesian inference uses a numerical estimate of the degree of belief in a hypothesis before evidence has been observed and calculates a numerical estimate of the degree of belief in the hypothesis after evidence has been observed.

BCF: Bioconcentration factor

Benthic: Associated with freshwater or saltwater substrata (upper layer of the sediment in rivers and ponds) at the sediment–water interface

Bioavailability: Rate and extent to which a pesticide or metabolite can be absorbed by an organism and is available for metabolism or interaction with biologically significant receptors. It involves both release from a medium (if present) and absorption by an organism.

Biocidal: Active substances, put up in the form in which they are supplied to destroy, deter, render harmless, prevent the action of, or otherwise exert a controlling effect on any harmful organism by chemical or biological means

Bioconcentration: Uptake of a pesticide residue from an environmental matrix, usually through partitioning across body surfaces to a concentration in the organism that is usually higher than in the environmental matrix

Biodegradation: Conversion or breakdown of the chemical structure of a pesticide catalyzed by enzymes in vitro or in vivo, often resulting in loss of biological activity. For hazard assessment, categories of chemical degradation include

1) Primary — loss of specific activity
2) Environmentally acceptable — loss of any undesirable activity (including any toxic metabolites)

3) Ultimate — mineralization to small molecules such as water and carbon dioxide

Biota: Ensemble of plant and animal life in an ecosystem

Buffer strips: Distance for environmental protection between the edge of an area where pesticide application is permitted and a sensitive nontarget area (e.g., watercourse)

Calibration: Testing of a model with known input and output information for adjustment or estimation of factors for which data are not available

CASM: Comprehensive aquatic systems model

CBR: Critical body residue

C-COSM: A model of the fate of nutrients and pesticides and their effects on biota in microcosms

Chronic: Responses occurring after an extended time relative to the life span of an organism. Long-term effects are related to changes in metabolism, growth, reproduction, or the ability to survive.

CLASSIC: Community-Level Aquatic System Studies–Interpretation Criteria (SETAC Guidance Document 2002)

Coenosis: An assemblage of organisms having similar ecological preferences

Community model: Model that addresses the relationships between any group of organisms belonging to a number of different species that cooccur in the same habitat and interact through trophic and spatial relationships

Conductivity: Property of conducting electricity expressed as $S \times m^{-1}$

COST: European Cooperation in the Field of Scientific and Technical Research

DAM: Damage assessment model

Data-mining model: Empirical models aim to use existing data for extrapolation (e.g., by regression), although they are not based on a mechanistic understanding. As almost all models use empirical data, those explicitly built on large data sets are separated here as data-mining models.

DEBtox: Dynamic energy budget model

Degradation: Process by which a pesticide is broken down to simpler structures through biological or abiotic mechanisms. Synonyms include "breakdown" and "decomposition."

Desorption: Decrease in the amount of adsorbed substance (e.g., pesticide) at the interphase of the soil colloids (clay or organic matter). Antonym: "adsorption."

Diapause: A resting phase; a period of suspended growth or development characterized by greatly reduced metabolic activity

Direct effect: Response directly caused by the stressor (in ecotoxicology, by the toxicant)

Dissipation: Loss of compound residues from an environmental compartment due to degradation and transfer to another environmental compartment

Diversity: A measure of the number of species and their relative abundance in a community

Dominant (organism): An organism exerting considerable influence on a community by its size, abundance, or coverage

Dormancy: A state of relative metabolic quiescence of an organism

Drift (of organisms): Lateral (downstream) displacement of organisms as a result of water flow

DT50/90: Period required for 50% or 90% disappearance

E_bC50: Median effective concentration on biomass

EC50: Median effective concentration that affects 50% of the test organisms

Ecological or biological trait: Any ecological or biological property of an organism

Ecological model: Any model that addresses ecological properties of a species, population, or community

Ecosystem: Assembly of populations of different species (often interdependent on and interacting with each other) together with nonliving components of their environment

Ecosystem model: Model that addresses relationships between components (biotic and abiotic) of an ecosystem

Ecotoxicology: The study of toxic effects of substance and physical agents in living organisms, especially on populations and communities within defined ecosystems. It includes transfer pathways of these agents and their interaction with the environment.

Effect assessment: Combination of analysis and inference of possible consequences of the exposure to a particular agent (e.g., pesticide) based on knowledge of the dose–effect relationship associated with that agent in a specific target organism, system, or (sub)population

Effect threshold: Concentration of a compound in an organism or environmental compartment below which an adverse effect is noted

EFSA: European Food Safety Authority

Empirical model: Model based on direct observation or experimentation rather than theory or preconception

Endemic species: Species native to, and restrictive to, a particular geographical region

Endpoint: Measurable ecological or toxicological characteristic or parameter of the test system (usually an organism) that is chosen as the most relevant assessment criterion (e.g., death in an acute test or tumor incidence in a chronic study)

Equational model: Equational formulation of an object-oriented data model

ERA: Ecological risk assessment. A process that evaluates the likelihood that adverse ecological effects may occur or are occurring as a result of exposure to one or more stressors

ERC: Ecotoxicologically relevant concentration. Concentration of a pesticide (active ingredient, formulations, or relevant metabolites) that is likely to affect a determinable ecological characteristic of an exposed system. The ERC describes the relevant type of concentration in terms of matrix, location, and time. The definition of the ERC has to include the following aspects: 1) the definition of the quantity itself, 2) the definition of the spatial scale of this quantity, and 3) the definition of the temporal scale of this quantity. Usually, the ERC is derived without applying an uncertainty factor. *See also* RAC.

E_rC50: Median effective concentration on growth rate

Eurythermic: Tolerant to a wide range of ambient temperatures

Eutrophic: Having high primary productivity; pertaining to waters rich in mineral nutrients

Exposure assessment: Evaluation of the exposure of an organism, system, or (sub) population to a pesticide or agent (and its derivatives). It includes estimation of transport, fate, and uptake.

Fecundity: The potential reproductive capacity of an organism or population

Field study: *Monitoring field study:* An investigation into the overall impact of pesticide use on a specific ecosystem through surveying or monitoring that consists of characterization of exposure (chemical monitoring and exposure modeling) and observations of effects (biological monitoring) occurring in the field or treated area as a consequence of use or misuse of pesticides *Experimental field study:* An experiment into the impact of a specific product or active substance applied under controlled conditions in the field. Such studies are performed in the natural environment within an agricultural context (and thus contrast with mesocosm studies).

First order: Pattern of decline that is the same pattern observed in a radioactive decay curve

First-tier test: Standardized protocol test in the initial phase of the research

FLC: Full life cycle

Flow-through toxicity test: Toxicity test in which the water (with toxicant) is constantly renewed to maintain the desired exposure concentration

FOCUS: Forum for the Co-ordination of Pesticide Fate Models and Their Use

Food web: The network of interconnected food chains of a community

Food web model: Model that addresses the network of interconnected food chains of a community

GLP: Good laboratory practice

Habitat: The locality, site, and particular type of local environment occupied by an organism

HARAP: Higher-Tier Aquatic Risk Assessment for Pesticides (SETAC Guidance Document 1999)

Hazard rate: Term used in toxicokinetic and toxicodynamic modeling to describe the probability of a particular effect (e.g., death) over time

HC5: Hazardous concentration to 5% of the tested taxa

Higher-tier test: Advanced test with a higher level of complexity that addresses remaining uncertainties

Hydrolysis: Reaction in which a chemical bond is cleaved and a new bond formed with the oxygen atom of a molecule of water

Hydrophobous: Intolerant of water or wet conditions; not thriving in aquatic matrix or medium

IBM: Individual-based model. Individual-based models are simulations based on the overall consequences of local interactions of members of a population and therefore include properties of individuals within populations.

IBMWP: Iberian Biomonitoring Working Party

Indirect effect: An effect resulting from the action of an agent on some components of the ecosystem, which in turn affect the assessment endpoint or other ecological component of interest. Indirect effects of substance contaminants include reduced abundance due to adverse effects on food species or on plants that provide habitat structure.

Keystone species: A species having a major influence on community structure or functioning, often in excess of that expected from its relative abundance

k-strategist: Species with superior competitive ability in stable predictable environments in which rapid population growth is less important as the population is maintained at or near the carrying capacity of the habitat

Latency of effects: Delayed effects that may become apparent after the stressor has disappeared

LC50: Median lethal concentration (concentration that is lethal to 50% of exposed test individuals)

Lentic: Pertaining to static, calm, or slow-moving aquatic habitats

Lithological: Description of mineral components of a water body, including the physical characteristics of rock, sediment, and water body edges

Littoral zone: Pertaining to the shore

LOEC: Lowest observed effect concentration

Logistic distribution: Distribution representing an exponential function (sigmoid curve)

Lognormal distribution: A distribution that is classically bell shaped and symmetrical only when the data are transformed to a logarithm

LTRE: Life table response experiment

MACRO: A model of water movement and solute transport in macroporous soils

MASTEP: Metapopulation model for assessing spatial and temporal effects of pesticides

Matrix model: A type of unstructured model focusing on temporal patterns, using, for example, an age–class matrix

Mechanistic model: A mathematical or functional representation of some component of an ecosystem with parameters that can be adjusted to closely describe a set of empirical data.

Mesocosm: *See* Model ecosystem

Mesosaprobic: Pertaining to a polluted aquatic habitat having reduced oxygen concentration and a moderately high level of organic decomposition

Mesoscale: Intermediate range

Metapopulation: A group of subpopulations of the same species coexisting in time but not in space

Metapopulation model: Model that addresses the interactions between metapopulations

Microcosm: *See* Model ecosystem

Model ecosystem: Human-made study system containing associated organism and abiotic components that is large enough to be representative of a natural ecosystem yet small enough to be experimentally manipulated. There is some subjective differentiation between larger, outdoor model ecosystems (mesocosms) and smaller, generally indoor model ecosystems (microcosms).

Model validation: Comparison of model results with numerical data independently derived from experiments or observations of the environment

Mode of action: Biochemical effect that occurs at the lowest dose or concentration or is the earliest among a number of biochemical effects that could, understandably, lead to the death of the pest. More precisely, this is the primary mode of action of a pesticide. However, there may also be other biochemical effects that occur later or at higher doses (i.e., secondary modes of action) that also may contribute to the death of the pest.

Monitoring: In the context of this document, survey or check of the status of an ecosystem being exposed to pesticides. The survey or monitoring implies observations and samplings for chemical or physical or biological indicators.

Narcotic: Any of a number of substances that have a depressant effect on the nervous system

NOEAEC: No observed ecologically adverse effect concentration

NOEC: No observed effect concentration

NOEC$_{community}$: NOEC based on a community level

Nominal concentration: Concentration in test medium based on the measured concentration in the dosing solution and the amount of dosing solution applied

Numerical model: Mathematical models that use a numerical time-stepping procedure to obtain the model's behavior over time

Oligotrophic: Having low primary productivity; pertaining to waters having low levels of mineral nutrients required by plants

PAF: Potentially affected fraction

PAT: Pesticide application timer. Software tool for finding the application time in a given time window of selected crop-growing conditions

PEC: Predicted environmental concentration. Predicted concentration of a pesticide within an environmental compartment based on estimates of quantities released, discharge patterns, and inherent disposition of the pesticide (fate and distribution) as well as the nature of the specific receiving ecosystems

PEC$_{act}$: Actual PEC

PEC$_{global\ max}$: Maximum PEC calculated for each single combination of a FOCUS scenario and type of water body

PEC$_{gw}$: PEC in groundwater

PEC$_{max}$: Maximum PEC

PEC$_{sed}$: PEC in sediment

PEC$_{sw}$: PEC in surface water

PEC$_{twa}$: Time-weighted average PEC

Pelagic zone: Pertaining to the water column; used for organisms inhabiting the open waters of a lake or the sea

Photolysis: Cleavage of one or more covalent bonds in a molecular entity resulting from absorption of light or a photochemical process in which such cleavage is an essential part

Phreatophyte communities: Plant communities dependent on groundwater

Phylogenetic: Pertaining to evolutionary relationships within and between groups

Phytosociological associations: Vegetation units and plant communities

Pioneer species: The first species to colonize a barren or disturbed area

Population: Assemblage of individual organisms of defined ages and growth stages belonging to one species within a specified location in space and time

Population model: Model that addresses the interactions between individuals of a population

Propagule: Any part of an organism, produced sexually or asexually, that is capable of giving rise to a new individual

PRZM: Pesticide root zone model

QBR: Riberian vegetation index

RAC: Regulatory acceptable concentration. Effects assessment endpoint, expressed in terms of a permissible concentration in the environment that is directly used in the risk assessment by comparing it with the appropriate field exposure estimate (e.g., PEC_{max}). Usually, the RAC is derived by applying an uncertainty factor. *See also* ERC.

Realistic worst-case exposure scenario: The scenarios should describe an overall vulnerability approximating the 90th percentile of all possible situations. *See also* Scenario.

Recovery: The extent of return of a population, community, or ecosystem function to a condition that existed before being affected by a stressor. Due to the complex and dynamic nature of ecological systems, the attributes of a "recovered" system must be carefully defined.

Refined exposure studies: Exposure studies with refined, usually more realistic, exposure profile

Refugia: Areas in which organisms may escape from or avoid stressors (e.g., toxicants and predators)

Resiliency: The degree to which a population of an organism or a community is able to tolerate a perturbation without the structure or function being affected

Resting stages: Periods in which organisms (or their propagules) are characterized by a state of relative metabolic quiescence

Rhithronic: Pertaining to the upper reaches of a river

Richness: Species diversity

RKM: Receptor kinetic model

r-strategist: Species favoring a rapid rate of population increase, typical for species that colonize short-lived environments or of species that undergo large fluctuations in population size

Scenario: A representative combination of crop, soil, climate, and agronomic parameters to be used in modeling

Semistatic: Condition in which the medium is refreshed at regular time intervals during the test

Sensitivity: The capacity of an organism to respond to stimuli (e.g., a stressor like a pesticide)

Shannon–Weaver index: An index of diversity; its minimum value occurs if all individuals belong to the same species and its maximum value if each individual belongs to a different species

SHM: Simple hazard model

Sorption: Removal of pesticide from solution by soil or sediment via mechanisms of adsorption and absorption

Spatially explicit model: Model that addresses the spatial distribution of organisms in their landscape (watershed)

Species traits: Any character or property of a species

SSD: Species sensitivity distribution. A function of the toxicity of a certain substance or mixture to a set of species that may be defined as a taxon, assemblage, or community. Empirically, an SSD is estimated from a sample of toxicity data for the specified species set.

Static: Condition in which the medium is not refreshed during the test

Sublethal endpoint: Ecotoxicological outcomes other than lethality that may be measured in ecotoxicology studies

SWASH: Surface Water Scenarios Help software

Taxon (*pl.*: taxa): Any group of organisms considered to be sufficiently distinct from other such groups to be treated as a separate unit

TDM: Threshold damage model

TD/TK: Toxicodynamic and toxicokinetic

Thermophilic: Thriving in warm environmental conditions

Threshold level: The highest dose of a substance to which an organism can be exposed without expecting the stated effect to occur

TOE: Time to onset of effects

Toxicodynamics: Description of the time course of injury and recovery in an organism in response to a chemical stressor

Toxicokinetics: Description of the time course of a chemical within an organism (e.g., rates of uptake and elimination)

TWA: Time-weighted average

UF: Uncertainty factor. A factor applied to an exposure or effect concentration or dose to correct for unidentified sources of uncertainty

Unstructured model: Unstructured models describe the population with state variables like population abundance or density N

Voltinism: Pertaining to the number of broods or generations per year

Vulnerability: The degree in which species or populations suffer from stressors and disturbances in their environment, including their rate of recovery

WFD: Water Framework Directive

Index

Other Titles from the Society of Environmental Toxicology and Chemistry (SETAC)

Freshwater Bivalve Ecotoxicology
Farris, Van Hassel, editors
2006

Estrogens and Xenoestrogens in the Aquatic Environment:
An Integrated Approach for Field Monitoring and Effect Assessment
Vethaak, Schrap, de Voogt, editors
2006

Assessing the Hazard of Metals and Inorganic Metal Substances
in Aquatic and Terrestrial Systems
Adams, Chapman, editors
2006

Perchlorate Ecotoxicology
Kendall, Smith, editors
2006

Natural Attenuation of Trace Element Availability in Soils
Hamon, McLaughlin, Stevens, editors
2006

Mercury Cycling in a Wetland-Dominated Ecosystem:
A Multidisciplinary Study
O'Driscoll, Rencz, Lean
2005

Atrazine in North American Surface Waters:
A Probabilistic Aquatic Ecological Risk Assessment
Giddings, editor
2005

Effects of Pesticides in the Field
Liess, Brown, Dohmen, Duquesne, Hart, Heimbach, Kreuger, Lagadic,
Maund, Reinert, Streloke, Tarazona
2005

Human Pharmaceuticals: Assessing the Impacts on Aquatic Ecosystems
Williams, editor
2005

SETAC

A Professional Society for Environmental Scientists and Engineers and Related
Disciplines Concerned with Environmental Quality

The Society of Environmental Toxicology and Chemistry (SETAC), with offices currently in North
America and Europe, is a nonprofit, professional society established to provide a forum for individuals and institutions engaged in the study of environmental problems, management and regulation of
natural resources, education, research and development, and manufacturing and distribution.

Specific goals of the society are

- Promote research, education, and training in the environmental sciences.
- Promote the systematic application of all relevant scientific disciplines to the evaluation
  of chemical hazards.
- Participate in the scientific interpretation of issues concerned with hazard assessment and
  risk analysis.
- Support the development of ecologically acceptable practices and principles.
- Provide a forum (meetings and publications) for communication among professionals in
  government, business, academia, and other segments of society involved in the use, protection, and management of our environment.

These goals are pursued through the conduct of numerous activities, which include:

- Hold annual meetings with study and workshop sessions, platform and poster papers, and
  achievement and merit awards.
- Sponsor a monthly scientific journal, a newsletter, and special technical publications.
- Provide funds for education and training through the SETAC Scholarship/Fellowship
  Program.
- Organize and sponsor chapters to provide a forum for the presentation of scientific data
  and for the interchange and study of information about local concerns.
- Provide advice and counsel to technical and nontechnical persons through a number of
  standing and ad hoc committees.

SETAC membership currently is composed of more than 5000 individuals from government,
academia, business, and public-interest groups with technical backgrounds in chemistry, toxicology,
biology, ecology, atmospheric sciences, health sciences, earth sciences, and engineering.

If you have training in these or related disciplines and are engaged in the study, use, or management
of environmental resources, SETAC can fulfill your professional affiliation needs.

All members receive a newsletter highlighting environmental topics and SETAC activities and
reduced fees for the Annual Meeting and SETAC special publications.

All members except Students and Senior Active Members receive monthly issues of Environmental
Toxicology and Chemistry (ET&C) and Integrated Environmental Assessment and Management
(IEAM), peer-reviewed journals of the Society. Student and Senior Active Members may subscribe
to the journal. Members may hold office and, with the Emeritus Members, constitute the voting
membership.

If you desire further information, contact the appropriate SETAC Office.

1010 North 12th Avenue	Avenue de la Toison d'Or 67
Pensacola, Florida 32501-3367 USA	B-1060 Brussels, Belgium
T 850 469 1500 F 850 469 9778	T 32 2 772 72 81 F 32 2 770 53 86
E setac@setac.org	E setac@setaceu.org

www.setac.org
Environmental Quality Through Science®